Nonstandard Analysis and Vector Lattices

Volume 525

Nonstandard Analysis and Vector Lattices

Edited by

S.S. Kutateladze

Sobolev Institute of Mathematics,
Siberian Division of the Russian Academy of Sciences,
Novosibirsk, Russia

SPRINGER-SCIENCE+BUSINESS MEDIA, B.V.

A C.I.P. Catalogue record for this book is available from the Library of Congress.

ISBN 978-0-7923-6619-5 ISBN 978-94-011-4305-9 (eBook)
DOI 10.1007/978-94-011-4305-9

This is an updated translation of the original Russian work.

Nonstandard Analysis and Vector Lattices,

A.E. Gutman, \'E.Yu. Emelyanov,
A.G. Kusraev and S.S. Kutateladze.
Novosibirsk, Sobolev Institute Press, 1999.
The book was typeset using AMS-TeX.

Printed on acid-free paper

Contents

Foreword

Nonstandard methods of analysis consist generally in comparative study of two interpretations of a mathematical claim or construction given as a formal symbolic expression by means of two different set-theoretic models: one, a "standard" model and the other, a "nonstandard" model. The second half of the twentieth century is a period of significant progress in these methods and their rapid development in a few directions.

The first of the latter appears often under the name coined by its inventor, A. Robinson. This memorable but slightly presumptuous and defiant term, *nonstandard analysis*, often swaps places with the term *Robinsonian* or *classical nonstandard analysis*. The characteristic feature of Robinsonian analysis is a frequent usage of many controversial concepts appealing to the actual infinitely small and infinitely large quantities that have resided happily in natural sciences from ancient times but were strictly forbidden in modern mathematics for many decades. The present-day achievements revive the forgotten term *infinitesimal analysis* which reminds us expressively of the heroic bygones of Calculus.

Infinitesimal analysis expands rapidly, bringing about radical reconsideration of the general conceptual system of mathematics. The principal reasons for this progress are twofold. Firstly, infinitesimal analysis provides us with a novel understanding for the method of indivisibles rooted deeply in the mathematical classics. Secondly, it synthesizes both classical approaches to differential and integral calculuses which belong to the noble inventors of the latter. Infinitesimal analysis finds newer and newest applications and merges into every section of contemporary mathematics. Sweeping changes are on the march in nonsmooth analysis, measure theory, probability, the qualitative theory of differential equations, and mathematical economics.

The second direction, *Boolean valued analysis*, distinguishes itself by ample usage of such terms as the technique of ascending and descending, cyclic envelopes and mixings, B-sets and representation of objects in $\mathbf{V}^{(B)}$. Boolean valued analysis originated with the famous works by P. J. Cohen on the continuum hypothesis.

Progress in this direction has evoked radically new ideas and results in many sections of functional analysis. Among them we list Kantorovich space theory, the theory of von Neumann algebras, convex analysis, and the theory of vector measures.

The book [3], printed by the Siberian Division of the Nauka Publishers in 1990 and translated into English by Kluwer Academic Publishers in 1994 (see [4]), gave a first unified treatment of the two disciplines forming the core of the present-day nonstandard methods of analysis.

The reader's interest as well as successful research into the field assigns a task of updating the book and surveying the state of the art. Implementation of the task has shown soon that it is impossible to compile new topics and results in a single book. Therefore, the Sobolev Institute Press decided to launch the series "Nonstandard Methods of Analysis" which will consist of monographs on various aspects of this direction in mathematical research.

The series started with the book [5] whose English edition [6] appeared quite simultaneously.

The present book continues the series and addresses applications to vector lattice theory. The latter stems from the early thirties and its rise is attributed primarily to the effort and contribution of H. Freudenthal, L. V. Kantorovich, and F. Riesz. Drifting in the general wake of functional analysis, the theory of vector lattices has studied those features of classical Banach spaces and operators between them which rest on the innate order relations.

The mid-seventies landmark a new stage of rapid progress in vector lattice theory. The reason behind this is an extraordinarily fruitful impact of the principal ideas of the theory on the mathematical research inspired by social sciences and, first of all, economics. The creative contribution of L. V Kantorovich has played a leading role in a merger between ordered vector spaces, optimization, and mathematical economics.

The next most important circumstance in the modern development of vector lattice theory is the discovery of a prominent place of Kantorovich spaces in Boolean valued models of set theory. Constructed by D. Scott, R. Solovay, and P. Vopěnka while interpreting the topical work of P. J. Cohen, these models turn out inseparable from vector lattices. The fundamental theorem by E. I. Gordon demonstrates that the members of each Dedekind complete vector lattice depict reals in an appropriate nonstandard model of set theory. This rigorously corroborates the *heuristic Kantorovich principle* which declares that the elements of every vector lattice are generalized numbers.

Some results of the development of vector lattice theory in the eighties were summarized in the book [1] published by the Siberian Division of the Nauka Publishers in 1992. It was in 1996 that Kluwer Academic Publishers printed a revised and enlarged English version of this book [2]. These articles, in particular, drafted some new synthetic approaches to vector lattice theory that use the modern nonstandard methods of analysis. The aim of the present monograph is to reveal the most recent results that were obtained along these lines in the last decade.

This book consists of five chapters which are closely tied by the scope of the problems addressed and the common methods involved. For the reader's convenience, exposition proceeds so that the chapters can be studied independently of one another. To this end, each chapter contains its own introduction and list of references, whereas the subject and notation indexes are common for the entire book.

Chapter 1 is a general introduction to the nonstandard methods of analysis applicable to vector lattice theory. That is why to study its first sections will do no harm to the reader, his or her further intentions notwithstanding. This chapter gives quite a few diverse applications among which we mention the technique for combining nonstandard models and the theory of cyclically compact operators. Chapter 1 is written by A. G. Kusraev and S. S. Kutateladze.

Chapters 2 and 3 belong to Boolean valued analysis. The former studies a new concept of continuous polyuniverse which is a continuous bundle of set-theoretic models. The class of continuous sections of such a polyuniverse maintains all principles of Boolean valued analysis. Furthermore, each of the similar algebraic systems is realizable as the class of sections of a suitable continuous polyuniverse. Chapter 2 was prepared by A. E. Gutman jointly with G. A. Losenkov.

Chapter 3 suggests a new approach to the definition of dual bundle which is motivated by studying the realization of dual Banach spaces in Boolean valued models. Chapter 3 is written by A. E. Gutman jointly with A. V. Koptev.

Chapter 4 by È. Yu. Emel′yanov deals mainly with adapting the methods of infinitesimal analysis for study of intrinsic problems of vector lattice theory. Incidentally, the author explicates some properties of the infinite dimensional analogs of the standard part operation over the hyperreals that are unpredictable in advance.

Chapter 5 is written by A. G. Kusraev and S. A. Malyugin and belongs to vector measure theory. Study of Banach-space-valued measures is well known to involve other tools than those of Boolean valued measures. The bulk of the chapter sets forth a principally new unified approach to both directions of research in measure theory which rests upon the concept of lattice normed space. It is worth observing that locally convex spaces and vector lattices are just some particular instances of lattice normed space. Also important is the fact that these spaces depict Banach spaces inside Boolean valued models.

Among particular topics of this chapter, we mention a criterion for a dominated operator to admit integral representation with respect to some quasi-Radon measure, a new version of the celebrated Fubini Theorem, and analysis of various statements of the Hausdorff and Hamburger moment problems.

The authors of the chapters and the editor tried to ensure the unity of style and level exposition, striving to avoid nauseating repetitions and verbosity. As it usually happens, the ideal remains intact and unreachable. The editor is the sole person to be blamed for this and other shortcomings of the book.

S. Kutateladze

References

1. Bukhvalov A. V. et al., Vector Lattices and Integral Operators [in Russian], Novosibirsk, Nauka (1992).
2. Kutateladze S. S. (ed.), Vector Lattices and Integral Operators, Dordrecht etc., Kluwer Academic Publishers (1996).
3. Kusraev A. G. and Kutateladze S. S., Nonstandard Methods of Analysis [in Russian], Novosibirsk, Nauka (1990).
4. Kusraev A. G. and Kutateladze S. S., Nonstandard Methods of Analysis, Dordrecht etc.: Kluwer Academic Publishers (1994).
5. Kusraev A. G. and Kutateladze S. S., Boolean Valued Analysis [in Russian], Novosibirsk, Sobolev Institute Press (1999).
6. Kusraev A. G. and Kutateladze S. S., Boolean Valued Analysis, Dordrecht etc., Kluwer Academic Publishers (1999).

CHAPTER 1

Nonstandard Methods and Kantorovich Spaces

BY

A. G. Kusraev and S. S. Kutateladze

S.S. Kutateladze (ed.), Nonstandard Analysis and Vector Lattices, 1–79.

It is universally recognized that the thirties of the 20th century play a special role in the development of the modern science. Outlined at the turn of the century, the tendency towards drastic reorganization of mathematics has revealed itself since these years. This led to the creation of a number of new mathematical subjects, functional analysis among the first. Nowadays we realize an exceptional place of the seventies framed sweeping changes both in volume and in essence of mathematical theories. In this period, a qualitative leap forward was registered in understanding interrelation and interdependence of mathematical topics; outstanding advances took place in working out new synthetic approaches and finding solutions to certain deep and profound problems intractable for a long time.

The processes we have indicated above are also characteristic of the theory of ordered vector spaces, one of the most actual and attractive branches of functional analysis.

This trend, stemming from the beginning of the thirties under the influence of contributions by F. Riesz, L. V. Kantorovich, H. Freudenthal, G. Birkhoff, et al., experiences a period of revival partly due to assimilation of the mathematical ideas of nonstandard models of set theory. Boolean valued interpretations, popular in connection with P. J. Cohen's final solution to the continuum hypothesis, open up new possibilities for interpreting and corroborating L. V. Kantorovich's heuristic transfer principle.

Robinsonian nonstandard analysis has in turn legitimized the resurrection and development of infinitesimal methods, substantiating G. W. Leibniz's logical dream and opening broad vistas to general monadology of vector lattices.

Brand-new nonstandard methods in Kantorovich space theory are under way. Expanding the well-known lines by N. S. Gumilëv [22, p. 309], a celebrated Russian poet of the "Silver Age," we may say that presently Kantorovich spaces "... are sloughing their skins to make room for souls to grow and mature ... ." Many of the arising lacunas remain unfilled in yet for lack of proper understanding rather than a short span of time for settling the corresponding problems. At the same time many principal and intriguing questions stand in line, waiting for comprehension and novel ideas.

This chapter presents the prerequisites for adaptation and application of the model-theoretic tools of nonstandard set theory to investigating Kantorovich spaces and classes of linear operators in them.

Sections 1.1–1.4 collect information on the formal set theories commonest to the contemporary research in functional analysis.

We start with recalling the axiomatics of the classical Zermelo–Fraenkel set theory. We then overview the Boolean valued models which stem from the works of D. Scott, R. Solovay, and P. Vopěnka. We also sketch Nelson's internal set theory and one of the most powerful and promising variants of external set theory which was recently proposed by T. Kawai and is widely used in modern infinitesimal analysis. We finish with brief exposition of the version of relative internal set theory due to E. I. Gordon and Y. Péraire.

Sections 1.5–1.8 deal with Boolean valued analysis of vector lattices. We are well aware by now that the most principal new nonstandard opportunity for Kantorovich spaces consists in formalizing the *heuristic transfer principle* by L. V. Kantorovich which claims that the members of every Dedekind complete vector lattice are generalized numbers. Boolean valued analysis demonstrates rigorously that the elements of each Kantorovich space depict reals in an appropriate nonstandard model of set theory. The formalism we reveal in this chapter belongs undoubtedly to the list of the basic and compulsory conceptions of the theory of ordered vector spaces.

Sections 1.9–1.12 treat infinitesimal constructions. A. Robinson's apology for infinitesimal has opened new possibilities in Banach space theory from scratch. The central place is occupied by the concept of *nonstandard hull* of a normed space X; i.e., the factor space of the external subspace of elements with limited norm by the *monad* of X which is the external set of members of X with infinitesimal norm. We discuss adaptation of nonstandard hulls to vector lattice theory in Section 1.9. We proceed to Section 1.10 with introducing another important construction of nonstandard analysis, the *Loeb measure.*

Sections 1.11 and 1.12 address the still-uncharted topic of combining Boolean valued and infinitesimal methods. Two approaches seem feasible theoretically: the first may consist in studying a Boolean valued model immersed into the inner universe of some external set theory. This approach is pursued in Section 1.11. The other approach consists in studying an appropriate fragment of some nonstandard set theory (for instance, in the form of ultraproduct or ultralimit) which lies inside a relevant Boolean valued universe. We take this approach in Section 1.12. It is worth emphasizing that the corresponding formalisms, in spite of their affinity, lead to the principally different constructions in Kantorovich space theory. We illustrate these particularities in technique by examining the "cyclic" topological notions of import for applied Boolean valued analysis.

Sections 1.13–1.16 deal with nonstandard analysis in operator theory. We first address positive linear operators which are listed among the central objects of the theory of ordered vector spaces. The main opportunity, offered by nonstandard methods, consists in the fact that the available formalism allows us to simplify essentially the analysis of operators and vector measures by reducing the environment to the case of functionals and scalar measures and sometimes even to ordinary numbers.

In Sections 1.13–1.16 we illustrate the general tricks of nonstandard analysis in operator theory in connection with the problems of extending and decomposing spaces and operators as well as representing homomorphisms and Maharam operators. We also distinguish a new class of cyclically compact operators. We leave some place for the problem of generating the fragments of a positive operator. The matter is that the complete description we suggest rests on successive application of nonstandard analysis in the Boolean valued and infinitesimal versions. The chapter is closed with Boolean valued analysis of one of the most important facts of the classical theory of operator equations, the Fredholm Alternative. We give its analog for a new class of equations with cyclically compact operators.

1.1. Zermelo–Fraenkel Set Theory

Zermelo–Fraenkel set theory, abbreviated to ZF, is commonly accepted as an axiomatic foundation for mathematics today. We will briefly recall some of the notions of ZF and introduce necessary notation. The details can be found in [9, 36].

1.1.1. The set-theoretic language of ZF uses the following symbols that comprise the *alphabet* of ZF: the symbols of variables $x, y, z, \dots$; the parentheses (,); the propositional connectives (= the signs of propositional algebra) $\wedge, \vee, \rightarrow, \leftrightarrow, \neg$; the quantifiers $\forall, \exists$; the equality sign $=$; and the symbol of the binary predicate of membership $\in$. Informally, the domain of variables of ZF is thought of as the world or *universe* of sets. The relation $\in (x, y)$ is written as $x \in y$ and read as "*x is an element of y*.

1.1.2. The formulas of ZF are defined by the usual recursive procedure. In other words, a *formula of* ZF is a finite text resulting from the atomic formulas such as $x = y$ and $x \in y$, where x and y are variables of ZF, with the help of reasonable arrangement of parentheses, quantifiers, and propositional connectives. So, the theorems of ZF form the least set of formulas which contains the axioms of ZF and is closed under the rules of inference (see 1.1.4 below).

1.1.3. Common mathematical abbreviations are convenient in working with ZF. Some of them follow:

$$(\forall x \in y)\, \varphi(x) := (\forall x)\, (x \in y \rightarrow \varphi(x));$$
$$(\exists x \in y)\, \varphi(x) := (\exists x)\, (x \in y \wedge \varphi(x));$$
$$\bigcup x := \{z : (\exists y \in x)\, (z \in y)\};$$
$$\bigcap x := \{z : (\forall y \in x)\, (z \in y)\};$$
$$x \subset y := (\forall z)\, (z \in x \rightarrow z \in y);$$
$$\mathscr{P}(x) := \text{"the class of subsets of } x\text{"} := \{z : z \subset x\};$$

$$\mathbf{V} := \text{“}\textit{the class of sets}\text{”} := \{x : x = x\};$$
$$\text{“}\textit{the class } A \textit{ is a set}\text{”} := A \in \mathbf{V} := (\exists x)\,(\forall y)\,(y \in A \leftrightarrow y \in x);$$
$$f : X \to Y := \text{“}f \textit{ is a function from } X \textit{ to } Y\text{”};$$
$$\mathrm{dom}(f) := \text{“}\textit{the domain of definition of } f\text{”};$$
$$\mathrm{im}(f) := \mathrm{rng}(f) := \text{“}\textit{the image of } f\text{”}.$$

1.1.4. The set theory ZF includes the conventional axioms and rules of inference of a first-order theory with equality. These axioms fix the standard ways of classical reasoning (syllogisms, the excluded middle, modus ponens, generalization, etc.). Moreover, the following six special or nonlogical axioms are accepted, written down with the standard abbreviations; cf. 1.1.3.

(1) AXIOM OF EXTENSIONALITY:
$(\forall x)\,(\forall y)\,((x \subset y \wedge y \subset x) \leftrightarrow x = y)$.

(2) AXIOM OF UNION:
$(\forall x)\,(\bigcup x \in \mathbf{V})$.

(3) AXIOM OF POWERSET:
$(\forall x)\,(\mathscr{P}(x) \in \mathbf{V})$.

(4) AXIOM SCHEMA OF REPLACEMENT:
$(\forall x)\,(\forall y)\,(\forall z)(\varphi(x, y) \wedge \varphi(x, z) \to y = z)$
$\to (\forall a)\,(\{v : (\exists u \in a)\,\varphi(u, v)\} \in \mathbf{V})$.

(5) AXIOM OF FOUNDATION:
$(\forall x)\,(x \neq \varnothing \to (\exists y \in x)\,(y \cap x = \varnothing))$.

(6) AXIOM OF INFINITY:
$(\exists \omega)\,((\varnothing \in \omega) \wedge (\forall x \in \omega)\,(x \cup \{x\} \in \omega))$.

The *theory* ZFC, *Zermelo–Fraenkel theory with choice*, results from ZF by adding the following

(7) AXIOM OF CHOICE:
$(\forall F)\,(\forall x)\,(\forall y)\,(x \neq \varnothing \wedge F : x \to \mathscr{P}(y))$
$\to ((\exists f)\,(f : x \to y) \wedge (\forall z \in x)\,f(z) \in F(z))$.

1.1.5. *Zermelo set theory* Z appears from ZFC by deleting the axiom of foundation 1.1.4 (5) and inserting instead of the axiom schema of replacement 1.1.4 (4) the following pair of its consequences:

(1) AXIOM SCHEMA OF COMPREHENSION:
$(\forall x)\,\{y \in x : \psi(y)\} \in \mathbf{V}$,
with ψ a formula of ZF.

(2) AXIOM OF PAIRING:
$(\forall x)\,(\forall y)\{x, y\} \in \mathbf{V}$.

Thus, the special axioms of Z are 1.1.4 (1–3, 6, 7) and 1.1.5 (1, 2).

So, all theories Z, ZF, and ZFC have the same language and logical axioms, differing only in the collections of their special axioms.

1.1.6. Comments.

(1) Zermelo–Fraenkel set theory slightly restricts a philistine mathematician by the axiom of foundation which, as a matter of fact, was proposed by J. von Neumann in 1925. However, this postulate gives a sound footing for the widely accepted set-theoretic view of the world of sets as the "von Neumann universe" growing up hierarchically from the empty set, the mathematical proatom.

(2) The axiomatics of Zermelo–Fraenkel set theory has never banned attempts at searching alternative set-theoretic foundations. In this regard, we refer in particular to [116].

1.2. Boolean Valued Set Theory

Here we sketch the theory of Boolean valued models of set theory in brief. More complete introductions are available elsewhere; see [6, 62, 109].

1.2.1. Let B stand for a fixed complete Boolean algebra. By a *Boolean valued interpretation* of an n-ary predicate P on a class X we mean any mapping $R : X^n \to B$ from X^n to B.

We suppose that $\mathscr{L}$ is a first-order language with the predicates $P_0, P_1, \dots, P_n$, and let $R_0, R_1, \dots, R_n$ stand for some fixed Boolean valued interpretations of these predicates on a class X. Given a formula $\varphi(u_1, \dots, u_m)$ of the language $\mathscr{L}$ and elements $x_1, \dots, x_m \in X$, we define the *truth value* $[\![\, \varphi(x_1, \dots, x_m) \,]\!] \in B$ by usual induction on the length of φ. Dealing with atomic formulas, we put

$$[\![\, P_k(x_1, \dots, x_m) \,]\!] := R_k(x_1, \dots, x_m).$$

The steps of induction use the following rules:

$$[\![\, \varphi \vee \psi \,]\!] := [\![\, \varphi \,]\!] \vee [\![\, \psi \,]\!],$$

$$[\![\, \varphi \wedge \psi \,]\!] := [\![\, \varphi \,]\!] \wedge [\![\, \psi \,]\!],$$

$$[\![\, \varphi \to \psi \,]\!] := [\![\, \varphi \,]\!] \Rightarrow [\![\, \psi \,]\!],$$

$$[\![\, \neg\varphi \,]\!] := [\![\, \varphi \,]\!]^*,$$

$$[\![\, (\forall x)\, \varphi \,]\!] := \bigwedge_{x \in X} [\![\, \varphi(x) \,]\!],$$

$$[\![(\exists x)\,\varphi]\!] := \bigvee_{x\in X} [\![\varphi(x)]\!],$$

with the symbols $\vee$, $\wedge$, $\Rightarrow$, $(\cdot)^*$, $\bigvee$, $\bigwedge$ on the right-hand sides of the equalities designating the conventional Boolean operations on B and $a \Rightarrow b := a^* \vee b$.

1.2.2. A proposition $\varphi(x_1,\dots,x_m)$, with $x_1,\dots,x_m \in X$ and $\varphi(u_1,\dots,u_m)$ a formula, is *valid* (*true*, *veritable*, etc.) in the system $\mathbb{X} := (X, R_0,\dots,R_n)$ if $[\![\varphi(x_1,\dots,x_m)]\!] = \mathbf{1}$. In this event we write $\mathbb{X} \models \varphi(x_1,\dots,x_m)$. All logically true statements are valid in $\mathbb{X}$. If a predicate P_0 symbolizes equality then we require that the B-system $\mathbb{X} := (X, =, R_1,\dots,R_n)$ satisfies the axioms of equality. If this requirement is fulfilled then all logically true statements of the first-order logic with equality, expressible in the language $\mathscr{L} := \{=, P_1,\dots,P_n\}$, are valid in the B-system $\mathbb{X}$.

1.2.3. We now consider a Boolean valued interpretation on a class X of the language $\mathscr{L} := \{=, \in\}$ of ZFC, i.e., the first-order language $\mathscr{L}$ with the two binary predicates: $=$ and $\in$. We denote the interpretations of these predicates by $[\![\cdot = \cdot]\!]$ and $[\![\cdot \in \cdot]\!]$, respectively. Thus, $[\![\cdot = \cdot]\!], [\![\cdot \in \cdot]\!] : X \times X \to B$, and

$$[\![= (x,y)]\!] = [\![x = y]\!], \quad [\![\in (x,y)]\!] = [\![x \in y]\!] \quad (x, y \in X).$$

Our nearest aim is to characterize B-systems $\mathbb{X} := (X, [\![\cdot = \cdot]\!], [\![\cdot \in \cdot]\!])$ that model ZFC so that $\mathbb{X} \models$ ZFC. The last condition amounts to the fact that all axioms of ZFC are valid in $\mathbb{X}$. So, for instance, by the rules of 1.2.1, the validity of the axiom of extensionality 1.1.4 (1) means that, for all $x, y \in X$,

$$[\![x = y]\!] = \bigwedge_{z\in X} ([\![z \in x]\!] \Leftrightarrow [\![z \in y]\!]),$$

where $a \Leftrightarrow b := (a \Rightarrow b) \wedge (b \Rightarrow a)$ for all $a, b \in B$.

1.2.4. A B-system $\mathbb{X}$ is called *separated* whenever for all $x, y \in X$ the statement $[\![x = y]\!] = \mathbf{1}$ implies $x = y$. An arbitrary B-system $\mathbb{X}$ becomes separated after taking the quotient modulo the equivalence relation $\sim := \{(x,y) \in X^2 : [\![x = y]\!] = \mathbf{1}\}$. (An equivalence class is defined with the help of the well-known Frege–Russell–Scott trick; see [62].)

A B-system $\mathbb{X}$ is said to be *isomorphic* to a B-system $\mathbb{X}' := (X', [\![\cdot = \cdot]\!]', [\![\cdot \in \cdot]\!]')$, if there is a bijection $\beta : X \to X'$ such that $[\![x = y]\!] = [\![\beta x = \beta y]\!]'$ and $[\![x \in y]\!] = [\![\beta x \in \beta y]\!]'$ for all $x, y \in X$.

1.2.5. Theorem. *There is a unique B-system $\mathbb{X}$ up to isomorphism such that*

(1) *$\mathbb{X}$ is separated* (see 1.2.4);

(2) *the axioms of equality are valid in* $\mathbb{X}$;

(3) *the axiom of extensionality* 1.1.4 (1) *and the axiom of foundation* 1.1.4 (5) *are true in* $\mathbb{X}$;

(4) *if a function* $f : \mathrm{dom}(f) \to B$ *satisfies* $\mathrm{dom}(f) \in \mathbf{V}$ *and* $\mathrm{dom}(f) \subset \mathbb{X}$, *then*

$$[\![y \in x]\!] = \bigvee_{z \in \mathrm{dom}(f)} (z) \wedge [\![z = y]\!] \quad (y \in \mathbb{X})$$

for some $x \in \mathbb{X}$;

(5) *for each* $x \in \mathbb{X}$, *there is a function* $f : \mathrm{dom}(f) \to B$ *with* $\mathrm{dom}(f) \in \mathbf{V}$, $\mathrm{dom}(f) \subset \mathbb{X}$, *such that equality holds in* (4) *for every* $y \in \mathbb{X}$.

1.2.6. A B-system enjoying 1.2.5 (1–5) is called a *Boolean valued model* of set theory and is denoted by the symbol $\mathbf{V}^{(B)} := (\mathbf{V}^{(B)}, [\![\cdot = \cdot]\!], [\![\cdot \in \cdot]\!])$. The class $\mathbf{V}^{(B)}$ is also called the *Boolean valued universe* over B. The basic properties of $\mathbf{V}^{(B)}$ are formulated as follows:

(1) Transfer Principle. *Every axiom, and hence every theorem, of* ZFC *is valid in* $\mathbf{V}^{(B)}$; *in symbols,* $\mathbf{V}^{(B)} \models \mathrm{ZFC}$.

(2) Mixing Principle. *If* $(b_\xi)_{\xi \in \Xi}$ *is a partition of unity in* B, *and* $(x_\xi)_{\xi \in \Xi}$ *is a family of elements of* $\mathbf{V}^{(B)}$, *then there is a unique element* $x \in \mathbf{V}^{(B)}$ *satisfying* $b_\xi \le [\![x = x_\xi]\!]$ *for all* $\xi \in \Xi$.

The element x is called the *mixing* of $(x_\xi)_{\xi \in \Xi}$ by $(b_\xi)_{\xi \in \Xi}$ and is denoted by $\mathrm{mix}_{\xi \in \Xi}\, b_\xi x_\xi$.

(3) Maximum Principle. *For every formula* $\varphi(u)$ *of* ZFC, *possibly with constants from* $\mathbf{V}^{(B)}$, *there is an element* $x_0 \in \mathbf{V}^{(B)}$ *satisfying*

$$[\![(\exists u)\varphi(u)]\!] = [\![\varphi(x_0)]\!].$$

It follows in particular that if $[\![(\exists ! x)\, \varphi(x)]\!] = \mathbf{1}$, then there is a unique x_0 in $\mathbf{V}^{(B)}$ satisfying $[\![\varphi(x_0)]\!] = \mathbf{1}$.

1.2.7. *There is a unique mapping* $x \mapsto x^\wedge$ *from* $\mathbf{V}$ *to* $\mathbf{V}^{(B)}$ *obeying the following conditions:*

(1) $x = y \leftrightarrow [\![x^\wedge = y^\wedge]\!] = \mathbf{1};\ x \in y \leftrightarrow [\![x^\wedge \in y^\wedge]\!] = \mathbf{1} \quad (x, y \in \mathbf{V})$,

(2) $[\![z \in y^\wedge]\!] = \bigvee_{x \in y} [\![x^\wedge = z]\!] \quad (z \in \mathbf{V}^{(B)},\ y \in \mathbf{V})$.

This mapping is called the *canonical embedding* of $\mathbf{V}$ into $\mathbf{V}^{(B)}$ and $x^\wedge$ is referred to as the *standard name* of x.

(3) ***Restricted Transfer Principle.*** *Let $\varphi(u_1, \dots, u_n)$ be some restricted formula, i.e., with quantifiers of the form $(\forall u)\,(u \in v \to \dots)$ or $(\exists u)\,(u \in v \wedge \dots)$ abbreviated to $(\forall u \in v)$ and $(\exists u \in v)$. Then for every $x_1, \dots, x_n \in \mathbf{V}$*

$$\varphi(x_1, \dots, x_n) \leftrightarrow \mathbf{V}^{(B)} \models \varphi(x_1^{\wedge}, \dots, x_n^{\wedge}).$$

1.2.8. Given an element $X \in \mathbf{V}^{(B)}$, we define its *descent* $X{\downarrow}$ as $X{\downarrow} := \{x \in \mathbf{V}^{(B)} : [\![x \in X]\!] = \mathbf{1}\}$. The descent of X is a *cyclic* set; i.e., $X{\downarrow}$ is closed under mixing. More precisely, if $(b_\xi)_{\xi \in \Xi}$ is a partition of unity in B and $(x_\xi)_{\xi \in \Xi}$ is a family of elements of $X{\downarrow}$, then the mixing $\mathrm{mix}_{\xi \in \Xi}\, b_\xi x_\xi$ lies in $X{\downarrow}$.

1.2.9. *Let F be a correspondence from X to Y inside $\mathbf{V}^{(B)}$, i.e., $X, Y, F \in \mathbf{V}^{(B)}$ and $[\![F \subset X \times Y]\!] = [\![F \neq \varnothing]\!] = \mathbf{1}$. There is a unique correspondence $F{\downarrow}$ from $X{\downarrow}$ to $Y{\downarrow}$ satisfying $F(A){\downarrow} = F{\downarrow}(A{\downarrow})$ for every set $A \subset X{\downarrow}$ inside $\mathbf{V}^{(B)}$. Furthermore, $[\![F$ is a mapping from X to $Y]\!] = \mathbf{1}$ if and only if $F{\downarrow}$ is a mapping from $X{\downarrow}$ to $Y{\downarrow}$.*

In particular, a function $f : Z^{\wedge} \to Y$ inside $\mathbf{V}^{(B)}$, where $Z \in \mathbf{V}$, defines its *descent* $f{\downarrow} : Z \to Y{\downarrow}$ by $f{\downarrow}(z) = f(z^{\wedge})$ for all $z \in Z$.

1.2.10. We suppose that $X \in \mathscr{P}(\mathbf{V}^{(B)})$. We then define a function $f : \mathrm{dom}(f) \to B$ by putting $\mathrm{dom}(f) = X$ and $\mathrm{im}(f) = \{\mathbf{1}\}$. By 1.2.5 (4) there is an element $X{\uparrow} \in \mathbf{V}^{(B)}$ satisfying

$$[\![y \in X{\uparrow}]\!] = \bigvee_{x \in X} [\![x = y]\!] \quad (y \in \mathbf{V}^{(B)}).$$

The element $X{\uparrow}$, unique by the axiom of extensionality, is called the *ascent* of X. Moreover, the following are true:

(1) $Y{\downarrow}{\uparrow} = Y \quad (Y \in \mathbf{V}^{(B)})$,

(2) $X{\uparrow}{\downarrow} = \mathrm{mix}(X) \quad (X \in \mathscr{P}(\mathbf{V}^{(B)}))$,

where $\mathrm{mix}(X)$ consists of all mixings of the form $\mathrm{mix}\, b_\xi x_\xi$, with $(x_\xi) \subset X$ and (b_ξ) a partition of unity in B.

1.2.11. *Assume that $X, Y \in \mathscr{P}(\mathbf{V}^{(B)})$ and let F be a correspondence from X to Y. The following are equivalent:*

(1) *there is a unique correspondence $F{\uparrow}$ from $X{\uparrow}$ to $Y{\uparrow}$ inside $\mathbf{V}^{(B)}$ such that $\mathrm{dom}(F{\uparrow}) = \mathrm{dom}(F){\uparrow}$ and*

$$F{\uparrow}(A{\uparrow}) = F(A){\uparrow}$$

for every subset A of $\mathrm{dom}(F)$;

(2) *the correspondence F is extensional, i.e.,*

$$y_1 \in F(x_1) \to [\![x_1 = x_2]\!] \leq \bigvee_{y_2 \in F(x_2)} [\![y_1 = y_2]\!].$$

A correspondence F is a mapping from X to Y if and only if $[\![F\uparrow : X\uparrow \to Y\uparrow]\!] = \mathbf{1}$.

In particular, a mapping $f : Z \to Y\downarrow$ generates a function $f\uparrow : Z^{\wedge} \to Y$ such that $[\![f\uparrow(x^{\wedge}) = f(x)]\!] = \mathbf{1}$ for all $x \in Z$.

1.2.12. We assume that a nonempty set X carries some *B-structure*; i.e., we assume fixed a mapping $d : X \times X \to B$ satisfying the "metric axioms":

(1) $d(x, y) = \mathbf{0} \leftrightarrow x = y$;

(2) $d(x, y) = d(y, x)$;

(3) $d(x, y) \leq d(x, z) \vee d(z, y)$.

Then there are an element $\mathscr{X} \in \mathbf{V}^{(B)}$ and an injection $\iota : X \to X' := \mathscr{X}\downarrow$ such that $d(x, y) = [\![\iota(x) \neq \iota(y)]\!]$ and every element $x' \in X'$ may be represented as $x' = \operatorname{mix} b_\xi \iota x_\xi$, with $(x_\xi) \subset X$ and (b_ξ) a partition of unity in B. This fact enables us to consider sets with B-structure as subsets of $\mathbf{V}^{(B)}$ and to handle them with means of the rules described above.

1.2.13. Comments.

(1) Boolean valued analysis (the term was coined by G. Takeuti) is a branch of functional analysis which uses Boolean valued models of set theory. Since recently this term has been treated in a broader sense implying the tools that rest on simultaneous use of two distinct Boolean valued models.

It is interesting to note that the invention of Boolean valued analysis was not connected with the theory of vector lattices. The necessary language and technique had already been available within mathematical logic by 1960. Nevertheless, the main idea was still absent for rapid progress in model theory and its applications. This idea emerged with P. J. Cohen's establishing undecidability in a rigorous mathematical sense of the classical continuum hypothesis. It was the Cohen method of forcing whose comprehension led to the invention of Boolean valued models of set theory which is attributed to the efforts by D. Scott, R. Solovay, and P. Vopěnka (see [6, 9, 36, 62, 109]).

(2) The method of forcing splits naturally into two parts, general and special. The general part contains the apparatus of Boolean valued models of set theory. The Boolean algebra B we use in construction is absolutely arbitrary here.

The special part consists in selecting a specific Boolean algebra B so as to provide the necessary (often pathological, even exotic) properties of the objects (e.g., of a Kantorovich space) sprouting from B. Either of the parts is of interest in its own right. However, the truly impressive results are obtained by combining both. Most works on Boolean valued analysis use only the general forcing. The future progress in Boolean valued analysis will surely be connected with forcing in full strength.

(3) A more detailed information on this section can be found in [6, 53, 60, 62, 109]; also see [36, 84]. Various modifications of the tools in 1.2.8–1.2.11 are widely used in research into the theory of Boolean valued models. In [52, 72] the machinery is framed as the technique of descending and ascending which suits the problems in analysis. The embedding 1.2.12 of the sets with Boolean structure into a Boolean valued universe is carried out in [52]. The motivation for such embedding is the Solovay–Tennenbaum method which was previously proposed for complete Boolean algebras [103].

1.3. Internal and External Set Theories

E. Nelson provided a convenient footing for nonstandard methods of analysis in the form of internal set theory, IST, in the end of the seventies. The formalism of this theory shortly gains popularity. The reason behind this is as follows: E. Nelson waived the preconceived notion of some especial "ideal" character of actual infinitely large and infinitely small quantities.

1.3.1. The alphabet of the formal theory IST results from supplementing the alphabet of ZFC by a sole new symbol, the symbol of the one-place predicate St expressing the property of a set to be *standard.* In other words, we admit into the feasible texts of IST the records like $\mathrm{St}(x)$ or, lengthily, "x is standard," or, finally, "x is a standard set." Thus, the semantic universe of discourse of the variables of IST is the world of Zermelo–Fraenkel set theory, the *von Neumann universe*, furnished with the possibility of distinguishing between standard and nonstandard sets.

The formulas of IST are defined by the routine procedure, but now we append to the list of atomic formulas the records $\mathrm{St}(x)$, with x a variable. Each formula of ZFC is a formula of IST; the converse is clearly false. To distinguish between formulas, we use the following terminology: We call the formulas of ZFC *internal*; whereas the epithet an *external formula* implies that this formula of IST is not a formula of ZFC. Therefore, the record "x is standard" exemplifies an external formula of IST.

This distinction between formulas of IST leads to specifying external and internal classes. If φ is an external formula of IST, then we verbalize the record $\varphi(y)$ as follows: "y is a member of the *external class* $\{x : \varphi(x)\}$." The term *internal class* implies the same as the term *class* in regard to Zermelo–Fraenkel theory. In case this does not lead to confusion, external and internal classes are referred to simply as classes. The external classes, consisting of members of some internal set, are called *external sets* or, amply, external subsets of this internal set.

It is worth observing once again that each internal class, consisting of some elements of an internal set, is again an internal set.

Alongside the abbreviations of ZFC, internal set theory uses extra conventions. We list a few:

$$x \in \mathbf{V}^{\mathrm{st}} := x \text{ is standard} := (\exists\, y)\,(\mathrm{St}(y) \wedge y = x);$$

$$(\forall^{\mathrm{st}} x)\,\varphi := (\forall\, x)\,(x \text{ is standard} \rightarrow \varphi);$$

$$(\exists^{\mathrm{st}} x)\,\varphi := (\exists\, x)\,(x \text{ is standard} \wedge \varphi);$$

$$(\forall^{\mathrm{st\,fin}} x)\,\varphi := (\forall^{\mathrm{st}} x)\,(x \text{ is finite} \rightarrow \varphi);$$

$$(\exists^{\mathrm{st\,fin}} x)\,\varphi := (\exists^{\mathrm{st}} x)\,(x \text{ is finite} \wedge \varphi);$$

$$^{\circ}x := \{y \in x : y \text{ is standard}\}.$$

The external set $^{\circ}x$ is often called the *standard core* of x.

1.3.2. The axioms of IST result from supplementing the list of axioms of ZFC with the following new axiom schemas called the *principles of nonstandard set theory*:

(1) TRANSFER PRINCIPLE:
$(\forall^{\mathrm{st}} x_1)\,(\forall^{\mathrm{st}} x_2) \dots (\forall^{\mathrm{st}} x_n)\,((\forall^{\mathrm{st}} x)\,\varphi(x, x_1, \dots, x_n)$
$\rightarrow (\forall\, x)\,\varphi(x, x_1, \dots, x_n))$
for every internal formula φ;

(2) IDEALIZATION PRINCIPLE:
$(\forall\, x_1)\,(\forall\, x_2)\,\dots(\forall\, x_n)\,((\forall^{\mathrm{st\,fin}} z)\,(\exists\, x)\,(\forall\, y \in z)\,\varphi(x, y, x_1, \dots, x_n)$
$\leftrightarrow (\exists\, x)\,(\forall^{\mathrm{st}} y)\,\varphi(x, y, x_1, \dots, x_n)),$
with φ an arbitrary internal formula;

(3) STANDARDIZATION PRINCIPLE:
$(\forall\, x_1) \dots (\forall\, x_n)$
$((\forall^{\mathrm{st}} x)\,(\exists^{\mathrm{st}} y)\,(\forall^{\mathrm{st}} z) z \in y \leftrightarrow z \in x \wedge \varphi(z, x_1, \dots, x_n))$
for every formula φ.

1.3.3. ***Powell Theorem.*** *The theory* IST *is conservative over the theory* ZFC.

This theorem means that all internal theorems of IST are theorems of Zermelo–Fraenkel theory. In other words, proving a "standard" theorem that is an internal theorem about some members of the von Neumann universe we may use the formalism of IST with the same degree of reliability as we enjoy in working within the realm of ZFC.

1.3.4. The expressive means of the axiomatic theory IST are rather effectual; however, there is a serious deficiency, that is, the lack of variables for external sets. This shortcoming prevents the possibility of dealing with such profound infinitesimal constructions as the nonstandard hull and Loeb measure.

At present, there are several versions of formal foundation for infinitesimal methods within the framework of axiomatic theories of external sets; see [5, 21, 30, 45, 46]. As regards applications, all these formalisms have practically the same power. We will give here one of the strongest versions of external set theory, the theory NST propounded by T. Kawai [45, 46].

The alphabet of NST results from enriching the alphabet of ZFC by the two constants $\mathbf{V}^S$ and $\mathbf{V}^I$. Semantically, we imagine $\mathbf{V}^S$ as the *universe of standard sets*; and $\mathbf{V}^I$, as the *universe of internal sets* (in any substantial interpretation).

It is worth observing that $\mathbf{V}^S$ and $\mathbf{V}^I$ viewed as particular external sets; i.e., $\mathbf{V}^S \in \mathbf{V}^E$ and $\mathbf{V}^I \in \mathbf{V}^E$, where $\mathbf{V}^E := \{x : x = x\}$ is the *class of external sets.* We sometimes write $\mathrm{St}(x)$ or "x is a standard set" instead of $x \in \mathbf{V}^S$. By analogy, we introduce the predicate $\mathrm{Int}(\,\cdot\,)$, expressing the property that a set is internal.

Formulas are defined as usual. Moreover, given a formula φ of ZFC we let the symbol φ^S (φ^I) stand for the *relativization* of φ to $\mathbf{V}^S$ ($\mathbf{V}^I$, respectively), i.e., the formula resultant from replacing all variables of φ with variables ranging over standard (internal) sets.

If φ is a formula of ZFC then, treating it as a formula of NST, we sometimes write φ^E and use the term an *E-formula.* By analogy we understand the concepts of *S-formula* and *I-formula.*

We use the routine abbreviations like $(\forall^{\mathrm{st}} x)\,\varphi := (\forall x \in \mathbf{V}^S)\,\varphi$; $(\exists^{\mathrm{Int}} x)\,\varphi := (\exists x \in \mathbf{V}^I)\,\varphi$; $\mathrm{fin}(x) := x$ is finite (= admits no one-to-one mapping onto a proper subset of x), etc.

1.3.5. The *special axioms* of NST split into three groups (the situation is similar with other versions of external set theory). The first group comprises the so-called *rules for constructing external sets.* The second group contains the *axioms of interplay between the universes of sets* $\mathbf{V}^S$, $\mathbf{V}^I$, and $\mathbf{V}^E$. Finally, the third group consists of the usual *postulates of nonstandard analysis*: the principles of transfer, idealization, and standardization.

1.3.6. We begin with the structure of the universe $\mathbf{V}^E$:

(1) SUPERRULE FOR INTRODUCING EXTERNAL SETS: *If φ is an axiom of* ZFC *other than the axiom of foundation then φ^E is an axiom of* NST.

Thus, the axioms of Zermelo theory Z act in NST, and also the axiom schema of replacement is valid. Moreover, we assume the following

(2) Restricted Axiom of Foundation:
$(\forall A)\,(A = \varnothing \vee A \cap \mathbf{V}^I = \varnothing) \to (\exists x \in A)\, x \cap A = \varnothing.$

In other words, regularity is postulated for external sets lacking in internal elements.

We note that $\mathbf{V}^S \in \mathbf{V}^E$. In other words, the usual *axiom of acceptability* is valid (see [62, 3.4.17]).

We recall in this connection that an external set A is *of acceptable size* (or *S-size*) if there is an external function that maps $\mathbf{V}^S$ onto A. In this case we write $A \in \mathbf{V}^{\text{a-size}}$.

1.3.7. The second group of axioms of NST contains the following

(1) Modeling Principle for Standard Sets: *the world* $\mathbf{V}^S$ *is the von Neumann universe*; i.e., for each axiom φ of ZFC the standardization φ^S is an axiom of NST;

(2) Axiom of Transitivity for Internal Sets: $(\forall x \in \mathbf{V}^I) x \subset \mathbf{V}^I$; i.e., internal sets are composed of internal elements;

(3) Axiom of Embedding: $\mathbf{V}^S \subset \mathbf{V}^I$; i.e., standard sets are internal.

1.3.8. The third group of postulates of NST consists of the following axiom schemas:

(1) Transfer Principle:
$(\forall^{\text{st}} x_1) \dots (\forall^{\text{st}} x_n) \varphi^S(x_1, \dots, x_n) \leftrightarrow \varphi^I(x_1, \dots, x_n)$
for every formula $\varphi = \varphi(x_1, \dots, x_n)$ of ZFC;

(2) Standardization Principle:
$(\forall A)\,(\exists^{\text{st}\,t})\,({}^\circ A \subset t) \to (\exists^{\text{st}\,a})\,(\forall^{\text{st}\,x})\,(x \in A \leftrightarrow x \in a),$
where ${}^\circ A := A \cap \mathbf{V}^S$ is the *standard core of* A.

The invoked set a is obviously unique. It is denoted by ${}^*A$ and called the *standardization* of A.

(3) Idealization Principle (the axiom schema of saturation):
$(\forall^{\text{Int}} x_1) \dots (\forall^{\text{Int}} x_n)\,(\forall A \in \mathbf{V}^{\text{a-size}})(((\forall z)\, z \subset A \wedge \text{fin}^E(z)$
$\to (\exists^{\text{Int}} x)\,(\forall y \in z)\,\varphi^I(x, y, x_1, \dots, x_n))$
$\to (\exists^{\text{Int}} x)\,(\forall^{\text{Int}} y \in A)\,\varphi^I(x, y, x_1, \dots, x_n))$
for every $\varphi = \varphi(x, y, x_1, \dots, x_n)$ of ZFC.

1.3.9. ***Kawai Theorem.*** *The theory* NST *is conservative over* ZFC.

1.3.10. As usual, working inside $\mathbf{V}^E$, we may construct the universe $\mathbf{V}^C$ which consists of *classical sets* (called *standard* or *ordinary* in Robinson's approach), by using the class of standard ordinals, $\mathrm{On}^{\mathrm{St}}$. Namely,

$$\mathbf{V}^C_\beta := \{x : (\exists^{\mathrm{st}} \alpha \in \beta)\, x \in \mathscr{P}(\mathbf{V}^C_\alpha)\},$$
$$\mathbf{V}^C := \bigcup_{\beta \in \mathrm{On}^{\mathrm{St}}} \mathbf{V}^C_\beta.$$

Robinson's standardization $* : \mathbf{V}^C \to \mathbf{V}^S$ appears in this situation by the recursion schema:

$$*\varnothing := \varnothing, \quad *A := {}^*\{*a : a \in A\}.$$

Robinson's standardization corroborates the *Leibniz Principle* in the form

$$(\forall x_1 \in \mathbf{V}^C) \ldots (\forall x_n \in \mathbf{V}^C)\, \varphi^C(x_1, \ldots, x_n) \leftrightarrow \varphi^S(x_1, \ldots, x_n)$$

for a formula $\varphi = \varphi(x_1, \ldots, x_n)$ of ZFC and its relativizations φ^C and φ^S to $\mathbf{V}^C$ and $\mathbf{V}^S$ respectively.

1.3.11. The world of the radical (as well as classical) stance of nonstandard analysis also admits an axiomatic description. We will describe UNST, a theory that was suggested by T. Kawai.

The variables of UNST stand for external sets. UNST contains the three constants $\mathbf{V}^C$, $\mathbf{V}^I$, and $*$. The corresponding external sets are naturally called the *classical universe, universe of internal sets,* and *Robinson's standardization.*

The special axioms of UNST resemble those of NST.

1.3.12. The structure of the universe of UNST is defined by the following postulates:

(1) Superrule for Introducing External Sets
similar to 1.3.6 (1).

(2) Restricted Axiom of Foundation (cf. 1.3.6 (2)).

1.3.13. Axioms of Interplay between the Worlds of Sets contain the following:

(1) Modeling Principle for classical sets: $\mathbf{V}^C$ *is the von Neumann universe*;

(2) Axiom of Transitivity for internal sets in the form of 1.3.7 (2);

(3) Axiom of Transitivity for classical sets: $(\forall x \in \mathbf{V}^C)\, x \subset \mathbf{V}^C$; i.e., classical sets are composed only of classical elements;

(4) Axiom of Superstructure:
External subsets of a classical set are classical;

(5) AXIOM OF ROBINSON'S STANDARDIZATION:
$*$ *is an external mapping from* $\mathbf{V}^C$ *into* $\mathbf{V}^I$.

According to 1.3.13 (5) there is obviously a unique set $\mathbf{V}^S$ consisting exactly of the standardizations of sets: $\mathbf{V}^S := *(\mathbf{V}^C)$. In UNST elements of $\mathbf{V}^S$ are called *standard sets.* In analogy with 1.3.6 (2), a set A has *classical size* or is of *c-size* whenever there is an external function from $\mathbf{V}^C$ onto A. In this case we write $A \in \mathbf{V}^{\text{c-size}}$.

1.3.14. The postulates of nonstandard analysis in UNST are as follows:

(1) TRANSFER PRINCIPLE in Leibniz's form, 1.3.10;

(2) IDEALIZATION PRINCIPLE in the form of the axiom schema of saturation for sets of classic size (cf. 1.3.8 (3)).

Finally, the *standardization* ${}^*A$ (which is a subset of an element of $\mathbf{V}^S$) in UNST of a set A presents the procedure

$$\dot{{}^*}A := *(*^{-1}(A \cap \mathbf{V}^S)).$$

The following proposition is immediate from 1.3.12.

1.3.15. *Theorem.* UNST *is conservative over* ZFC.

While working with analytical objects below we will adapt a free stance close to the neoclassic and radical credos of nonstandard analysis. In particular, the reals will be considered as a standard element of the world of internal sets, whereas the classic realization of $\mathbb{R}$ will be identified with the standard core ${}^\circ\mathbb{R}$. The symbols we use in nonstandard analysis for infinitesimals, monads, etc. agree with those in [62].

1.3.16. Comments.

(1) The axiomatic approach to nonstandard analysis has started gaining popularity after the papers of E. Nelson [87, 88] who evoked an axiomatics of internal set theory. As a result, views of the essence of infinitesimal methods have changed drastically (see [62, 76]). The most distinctive feature of changes undergone is a refusal to take a "shy" approach to infinitesimals as some bizarre monsters.

(2) The axiomatic theories of external sets were propounded by K. Hrbáček [30] and T. Kawai [45]. Our presentation follows [46]. Among the most recent works we mention [18, 95] which expose, as a matter of fact, some convenient formalisms for a "graded" theory of external sets connected with the conception of relative standardness. E. I. Gordon has suggested a nonstandard theory of classes which generalizes Gödel–Bernays theory; cf. [21].

(3) V. Kanoveĭ and M. Reeken [37] suggested the theory of bounded sets BST which differs from IST in a supplementary *boundedness axiom* $(\forall x)\,(\exists y)\,(x \in y)$

and an appropriate modification of the idealization principle (note that the idealization principle of IST is in an outright contradiction with the boundedness axiom). Clearly, BST suffices for applications; moreover, it simplifies some constructions of nonstandard analysis.

1.4. Relative Internal Set Theory

In this section we consider the theory of relative internal sets within Nelson's internal set theory.

1.4.1. The presence of infinitesimals in nonstandard analysis opens a way of constructing new concepts (and in fact of legitimizing the concepts that were refuted long ago) in order to study the classical objects of mathematical analysis. In particular, our new attractive acquisition is new mathematical concepts such as a microlimit of a finite sequence or microcontinuity of a function at a point. A number a is a *microlimit* of a sequence $a[N] := (a_1, \dots, a_N)$, with N an illimited natural number, in case for any infinitely large M less than N, we have $a_M \approx a$. A function $f : \mathrm{dom}(f) \to \mathbb{R}$ is *microcontinuous at a point* x in $\mathrm{dom}\, f$ whenever for $x' \in \mathrm{dom}\, f$ and $x' \approx x$ we have $f(x') \approx f(x)$. These definitions are justified by the following criteria:

(1) *A standard number $a \in \mathbb{R}$ is the limit of a standard sequence (a_n) if and only if a is a microlimit of $a[N]$, with N an arbitrary illimited natural number;*

(2) *A standard numerical function f is continuous at a standard point x of a standard point of the domain* $\mathrm{dom}(f)$ *of f if and only if f is microcontinuous at x.*

Here, as well as elsewhere in any equivalences of the same sort, it is essential that (a_n), a, f, x, and $\mathrm{dom}(f)$ are all standard. The question arises as whether there are some simple nonstandard criteria in the general case when we may encounter arbitrary nonstandard elements.

The most simple example reveals itself in attempts at nonstandardly defining the relation $\lim_{n\to\infty} \lim_{n\to\infty} f(x_n, y_n) = a$ even in the case of f and a standard. Indeed, (1) implies that $(\forall N \approx +\infty)(\lim_{n\to+\infty} f(x_N, y_n) \approx a)$. However, $f(x_N, \cdot)$ is in general a nonstandard function for $N \approx +\infty$ and the equivalence (1) is not applicable in this case. This example prompts us to introduce the infinitesimals that are essentially higher in order than the original $\alpha := x_N$; i.e., that remain infinitely small even on assuming α finite.

1.4.2. In the sequel, meeting the predicates "f is a function," "f is a finite set," the domain and target of f, and also the formulas "f is a function," and $\wedge(\forall x \in \mathrm{rng}(f))$ (x is finite), we use the designations $\mathrm{Fn}(f)$, $\mathrm{Fin}(f)$, $\mathrm{dom}(f)$, $\mathrm{rng}(f)$, and Ffin(f) respectively. We recall that Fin(x) means only that the cardinality of x is an element of ω, i.e. a natural number, possibly, illimited if x is nonstandard.

We call an element x *feasible,* in symbols, (Su(x)) on condition that $(\exists^{\mathrm{st}} X)\,(x \in X)$. We introduced the predicate "x is standard relative to y" by the formula

$$x \operatorname{st} y := (\exists^{\mathrm{st}} \varphi)(F \operatorname{fin}(\varphi) \wedge y \in \operatorname{dom}(\varphi) \wedge x \in \varphi(y)).$$

The two-place predicate $x \operatorname{st} y$ possesses the following properties:

(1) $x \operatorname{st} y \ \rightarrow\ \operatorname{Su}(x) \wedge \operatorname{Su}(y)$;

(2) $x \operatorname{st} y \wedge y \operatorname{st} z \ \rightarrow\ x \operatorname{st} z$;

(3) $x \operatorname{st} y \wedge \operatorname{Fin}(x) \ \rightarrow\ (\forall z \in x)\,(z \operatorname{st} y)$;

(4) $\operatorname{Su}(y) \wedge \operatorname{St}(x) \ \rightarrow\ x \operatorname{st} y$.

In the last claim St is a one-place predicate expressing the property of "standardness" in the theory of internal sets; cf. 1.3.1.

1.4.3. In analogy with 1.3.1, we abbreviate the predicates as follows:

$$(\forall^{\mathrm{st}\, y} x)\, \varphi := (\forall x)\, ((x \text{ is relative to } y) \rightarrow \varphi);$$
$$(\exists^{\mathrm{st}\, y} x)\, \varphi := (\exists x)\, ((x \text{ is relative to } y) \wedge \varphi);$$
$$(\forall^{\mathrm{st\,fin}\, y} x)\, \varphi := (\forall^{\mathrm{st}\, y} x)\, (x \text{ is finite} \rightarrow \varphi);$$
$$(\exists^{\mathrm{st\,fin}\, y} x)\, \varphi := (\exists^{\mathrm{st}\, y} x)\, (x \text{ is finite} \wedge \varphi).$$

1.4.4. *Relativized Transfer Principle.* *If φ is an internal formula with the only free variables $x, t_1, \dots, t_k$ $(k \geq 1)$, then the following holds*

$$(\forall^{\mathrm{st}\,\tau} t_1) \dots (\forall^{\mathrm{st}\,\tau} t_k)\, ((\forall^{\mathrm{st}\,\tau} x)\, \varphi(x, t_1, \dots, t_k) \ \rightarrow\ (\forall x)\, \varphi(x, t_1, \dots, t_k))$$

for an arbitrary feasible τ.

1.4.5. *Relativized Idealization Principle.* *If ϕ is an internal formula such that $\psi(x, y)$ may possess free variables other than x, y; then, given an admissible τ, the following holds*

$$(\forall^{\mathrm{st}\,\tau\,\mathrm{fin}}\ z)\, (\exists x)\, (\forall y \in z)\, \psi(x, y) \ \leftrightarrow\ (\exists x)\, (\forall^{\mathrm{st}\,\tau} y)\, \psi(x, y).$$

1.4.6. It is possible to show that the relativized standardization principle fails now. However, the already-established principles 1.4.4 and 1.4.5 suffice for solving the class of problems we have discussed in 1.4.1. We state a few results in this direction.

Let $x \in \mathbb{R}$ be an arbitrary, not necessarily standard, number. We say that x is *τ-infinitesimal* and write $x \overset{\tau}{\sim} 0$ if $(\forall^{\mathrm{st}\,\tau} y \in \mathbb{R}_+)\, |x| < y$. The following definitions are also natural: x is τ-infinite whenever $1/x$ is τ-infinitesimal; x is *τ-limited* provided that x is not τ-infinite.

1.4.7. Theorem. *If $f : \mathbb{R} \to \mathbb{R}$ and $a, b \in \mathbb{R}$ are arbitrary, not necessarily standard, elements and $\tau = (f, a, b)$, then*

$$\lim_{x\to a} f(x) = b \leftrightarrow (\forall \alpha \overset{\tau}{\sim} 0)\,(f(a+\alpha) - b \overset{\tau}{\sim} 0).$$

$\lhd$ By 1.4.2 f, a, and b are standard relative to τ. By the transfer principle 1.4.4 we now have

$$\lim_{x\to a} f(x) = b \leftrightarrow (\forall^{\mathrm{st}\,\tau}\varepsilon)(\exists^{\mathrm{st}\,\tau}\delta)(|x-a| < \delta \to |f(x) - b| < \varepsilon).$$

If $\alpha \overset{\tau}{\sim} 0$ and $x = a + \alpha$ then $(\forall^{\mathrm{st}\,\tau}\delta)\,|x - a| < \delta$; i.e., $(\forall^{\mathrm{st}\,\tau}\varepsilon)\,|f(x) - b| < \varepsilon$ and $f(x) - b \overset{\tau}{\sim} 0$.

Conversely, fix an arbitrary standard $\varepsilon\,\mathrm{st}\,\tau$ and consider the internal set M of α satisfying $|f(a+\alpha) - b| < \varepsilon$. By hypothesis M contains all τ-infinitesimals. Consider the set $M_1 := \{\delta : (0, \delta] \subset M\}$. This is also an internal set containing all τ-infinitesimals. Consequently, $\sup M_1$ cannot be τ-infinitesimal and so there is a τ-standard $\delta \in M_1$. It suffices to use the transfer principle. $\rhd$

1.4.8. Theorem. *Assume that $f : \mathbb{R}^2 \to \mathbb{R}$ and $a \in \mathbb{R}$ are standard and the limit exists $\lim_{y\to 0} f(x, y)$ for all x in some neighborhood of zero. Then*

$$\lim_{x\to 0}\lim_{y\to 0} f(x,y) = a \leftrightarrow (\forall \alpha \sim 0)\,(\forall \beta \overset{\alpha}{\sim} 0)\, f(\alpha, \beta) - a \sim 0.$$

$\lhd$ Assign $a := \lim_{x\to 0}\lim_{y\to 0} f(x, y)$ and put $g(x) := \lim_{y\to 0} f(x, y)$. Then $g(\alpha) \sim a$ for all $\alpha \sim 0$. Note that g is a standard function; consequently, $g(\alpha)\,\mathrm{st}\,\alpha$. Now, by Theorem 1.4.7 and 1.4.2 (2) the equality $g(\alpha) = \lim_{y\to 0} f(\alpha, y)$ amounts to $(\forall \beta \overset{\alpha}{\sim} 0)\,(f(\alpha,\beta) \overset{\alpha}{\sim} g(\alpha))$. By 1.4.2 (4) $f(\alpha, \beta) \overset{\alpha}{\sim} \varphi(\alpha) \to f(\alpha,\beta) \sim g(\alpha)$. However, $g(\alpha) \sim a$ and so $f(\alpha, \beta) \sim a$.

Prove the converse. To this end, it suffices to check that

$$(\forall \varepsilon > 0)\,(\exists \delta)(\forall x)(|x| < \delta \to (\exists \gamma)\,(\forall y)\,(|y| < \gamma \to |f(x,y) - a| < \varepsilon)).$$

Take an arbitrary standard ε and consider the internal set

$$M := \{\delta > 0 : (\forall x)\,(|x| < \delta \to (\exists \gamma)\,(\forall y)(|y| < \gamma \to |f(x,y) - a| < \varepsilon))\}.$$

Clearly, M contains all infinitesimals. Indeed, if $\delta \sim 0$ and $|x| < \delta$ then $x \sim 0$. If $\gamma \overset{x}{\sim} 0$ then $(\forall y)\,(|y| < \gamma \to y \overset{x}{\sim} 0)$. Therefore, $|f(x, y) - a| < \varepsilon$. It is obvious now that M contains some standard element as well. $\rhd$

1.4.9. The argument of the above two subsections is easy to translate to the case of an arbitrary topological space.

We now assume that X is a topological space, τ is a feasible element, and $X \operatorname{st} \tau$. Given an element $x \in X$ standard relative to τ, we define the *τ-monad* $\mu^{\tau}(x)$ as the intersection of τ-standard neighborhoods about x, i.e., $\mu^{\tau}(x) := \{y : (\forall^{\operatorname{st} \tau} u)\,((u \text{ is open} \wedge x \in u) \to y \in u\})$.

(1) *In these circumstances, a set $U \subset X$ standard relative to τ is open if and only if $\mu^{\tau}(x) \subset U$ for all $x \in U$ standard relative to τ.*

(2) *Let X and Y be feasible topological spaces, $f : X \to Y$, $a \in X$, and $b \in Y$. If $\tau := (X, Y, f, a, b)$ then the equivalence holds:*

$$\lim_{x \to a} f(x) = b \leftrightarrow (\forall x \in \mu^{\tau}(a))\, f(x) \in \mu^{\tau}(b).$$

1.4.10. In closing we briefly present the axiomatic theory RIST of relative internal sets. The language of this theory results from the language of Zermelo–Fraenkel set theory by supplementing a sole two-place predicate st. As before, we read the expression $x \operatorname{st} y$ as "x is standard relative to y." A formula of RIST is internal if it contains no occurrences of the predicate st. Like in 1.4.3 we define the external quantifiers $\forall^{\operatorname{st} \alpha}$, $\exists^{\operatorname{st} \alpha}$, $\forall^{\operatorname{st\,fin} \alpha}$, and $\exists^{\operatorname{st\,fin} \alpha}$.

The axioms of RIST contain all axioms of Zermelo–Fraenkel theory. Moreover, the predicate st obeys the following three axioms:

(1) $(\forall x)\,(x \operatorname{st} x)$;

(2) $(\forall x)\,(\forall y)\,(x \operatorname{st} y \vee y \operatorname{st} x)$;

(3) $(\forall x)\,(\forall y)\,(\forall z)\,(x \operatorname{st} y \wedge y \operatorname{st} z \to x \operatorname{st} y)$.

In addition, the theory RIST, like IST, includes three new axiom schemas. The axiom schemas of transfer and idealization are the same as in 1.4.4 and 1.4.5, while we must restrict the class of formulas in the axiom schema of standardization in accord with Remark 1.4.6.

1.4.11. ***Axiom Schema of Transfer.*** *If $\varphi(x, t_1, \dots, t_k)$ is an internal formula with free variables $x, t_1, \dots, t_k$ and τ a fixed set, then*

$$(\forall^{\operatorname{st} \tau} t_1) \dots (\forall^{\operatorname{st} \tau} t_k)\,((\forall^{\operatorname{st} \tau} x)\,\varphi(x, t_1, \dots, t_k) \to (\forall x)\,\varphi(x, t_1, \dots, t_k)).$$

1.4.12. ***Axiom Schemata of Idealization.*** Let $\varphi(x_1, \dots, x_k, y)$ be an internal formula with free variables $x_1, \dots, x_k, y$ and possibly other variables. Assume that $\tau_1, \dots, \tau_k$ are fixed sets and β is not standard relative to $(\tau_1, \dots, \tau_k)$. Then the following hold:

(1) RESTRICTED IDEALIZATION PRINCIPLE:
$(\forall^{\text{st}\,\tau_1\,\text{fin}} z_1)\dots(\forall^{\text{st}\,\tau_k\,\text{fin}} z_k)$
$(\exists^{\text{st}\,\beta} y)\,(\forall x_1 \in z_1)\dots(\forall x_k \in z_k)\,\psi(x_1,\dots,x_k,y)$
$\leftrightarrow\ (\exists^{\text{st}\,\beta} y)\,(\forall^{\text{st}\,\tau_1} x_1)\dots(\forall^{\text{st}\,\tau_k} x_k)\,\psi(x_1,\dots,x_k,y).$

(2) UNRESTRICTED IDEALIZATION PRINCIPLE:
$(\forall^{\text{st}\,\tau\,\text{fin}} z_1)\dots(\forall^{\text{st}\,\tau\,\text{fin}} z_k)$
$(\exists y)\,(\forall x_1 \in z_1)\dots(\forall x_k \in z_k)\,\psi(x_1,\dots,x_k,y)$
$\leftrightarrow\ (\exists y)\,(\forall^{\text{st}\,\tau_1} x_1)\dots(\forall^{\text{st}\,\tau_k} x_k)\,\psi(x_1,\dots,x_k,y).$

1.4.13. To formulate the axiom schema of standardization, we introduce the class $\mathscr{F}_\tau$ of *τ-external formulas*, with τ a fixed set. If $\mathscr{F}$ is a class of formulas of the theory RIST then $\mathscr{F}_\tau$ is defined as the least subclass of $\mathscr{F}$ meeting the conditions:

(1) Each atomic formula $x \in y$, with x and y variables or constants, belongs to $\mathscr{F}_\tau$;

(2) If some formulas φ and ψ belong to $\mathscr{F}_\tau$ then the formulas $\neg\varphi$ and $\varphi \to \psi$ belong to $\mathscr{F}_\tau$ too;

(3) If a formula $\varphi(x,y)$ belongs to $\mathscr{F}_\tau$ then the formula $(\exists y)\,\varphi(x,y)$ belongs to $\mathscr{F}_\tau$ as well;

(4) If a formula $\varphi(x,y)$ belongs to $\mathscr{F}_\tau$ and β is a set such that the set τ is standard relative to β, then the formula $(\exists^{\text{st}\,\beta} y)\,\varphi(x,y)$ belongs to $\mathscr{F}_\tau$.

1.4.14. *Axiom Schema of Standardization.* *If τ is a fixed set and φ is some τ-external formula then*

$$(\forall^{\text{st}\,\tau} y)\,(\exists^{\text{st}\,\tau} z)\,(\forall^{\text{st}\,\tau} t)\,(t \in z \leftrightarrow (t \in y \wedge \varphi(t))).$$

1.4.15. *Theorem.* *The theory* RIST *is conservative over* ZFC.

1.4.16. Comments.

(1) The content of Sections 1.4.2–1.4.9 is taken from the E. I. Gordon's article [18]; also see [21]. This article demonstrates that there are an infinitely large natural N and some $x \in [0, 1]$ for which no number N-infinitely close to x is N-standard. Since existence of the standard part of a real number is a consequence of the standardization principle; therefore, the relativized standardization principle is not valid. In particular, we may conclude that the standardization principle of IST is not a consequence of the other axioms of this theory (for details, see [18, 21]).

(2) The axiomatic theory RIST, as presented in 1.4.10–1.4.14, was propounded by Y. Péraire [96]. The same article contains Theorem 1.4.15. Prior to this,

Y. Péraire carried out an extension of IST (consistent with ZFC) by appending a sequence of the undefined predicates $\mathrm{St}_p(x)$ (read: x is standard to the power of $1/p$); cf. [94]. The articles [95] and [97] contain other results in this direction.

1.5. Kantorovich Spaces

The theory of vector lattices resides in many excellent monographs; see, for instance, [2, 4, 40, 41, 82, 98, 99, 117]. Vector lattices are also called *Riesz spaces.* Here we will briefly introduce Dedekind complete vector lattices.

1.5.1. Let $\mathbb{F}$ be a linearly ordered field. An *ordered vector space* over $\mathbb{F}$ is a pair $(E, \leq)$, with E a vector space over $\mathbb{F}$ and $\leq$ a *vector order* on E, i.e. an order relation on E compatible with the vector structure. The last expression means that we may add inequalities in E and multiply them by positive elements of $\mathbb{F}$. Furnishing a vector space E over $\mathbb{F}$ with a vector order amounts to distinguishing a subset $E_+ \subset E$, the *positive cone* of E, such that $E_+ + E_+ \subset E_+$; $\lambda E_+ \subset E_+$ $(0 \leq \lambda \in \mathbb{F})$; and $E_+ \cap E_+ = \{0\}$. In this event the order $\leq$ and cone E_+ are interrelated as follows: $x \leq y \leftrightarrow y - x \in E_+ \quad (x, y \in E)$.

An ordered vector space that is a lattice is called a *vector lattice.* Given members x and y of a vector lattice E, we write $x \vee y := \sup\{x, y\}$, $x \wedge y := \inf\{x, y\}$, $|x| := \sup\{x, -x\}$, $x^+ := \sup\{x, 0\}$, and $x^- := (-x)^+$.

A *Kantorovich space* or, briefly, a *K-space* is a vector lattice whose every order bounded nonempty subset has a supremum and an infimum; i.e., a Dedekind complete vector lattice. If each countable order bounded nonempty subset of a vector lattice has a supremum and an infimum then this vector lattice is called a *K_σ-space.* Let E designate a Kantorovich space in the sequel.

Elements $x, y \in E$ are *disjoint,* in symbols $x \perp y$, provided that $|x| \wedge |y| = 0$. The set

$$M^\perp := \{x \in E : (\forall y \in M)\, x \perp y\},$$

with $M \subset E$, is the *disjoint complement* of M.

We note several simple properties of disjointness:

(1) $M \subset N \rightarrow N^\perp \subset M^\perp$;

(2) $M \subset M^{\perp\perp}$;

(3) $M^\perp = M^{\perp\perp\perp}$;

(4) $\left(\bigcup_\alpha M_\alpha\right)^\perp = \bigcap_\alpha M_\alpha^\perp$.

A *band* (or *component* in the Russian literature) of E is a set of the form $M^\perp$, with $M \subset E$ and $M \neq \varnothing$. The collection $\mathfrak{B}(E)$ of bands of E, ordered by inclusion, is a complete Boolean algebra with the Boolean operations as follows:

$$L \wedge K = L \cap K, \quad L \vee K = (L \cup K)^{\perp\perp}, \quad L^* = L^\perp \quad (L, K \in \mathfrak{B}(E)).$$

This algebra $\mathfrak{B}(E)$ is called the *base* of E.

1.5.2. Every band K of a Kantorovich space E gives rise to the decomposition $E = K \oplus K^{\perp}$. This uniquely defines the projection operator $[K]$ to K along $K^{\perp}$ which is called the *band projection* to K (or, simply, projection to K if the context excludes any possibility of confusion). In this event we have the inequalities $0 \le [K]x \le x$ for all $0 \le x \in E$. Conversely, if a linear projection π in E satisfies the inequalities $0 \le \pi x \le x$ for all $0 \le x \in E$ then $K := \pi(E)$ is a band and π serves as the band projection to K. The set of band projections $\mathfrak{P}(E)$ is ordered by putting $\rho \le \pi \leftrightarrow \operatorname{im}(\rho) \subset \operatorname{im}(\pi)$. The following equivalent definition is worth bearing in mind: $\rho \le \pi \leftrightarrow \rho\pi = \pi\rho = \rho$. The ordered set $\mathfrak{P}(E)$ is a complete Boolean algebra with the operations

$$\pi \wedge \rho = \pi\rho = \rho\pi, \quad \pi \vee \rho = \pi + \rho - \pi\rho, \quad \pi^* = I_E - \pi \quad (\pi, \rho \in \mathfrak{P}(E)).$$

Let $\mathbf{1}$ be a (*weak*) *order unity* in E; i.e., $\{\mathbf{1}\}^{\perp\perp} = E$. An element $e \in E$ is a *unit element* or a *fragment* of $\mathbf{1}$ provided that $e \wedge (\mathbf{1} - e) = 0$. The set $\mathfrak{E}(E) := \mathfrak{E}(\mathbf{1})$ of unit elements is endowed with the order induced from E. The ordered set $\mathfrak{E}(E)$ is a complete Boolean algebra whose Boolean complementation takes the form $e^* := \mathbf{1} - e$ for $e \in \mathfrak{E}(\mathbf{1})$.

1.5.3. ***Theorem.*** *The mapping $K \mapsto [K]$ is an isomorphism between the Boolean algebras $\mathfrak{B}(E)$ and $\mathfrak{P}(E)$. If E has an order unity then the mappings $\pi \mapsto \pi\mathbf{1}$ from $\mathfrak{P}(E)$ to $\mathfrak{E}(E)$ and $e \mapsto \{e\}^{\perp\perp}$ from $\mathfrak{E}(E)$ to $\mathfrak{B}(E)$ are also Boolean isomorphisms.*

1.5.4. A Kantorovich space E is called *universally complete* or *extended* in the Russian terminology if every nonempty set of disjoint elements of E has a supremum. We will list the most important examples of universally complete Kantorovich spaces. For the sake of brevity we restrict exposition to the case of real scalars, except for the example (4).

(1) The space $M(\Omega, \Sigma, \mu) := L^0(\Omega, \Sigma, \mu)$ of cosets of measurable functions, where (Ω, Σ, μ) is a measure space, and μ is σ-finite (or, more generally, μ possesses the direct sum property; cf. [40]). The base of the Kantorovich space $M(\Omega, \Sigma, \mu)$ is isomorphic to the Boolean algebra $\Sigma/\mu^{-1}(0)$ of measurable sets modulo negligible sets.

(2) The space $C_\infty(Q)$ of continuous functions defined on an extremally disconnected compact space Q with values in the extended real line and taking the values $\pm\infty$ only on a rare (= nowhere dense) set [2, 41]. The base of this Kantorovich space is isomorphic to the Boolean algebra of clopen (= closed and open) subsets of Q.

(3) The space $\operatorname{Bor}(Q)$ of the cosets of Borel functions defined on a topological space Q. Two functions are *equivalent* if they agree on the complement of a meager set. The base of the Kantorovich space Bor(Q) is isomorphic to the Boolean algebra of Borel subsets of Q modulo meager sets.

(4) The space $\overline{\mathfrak{A}}$ of hermitian (not necessarily bounded) operators in a Hilbert space which are adjoint to a commutative von Neumann algebra $\mathfrak{A}$ (see [115]). The base of the Kantorovich space $\overline{\mathfrak{A}}$ is isomorphic to the Boolean algebra of projections in $\mathfrak{A}$.

1.5.5. Let E and F be vector lattices. An operator $T : E \to F$ is *positive* provided that $Tx \geq 0$ for every $0 \leq x \in E$; an operator $T : E \to F$ is *regular* provided that $T = T_1 - T_2$, with T_1 and T_2 positive operators.

An operator T is *order bounded* or *o-bounded* if $T(M)$ is an order bounded set in F for every order bounded $M \subset E$. If F is a Kantorovich space then the sets of regular and order bounded operators coincide. Moreover, we have

1.5.6. *Riesz–Kantorovich Theorem.* *If E is a vector lattice and F is a Kantorovich space then the space $L^{\sim}(E, F)$ of regular operators from E into F is also a Kantorovich space.*

1.5.7. We say that a net $(x_\alpha)_{\alpha \in \mathrm{A}}$ in E converges in order (o-converges) to $x \in E$ if there is a decreasing net $(e_\beta)_{\beta \in B}$ in E such that $\inf_{\beta \in B} e_\beta = 0$ and for each $\beta \in B$ there exists $\alpha_0 \in \mathrm{A}$ with $|x_\alpha - x| \leq e_\beta$ $(\alpha \geq \alpha_0)$. An operator $T : E \to F$ is *order continuous* or *o-continuous* (*sequentially o-continuous* or *order σ-continuous*) provided that $Tx_\alpha \xrightarrow{(o)} 0$ in F for every net (x_α) order convergent to zero in E ($Tx_n \xrightarrow{(o)} 0$ in F for every sequence (x_n) order convergent to zero in E, respectively). The sets of order continuous and sequentially order continuous operators from E to F are denoted by $L_n^{\sim}(E, F)$ and $L_\sigma^{\sim}(E, F)$.

Theorem. *Let E and F be vector lattices, with F Dedekind complete. Then the sets $L_n^{\sim}(E, F)$ and $L_\sigma^{\sim}(E, F)$ are bands in $L^{\sim}(E, F)$.*

1.5.8. A *Kantorovich–Pinsker space* is a Kantorovich space having an order dense ideal with point separating set of order continuous functionals (or, which is the same, whose base admits an essentially positive locally finite completely additive measure).

Theorem. *If a measure space $(\Omega, \mathscr{A}, \mu)$ possesses the direct sum property then $L^0(\Omega, \mathscr{A}, \mu)$ is a Kantorovich–Pinsker space. Conversely, each Kantorovich–Pinsker space is linearly and order isomorphic to an order dense ideal of $L^0(\Omega)$ for some measure space (Ω, Σ, μ) with the direct sum property.*

We note in addition that if some order unity $\mathbf{1}$ is fixed in E then there is a unique isomorphism claimed by the above theorem which sends $\mathbf{1}$ to the coset of the identically one function on Ω. This E is universally complete if and only if its image under the above isomorphism coincides with $L^0(\Omega)$.

1.5.9. Comments.

(1) The invention of the theory of vector lattices is customarily attributed to research by G. Birkhoff, L. V. Kantorovich, M. G. Kreĭn, H. Nakano, F. Riesz, H. Freudenthal, et al. Nowadays the theory and applications of vector lattices form a vast area of mathematics. It is painstakingly charted in the monographs [31, 40, 41, 82, 98, 99, 115, 117].

For the prerequisites of Boolean algebras, see [23, 100, 111].

(2) The class of Dedekind complete vector lattices, i.e. of Kantorovich spaces, was introduced by L. V. Kantorovich in his first fundamental paper [38]. Therein he also suggested the *heuristic transfer principle* for Kantorovich spaces, claiming that the members of each Kantorovich space serve as generalized numbers.

The further research of the author himself and his disciples corroborated this principle. As a matter of fact, the Kantorovich heuristic principle became one of the key ideas leading to a deeper and more elegant theory of Kantorovich space abundant in versatile applications.

(3) Even at the first stage of the theory some attempts were made at formalizing the Kantorovich heuristic principle. This resulted in the so-called theorems on preservation of relations (sometimes a less exact term "conservation" is employed). These theorems assert that if a formal expression with finitely many function relations is proved for the reals then a similar fact holds for elements of every Kantorovich space (see [41, 115]). Unfortunately, there was no satisfactory explanation for the intrinsic mechanism controlling the phenomenon of preservation of relations. Limits to applying the above assertions and the general background for affinity and parallelism between them and their analogs in the classical function theory had not been sufficiently clarified. The depth and universality of Kantorovich's heuristic principle were fully explicated only within Boolean valued analysis (see 1.6, 1.7 and [8, 53, 60]).

(4) The definitions of order continuous and sequentially order continuous operators, as well as Theorem 1.5.7, belong to T. Ogasawara.

1.6. Reals Inside Boolean Valued Models

Boolean valued analysis began with representing the "genuine" reals in a Boolean valued model. Such representation happened to be a universally complete Kantorovich space. On varying a Boolean algebra B (the algebra of measurable sets, or regular open sets, or projections in a Hilbert space) and generating the Boolean valued model $\mathbf{V}^{(B)}$, we come to different universally complete Kantorovich spaces (the space of measurable functions, or semicontinuous functions, or selfadjoint operators). Thus opens a marvelous vista for transferring knowledge about numbers to other classical objects of modern analysis.

1.6.1. By a field of reals we mean an algebraic system that satisfies the axioms of an Archimedean ordered field (with distinct zero and unity) and the axiom of completeness. We recall two well-known propositions:

(1) *There is a unique field of reals* $\mathbb{R}$ *up to isomorphism.*

This fact allows us to speak of the *reals*, a particular instance of a field of reals "specified once and forever."

(2) *If* $\mathbb{P}$ *is an Archimedean ordered field then there exists an isomorphic embedding* h *of* $\mathbb{P}$ *into* $\mathbb{R}$ *such that the image* $h(\mathbb{P})$ *is a subfield of* $\mathbb{R}$ *including the subfield of rationals. In particular,* $h(\mathbb{P})$ *is dense in* $\mathbb{R}$.

1.6.2. On successively applying the transfer principle and the maximum principle to 1.6.1 (1), we find an element $\mathscr{R} \in \mathbf{V}^{(B)}$ such that $[\![\mathscr{R}$ is the reals$]\!] = \mathbf{1}$.

Moreover, for every $\mathscr{R}' \in \mathbf{V}^{(B)}$, satisfying the condition $[\![\mathscr{R}'$ is a field of real numbers$]\!] = \mathbf{1}$, the equality $[\![$the ordered fields $\mathscr{R}$ and $\mathscr{R}'$ are isomorphic$]\!] = \mathbf{1}$ also holds.

In other words, in the model $\mathbf{V}^{(B)}$ there is located a unique field of real numbers $\mathscr{R}$ up to isomorphism. We call $\mathscr{R}$ the *reals inside* $\mathbf{V}^{(B)}$.

1.6.3. We note also that the formula $\varphi(\mathbb{R})$, recording the axioms of an Archimedean ordered field, is restricted. So, $[\![\varphi(\mathbb{R}^\wedge)]\!] = \mathbf{1}$, i.e. $[\![\mathbb{R}^\wedge$ is an Archimedean ordered field$]\!] = \mathbf{1}$.

"Pulling" the statement 1.6.1 (2) through the transfer principle, we obtain the fact that $[\![\mathbb{R}^\wedge$ is isomorphic to a dense subfield of $\mathscr{R}]\!] = \mathbf{1}$. On these grounds we presume that $\mathscr{R}$ is the reals while $\mathbb{R}^\wedge$ is a dense subfield of the reals inside $\mathbf{V}^{(B)}$.

We now turn to the descent $\mathscr{R}{\downarrow}$ of the algebraic system $\mathscr{R}$. In other words, we look at the descent of the carrier set of $\mathscr{R}$ equipped with the descended operations and order. For the sake of simplicity, the operations and order in $\mathscr{R}$ and $\mathscr{R}{\downarrow}$ are always denoted by the same symbols $+, \cdot, \leq$.

1.6.4. Gordon's Theorem. *Let* $\mathscr{R}$ *be the reals inside* $\mathbf{V}^{(B)}$. *Then* $\mathscr{R}{\downarrow}$, *with the descended operations and order, is a universally complete Kantorovich space with unity* $\mathbf{1} := 1^\wedge$. *Furthermore, there is an isomorphism* χ *of the Boolean algebra* B *onto the base* $\mathfrak{P}(\mathscr{R}{\downarrow})$ *such that the equivalences*

$$\chi(b)x = \chi(b)y \leftrightarrow b \leq [\![x = y]\!],$$
$$\chi(b)x \leq \chi(b)y \leftrightarrow b \leq [\![x \leq y]\!]$$

hold for all $x, y \in \mathscr{R}$ *and* $b \in B$.

1.6.5. The universally complete Kantorovich space $\mathscr{R}{\downarrow}$ is at the same time a faithful f-algebra with ring unity $\mathbf{1} := 1^{\wedge}$, and for every $b \in B$ the projection $\chi(b)$ acts as multiplication by the unit element $\chi(b)\mathbf{1}$. It is clear therefore that the mapping $b \mapsto \chi(b)\mathbf{1}$ $(b \in B)$ is a Boolean isomorphism of B and the algebra of unit elements $\mathfrak{E}(\mathscr{R}{\downarrow})$. This isomorphism is denoted by the same letter χ.

1.6.6. We recall that if E is a Kantorovich space with unity and $x \in E$ then the projection of the unity to the band $\{x\}^{\perp\perp}$ is called the *trace* of x and is denoted by the symbol e_x. Given a real λ, denote by e_λ^x the trace of the positive part of $\lambda\mathbf{1} - x$, i.e., $e_\lambda^x := e_{(\lambda\mathbf{1}-x)^+}$. The mapping $\lambda \mapsto e_\lambda^x$ $(\lambda \in \mathbb{R})$ is called the *spectral function* or *characteristic* of x.

For every element $x \in \mathscr{R}{\downarrow}$ the following holds:

$$e_x = \chi([\![x \neq 0]\!]), \quad e_\lambda^x = \chi([\![x < \lambda^{\wedge}]\!]) \quad (\lambda \in \mathbb{R}).$$

The next result states that every Archimedean vector lattice is realizable as a sublattice of $\mathscr{R}$ in a suitable Boolean valued model.

1.6.7. Theorem. *Let E be an Archimedean vector lattice, with $\jmath$ an isomorphism of B onto the base $\mathfrak{B}(E)$, and let $\mathscr{R}$ be the reals inside $\mathbf{V}^{(B)}$. There is an element $\mathscr{E} \in \mathbf{V}^{(B)}$ satisfying the following:*

(1) *$\mathbf{V}^{(B)} \models \mathscr{E}$ is a vector sublattice of $\mathscr{R}$ considered as a vector lattice over $\mathbb{R}^{\wedge}$;*

(2) *$E' := \mathscr{E}{\downarrow}$ is a vector sublattice of $\mathscr{R}{\downarrow}$; moreover, E' is invariant under every projection $\chi(b)$ $(b \in B)$ and each set of disjoint elements of E' has a supremum;*

(3) *there is an o-continuous lattice isomorphism $\iota : E \to E'$ such that $\iota(E)$ is a coinitial sublattice of $\mathscr{R}{\downarrow}$, i.e., $(\forall 0 < x \in \mathscr{R}{\downarrow})(\exists y \in E)(0 < \iota(y) \leq x)$;*

(4) *for every $b \in B$ the band projection to the band generated in $\mathscr{R}{\downarrow}$ by the set $\iota(\jmath(b))$ coincides with $\chi(b)$.*

1.6.8. The element $\mathscr{E} \in \mathbf{V}^{(B)}$ in Theorem 1.6.7 is called a *Boolean valued realization* of E. Thus, Boolean valued realizations of Archimedean vector lattices are vector sublattices of the reals $\mathscr{R}$ inside $\mathbf{V}^{(B)}$ viewed as a vector lattice over $\mathbb{R}^{\wedge}$.

We now state a few corollaries to 1.6.4 and 1.6.7, keeping the previous notations.

(1) *If E is a Kantorovich space then $\mathscr{E} = \mathscr{R}$, $E' = \mathscr{R}{\downarrow}$ and $\iota(E)$ is an order dense ideal of the Kantorovich space $\mathscr{R}{\downarrow}$. Furthermore, $\iota^{-1} \circ \chi(b) \circ \iota$ is the band projection to $\jmath(b)$ for every $b \in B$.*

(2) *The image $\iota(E)$ coincides with the whole of $\mathscr{R}\!\downarrow$ if and only if E is a universally complete Kantorovich space.*

(3) *Universally complete Kantorovich spaces are isomorphic if and only if their bases are isomorphic.*

(4) *Let E be a universally complete Kantorovich space with unity* **1**. *Then in E admits a unique multiplication such that E becomes a faithful f-algebra with* **1** *as a ring unity.*

1.6.9. Boolean valued realization of Archimedean vector lattices is not the only way leading to subsystems of the reals $\mathscr{R}$; cf. 1.6.7. By way of example, we formulate several propositions from [55].

Theorem. *The following hold:*

(1) *Every Boolean valued realization of an Archimedean lattice ordered group is a subgroup of the additive group of the reals $\mathscr{R}$.*

(2) *An Archimedean f-ring splits into the sum of complementary bands: one, a group with zero multiplication realizable as in* (1) *and the other realizable as a subring of $\mathscr{R}$.*

(3) *An Archimedean f-algebra splits into the sum of complementary bands: one, a vector lattice with zero multiplication realizable as in* 1.6.7, *and the other realizable as a subring and sublattice of $\mathscr{R}$, the latter considered as an f-algebra over $\mathbb{R}^\wedge$.*

1.6.10. A *complex vector lattice* is the *complexification* $E \oplus iE$ of a real vector lattice E, with i the *imaginary unity*. It is often required additionally that each element $z \in E \oplus iE$ has the *modulus*

$$|z| := \sup\{Re(e^{i\theta}z) : 0 \leq \theta \leq \pi\}.$$

In the case of Kantorovich space, this requirement is excessive so that a *complex Kantorovich space* is the complexification of a real Kantorovich space. Speaking about order properties of a complex vector lattice $E \oplus iE$, we always bear in mind its real part E. The concepts of sublattice, ideal, band, etc. are naturally abstracted to the case of a complex vector lattice by complexification.

1.6.11. Comments.

(1) The Boolean valued status of the notion of Kantorovich space was first demonstrated by Gordon's Theorem 1.6.4 in [14]. This fact can be reformulated as follows: A universally complete Kantorovich space serves as interpretation of the reals in a suitable Boolean valued model.

Furthermore, every theorem (within ZFC) about real numbers has a full analog for the corresponding Kantorovich space. Translation of one theorem into the other is fulfilled by precisely-defined procedures: ascent, descent, canonical embedding, etc., i.e., by algorithm, as a matter of fact.

Thus, Kantorovich's motto: "Elements of a Kantorovich space are generalized numbers" acquires a rigorous mathematical formulation within Boolean valued analysis. On the other hand, the heuristic transfer principle which played an auxiliary role in many investigations of the pre-Boolean valued theory of Kantorovich spaces becomes a powerful and precise method of research in Boolean valued analysis.

(2) Assuming in 1.6.4 that B is the σ-algebra of measurable sets modulo sets of measure zero, we see that $\mathscr{R}{\downarrow}$ is isomorphic to the universally complete Kantorovich space of measurable functions $M(\Omega, \Sigma, \mu)$. This fact (in regard to Lebesgue measure for an interval) was known as far back as Scott and Solovay's article (see [102]). If B is a complete Boolean algebra of projections in a Hilbert space then $\mathscr{R}{\downarrow}$ is isomorphic to the space of selfadjoint operators whose spectral function acts to B.

These two particular cases of Gordon's Theorem were intensively and fruitfully exploited by G. Takeuti; see [106] and the references in [62]. T. Jech [33, 34] also considered the object $\mathscr{R}{\downarrow}$ for general Boolean algebras and rediscovered Gordon's Theorem. The notable distinction is the fact that in [33] a (complex) universally complete Kantorovich space with unity is defined by another system of axioms and is called a complete Stone algebra.

(3) The realization theorem 1.6.7 was obtained by A. G. Kusraev [55]. A close result in somewhat different terms appeared in the paper [35] which develops the Boolean valued interpretation of the theory of linearly ordered sets. Corollaries 1.6.8 (3, 4) are well known; see [41, 115].

The notion of a universal completion of a Kantorovich space was introduced by A. G. Pinsker. He also proved the existence of a unique universal completion up to an isomorphism for an arbitrary Kantorovich space. The existence of a Dedekind completion of an Archimedean vector lattice was established by A. I. Yudin. The corresponding references are in [41, 115]. All these facts can be easily derived from 1.6.4 and 1.6.7 (see [60] for more details).

(4) As was already mentioned in 1.6.11 (1), the initial attempts at formalizing the heuristic Kantorovich principle led to the theorems on preservation of relations (see [41, 115]). The modern forms of these theorems are presented in [16, 34]; also see [62].

1.7. Functional Calculus in Kantorovich Spaces

The most important structural properties of vector lattices such as representability by function spaces, the spectral theorem, the functional calculus, etc., replicate analogous properties of the reals inside a suitable Boolean valued model. We will briefly outline the Boolean valued approach to the functional calculus in Kantorovich spaces.

1.7.1. We need below the notion of integral with respect to a spectral measure.

Let (Ω, Σ) stand for a *measurable space*; i.e., Ω is a nonempty set, and Σ is a fixed σ-algebra of subsets of Ω. A mapping $\mu : \Sigma \to B$ is said to be a *spectral measure* if $\mu(\Omega \setminus A) = \mathbf{1} - \mu(A)$ and

$$\mu\Big(\bigcup_{n=1}^{\infty} A_n\Big) = \bigvee_{n=1}^{\infty} \mu(A_n)$$

for every sequence (A_n) of elements of the σ-algebra Σ.

Let $B := \mathfrak{C}(E)$ be the Boolean algebra of unit elements of a Kantorovich space E with unity $\mathbf{1}$. Consider a measurable function $f : \Omega \to \mathbb{R}$. Given an arbitrary partition of the real line $\beta := (\lambda_k)_{k\in\mathbb{Z}}$, $\lambda_k < \lambda_{k+1}$ $(k \in \mathbb{Z})$, $\lim_{n\to\pm\infty} \lambda_n = \pm\infty$, we denote by A_k the inverse image $f^{-1}([\lambda_k, \lambda_{k+1}))$ and arrange the integral sums

$$\underline{\sigma}(f, \beta) := \sum_{-\infty}^{\infty} \lambda_k \mu(A_k), \quad \overline{\sigma}(f, \beta) := \sum_{-\infty}^{\infty} \lambda_{k+1} \mu(A_k)$$

where summation is performed in E.

If there is an element $x \in E$ such that $\sup\{\underline{\sigma}(f, \beta)\} = x = \inf\{\overline{\sigma}(f, \beta)\}$, where the supremum and infimum are taken over all partitions $\beta := (\lambda_k)$ of the real line; then we call f *integrable with respect to the spectral measure* μ, say that the *spectral integral* $I_\mu(f)$ exists, and write

$$I_\mu(f) := \int_\Omega f\, d\mu := \int_\Omega f(t)\, d\mu(t) := x.$$

1.7.2. Theorem. *Put $E := \mathscr{R}{\downarrow}$ and let μ be a spectral measure with values in $B := \mathfrak{C}(E)$. Then for every measurable function f, the integral $I_\mu(f)$ is the unique element of the Kantorovich space E satisfying the following condition*

$$[\![I_\mu(f) < \lambda^\wedge]\!] = \mu(\{f < \lambda\}) \quad (\lambda \in \mathbb{R}),$$

where $\{f < \lambda\} := \{t \in \Omega : f(t) < \lambda\}$.

It is clear from this theorem that if the integral $I_\mu(f) \in E$ exists then the mapping $\lambda \mapsto \mu(\{f < \lambda\})$ coincides with the spectral function of $I_\mu(f)$.

In particular, if E is universally complete then $I_\mu(f)$ exists for every measurable function f. Moreover, on using elementary properties of the reals $\mathscr{R}$, the next result follows easily from Theorems 1.6.4 and 1.7.2.

1.7.3. Theorem. *Let E be a universally complete Kantorovich space, and let $\mu : \Sigma \to B := \mathfrak{C}(E)$ be a spectral measure. The spectral integral $I_\mu(\cdot)$ is a sequentially o-continuous (linear, ring, and lattice) homomorphism from the f-algebra of measurable functions $M(\Omega, \Sigma)$ to E.*

1.7.4. Let $e_1, \dots, e_n : \mathbb{R} \to B$ be a finite collection of spectral functions with values in a σ-algebra B. Then there is a unique B-valued spectral measure μ on the Borel σ-algebra $\mathscr{B}(\mathbb{R}^n)$ of the space $\mathbb{R}^n$ for which

$$\mu\left(\prod_{k=1}^{n}(-\infty, \lambda_k)\right) = \bigwedge_{k=1}^{n} e_k(\lambda_k)$$

whenever $\lambda_1, \dots, \lambda_n \in \mathbb{R}$.

1.7.5. We now consider an ordered collection of elements (n-tuple) $x_1, \dots, x_n$ of a Kantorovich space E with unity $\mathbf{1}$. Let $e^{x_k} : \mathbb{R} \to B := \mathfrak{C}(E)$ be the spectral function of x_k. According to the proposition above, there is a spectral measure $\mu : \mathscr{B}(\mathbb{R}^n) \to B$ such that

$$\mu\left(\prod_{k=1}^{n}(-\infty, \lambda_k)\right) = \bigwedge_{k=1}^{n} e^{x_k}(\lambda_k).$$

It is clear that the measure μ is uniquely determined from the n-tuple $\mathfrak{X} := (x_1, \dots, x_n) \in E^n$. This allows us to write $\mu_{\mathfrak{X}} := \mu$ and to say that $\mu_{\mathfrak{X}}$ is the spectral measure of $\mathfrak{X}$. For the integral of a measurable function $f : \mathbb{R}^n \to \mathbb{R}$ with respect to the spectral measure $\mu_{\mathfrak{X}}$ the following notations are convenient

$$\widehat{\mathfrak{X}}(f) := f(\mathfrak{X}) := f(x_1, \dots, x_n) := I_\mu(f).$$

If $\mathfrak{X} := (x)$ then we also write $\widehat{x}(f) := f(x) := I_\mu(f)$; in addition, the measure $\mu_x := \mu_{\mathfrak{X}}$ is said to be the *spectral measure* of x. For a function $f(t) = t$ $(t \in \mathbb{R})$ the *Freudenthal Spectral Theorem* follows from 1.7.2:

$$x = \int_{\mathbb{R}} t \, d\mu_x(t) = \int_{-\infty}^{\infty} \lambda \, de_\lambda^x.$$

We recall that the space $\mathscr{B}(\mathbb{R}^n, \mathbb{R})$ of Borel functions in $\mathbb{R}^n$ is a universally complete K_σ-space and a faithful f-algebra.

1.7.6. Theorem. *The spectral measures of an n-tuple $\mathfrak{X} := (x_1, \dots, x_n)$ and an element $f(x_1, \dots, x_n)$ are interrelated as follows*

$$\mu_{f(\mathfrak{X})} = \mu_{\mathfrak{X}} \circ f^{\leftarrow},$$

with $f^{\leftarrow} : \mathscr{B}(\mathbb{R}) \to \mathscr{B}(\mathbb{R}^n)$ the homomorphism acting as $A \mapsto f^{-1}(A)$.

In particular, for arbitrary measurable functions, $f \in \mathscr{B}(\mathbb{R}^n, \mathbb{R})$ and $g \in \mathscr{B}(\mathbb{R}, \mathbb{R})$, the identity $(g \circ f)(\mathfrak{X}) = g(f(\mathfrak{X}))$ holds provided that $f(\mathfrak{X})$ and $g(f(\mathfrak{X}))$ both exist.

$\lhd$ By 1.7.2, for every $\lambda \in \mathbb{R}$, we have

$$\mu_{\mathfrak{X}}(-\infty, \lambda) = e_{\lambda}^{f(\mathfrak{X})} = [\![f(\mathfrak{X}) < \lambda^{\wedge}]\!] = \mu_{\mathfrak{X}} \circ f^{-1}(-\infty, \lambda).$$

Hence, the spectral measures $\mu_{f(\mathfrak{X})}$ and $\mu_{\mathfrak{X}} \circ f^{-1}$ defined on $\mathscr{B}(\mathbb{R})$, agree on intervals of the form $(-\infty, \lambda)$. Using the standard arguments of measure theory, derive that these measures agree everywhere. To prove the second part, it suffices to note that $(g \circ f)^{\leftarrow} = f^{\leftarrow} \circ g^{\leftarrow}$ and to use twice the statement obtained. $\rhd$

The next fact follows from 1.7.3 and 1.7.6.

1.7.7. Theorem. *For each n-tuple $\mathfrak{X} := (x_1, \dots, x_n)$ of elements of a universally complete Kantorovich space E, the mapping*

$$\widehat{\mathfrak{X}} : f \mapsto \widehat{\mathfrak{X}}(f) \quad (f \in \mathscr{B}(\mathbb{R}^n, \mathbb{R}))$$

is a unique sequentially o-continuous homomorphism of the f-algebra $\mathscr{B}(\mathbb{R}^n, \mathbb{R})$ to E such that the following holds

$$\widehat{\mathfrak{X}}(d\lambda_k) = x_k \quad (k := 1, \dots, n),$$

with $d\lambda_k : (\lambda_1, \dots, \lambda_n) \mapsto \lambda_k$ a coordinate function in $\mathbb{R}^n$.

1.7.8. We will briefly discuss two realizations of the universally complete Kantorovich space $\mathscr{R}{\downarrow}$ which can be obtained with the help of 1.6.4. We recall the necessary definitions. Given a compact space Q, we let the symbol $C_{\infty}(Q)$ stand for the set of continuous functions from Q to $\overline{\mathbb{R}} := \mathbb{R} \cup \{-\infty, +\infty\}$ each of which takes the values $\pm\infty$ only on a rare (= nowhere dense) set (cf. 1.5.4 (2)).

Let $\mathfrak{R}(B)$ be the set of resolutions of unity in B.

1.7.9. Theorem. *Let B be a complete Boolean algebra. The set $\mathfrak{R}(B)$ with suitable operations and order is a universally complete Kantorovich space. The function, sending an element $x \in \mathscr{R}{\downarrow}$ into the resolution of unity $\lambda \mapsto [\![x < \lambda^{\wedge}]\!]$ $(\lambda \in \mathbb{R})$, is an isomorphism between the Kantorovich spaces $\mathscr{R}{\downarrow}$ and $\mathfrak{R}(B)$.*

1.7.10. Theorem. *Let Q be the Stone space of a complete Boolean algebra B, and let $\mathscr{R}$ be the reals in the model $\mathbf{V}^{(B)}$. The vector lattice $C_\infty(Q)$ is isomorphic to the universally complete Kantorovich space $\mathscr{R}{\downarrow}$. This isomorphism is defined by sending an element $x \in \mathscr{R}{\downarrow}$ to the function $\widehat{x} : Q \to \overline{\mathbb{R}}$ as follows:*

$$\widehat{x}(q) = \inf\{\lambda \in \mathbb{R} : [\![x < \lambda^\wedge]\!] \in q\}.$$

1.7.11. Comments.

(1) The notions of unity and unit element, as well as the characteristic or spectral function of an element, were introduced by H. Freudenthal. He also established the Spectral Theorem; see 1.7.5 and [41, 115].

It follows from Theorem 1.7.9 that, for a complete Boolean algebra B, the set of resolutions of unity is a universally complete Kantorovich space with base isomorphic to B. This fact belongs to L. V. Kantorovich [41].

The realization of an arbitrary Kantorovich space as a foundation of a universally complete Kantorovich space was implemented by A. G. Pinsker [41].

The possibility of realizing an arbitrary Kantorovich space as an order dense ideal of $C_\infty(Q)$ follows from 1.6.8 (1) and 1.7.3. This possibility was first established by B. Z. Vulikh and T. Ogasawara independently of each other [41, 115].

(2) It follows from 1.7.4 that every spectral function with values in a σ-algebra defines a spectral measure on the Borel σ-algebra of the real line. This fact was first mentioned by V. I. Sobolev in [101]. Nevertheless, it was supposed in [101] that such a measure can be obtained by the Carathéodory extension. As was shown by D. A. Vladimirov, for a complete Boolean algebra of countable type the Carathéodory extension is possible only if the algebra is regular. Thus, the extension method of 1.7.4 differs essentially from the Carathéodory extension and rests on the Loomis–Sikorski representation of Boolean σ-algebras. J. D. M. Wright derived 1.7.4 as a consequence of an abstraction of the Riesz Theorem for operators with values in a Kantorovich space.

(3) Apparently, Borel functions of elements of an arbitrary Kantorovich space with unity were first considered by V. I. Sobolev (see [101, 115]). Theorem 1.7.6 in full generality is presented in [65]. The Borel functional calculus of (countable and uncountable) collections of elements of an arbitrary Kantorovich space is constructed in [65] as well. A Boolean valued proof of Theorem 1.7.7 is also available (see [33]).

(4) For other aspects of Boolean valued analysis of vector lattices, see [15, 16, 33, 35, 53, 62, 84, 106, 107].

1.8. Lattice Normed Spaces

A function space X often admits a natural abstraction of a norm. Namely, we may assume that to each member of X there corresponds some member of another vector lattice called the *norm lattice* of X. The availability of a *lattice norm* on X is sometimes decisive in studying various structural properties of X. Furthermore, a norm taking values in a vector lattice makes it possible to distinguish an interesting class of the so-called dominated operators. The current section recalls preliminaries. For details, see [53, 58, 60].

1.8.1. Consider a vector space X and a real vector lattice E. We will assume each vector lattice Archimedean without further stipulations. A mapping $p : X \to E_+$ is called an (*E-valued*) *vector norm* or simply *lattice norm* if p satisfies the following axioms:

(1) $p(x) = 0 \leftrightarrow x = 0 \quad (x \in X)$;

(2) $p(\lambda x) = |\lambda| p(x) \quad (x \in X,\ \lambda \in \mathbb{R})$;

(3) $p(x + y) \leq p(x) + p(y) \quad (x, y \in X)$.

A vector norm p is a *decomposable* or *Kantorovich norm* if

(4) for arbitrary $e_1, e_2 \in E_+$ and $x \in X$ the equality $p(x) = e_1 + e_2$ implies the existence of $x_1, x_2 \in X$ such that $x = x_1 + x_2$ and $p(x_l) = e_l$ for $l := 1, 2$.

A triple (X, p, E) (simpler, X or (X, p) with the implied parameters omitted) is called a *lattice normed space* provided that p is an E-valued norm on X. If p is a decomposable norm then the space (X, p) itself is called *decomposable.*

If (X, p, E) is a lattice normed space, with E the norm lattice of X; we may equip X with the *mixed norm:*

$$|||x||| := \|p(x)\| \quad (x \in X).$$

In this event the normed space $X := (X, |||\cdot|||)$ is also called a *space with mixed norm.* By the inequality $|p(x) - p(y)| \leq p(x - y)$ and monotonicity of the norm of E, the vector norm p is a continuous operator from $(X, |||\cdot|||)$ to E.

1.8.2. Take a net $(x_\alpha)_{\alpha \in \mathrm{A}}$ in X. We say that (x_α) *bo-converges* to an element $x \in X$ and write $bo\text{-}\lim x_\alpha = x$ provided that there exists a decreasing net $(e_\gamma)_{\gamma \in \Gamma}$ in E such that $\inf_{\gamma \in \Gamma} e_\gamma = 0$ and, to every $\gamma \in \Gamma$, there exists an index $\alpha(\gamma) \in \mathrm{A}$ such that $p(x - x_\alpha) \leq e_\gamma$ for all $\alpha \geq \alpha(\gamma)$. A net (x_α) is *bo-fundamental* if the net $(x_\alpha - x_\beta)_{(\alpha,\beta) \in \mathrm{A} \times \mathrm{A}}$ *bo*-converges in order to zero. A lattice normed space X is *bo-complete* if every *bo*-fundamental net in it *bo*-converges to some element of X.

By analogy we define *relative uniform completeness.* It is easy that if E is a Banach lattice then a space $(X, |||\cdot|||)$ with mixed norm is Banach if and only if (X, p, E) is relatively uniformly complete.

We call a decomposable *bo*-complete lattice normed space a *Banach–Kantorovich space.*

Assume that (Y, q, F) is a Banach–Kantorovich space and $F = q(Y)^{\perp\perp}$. We say that Y is *universally complete,* if $mF = F$; i.e., if the norm space F is universally complete. This amounts to the fact that Y is a decomposable *bo*-complete space in which every disjoint family is *bo*-summable. A space Y is a *universal completion* of a lattice normed space (X, p, E) provided that

(1) $F = mE$ (in particular, Y is universally complete);

(2) there is a linear isometry $\imath : X \to Y$;

(3) if Z is a decomposable *bo*-complete subspace of Y and $\imath(X) \subset Z$ then $Z = Y$.

1.8.3. Theorem. *Let $(\mathscr{X}, \rho)$ be a Banach space inside $\mathbf{V}^{(B)}$. Put $X := \mathscr{X}{\downarrow}$ and $p := \rho{\downarrow}$. Then*

(1) *$(X, p, \mathscr{R}{\downarrow})$ is a universally complete Banach–Kantorovich space;*

(2) *the space X admits the structure of a faithful unitary module over the ring $\mathscr{C}{\downarrow}$ so that*

(a) $(\lambda \mathbf{1})x = \lambda x \quad (\lambda \in \mathbb{C},\ x \in X)$;

(b) $p(ax) = |a| p(x) \quad (a \in \mathscr{C}{\downarrow},\ x \in X)$;

(c) $b \le [\![x = \mathbf{0}]\!] \leftrightarrow \chi(b)x = \mathbf{0} \quad (b \in B,\ x \in X)$ *where χ is an isomorphism from B onto $\mathfrak{E}(\mathscr{R}{\downarrow})$.*

We call the resultant universally complete Banach–Kantorovich space $\mathscr{X}{\downarrow} := (\mathscr{X}, \rho){\downarrow} := (\mathscr{X}{\downarrow}, \rho{\downarrow}, \mathscr{R}{\downarrow})$ the *descent* of a Banach space $(\mathscr{X}, \rho)$.

1.8.4. Theorem. *To each lattice normed space (X, p, E) there is a Banach space $\mathscr{X}$ inside $\mathbf{V}^{(B)}$, with $B \simeq \mathscr{B}(p(X)^{\perp\perp})$, such that the descent $\mathscr{X}{\downarrow}$ of $\mathscr{X}$ is a universal completion of (X, p, E). Moreover, $\mathscr{X}$ is unique up to linear isometry inside $\mathbf{V}^{(B)}$.*

1.8.5. A Banach space $\mathscr{X}$ inside $\mathbf{V}^{(B)}$ is said to be a *Boolean valued representation* for a lattice normed space X if $\mathscr{X}{\downarrow}$ is a universal completion of X.

We suppose that $\mathscr{X}$ and $\mathscr{Y}$ are some Boolean valued representations of Banach –Kantorovich spaces X and Y normed by some universally complete Kantorovich space E. We further let $\mathscr{L}^B(\mathscr{X}, \mathscr{Y})$ stand for the space of bounded linear operators from $\mathscr{X}$ to $\mathscr{Y}$ inside $\mathbf{V}^{(B)}$, where $B := \mathfrak{B}(E)$.

We denote by $\mathscr{L}_b(X, Y)$ the space of all linear operators bounded in the following sense: there is some $\pi \in \operatorname{Orth}(E)$ such that $|Tx| \le \pi |x|$ for all $x \in X$.

The descent of operators $\mathscr{T} \mapsto \mathscr{T}{\downarrow}$ is a linear isometry between the lattice normed spaces $\mathscr{L}^{(B)}(\mathscr{X}, \mathscr{Y}){\downarrow}$ and $\mathscr{L}_b(X, Y)$.

1.8.6. Let X be a normed space. We suppose that $\mathscr{L}(X)$ has a complete Boolean algebra of norm one commuting projections $\mathscr{B}$ which is isomorphic to B. Note that we always imply the following Boolean operations in $\mathscr{B}$ (cf. 1.5.2): $\pi \wedge \rho := \pi\rho = \rho\pi$, $\pi \vee \rho := \pi + \rho - \pi\rho$, $\pi^* := I_X - \pi$ $(\pi, \rho \in \mathscr{B})$. In this event we will identify the Boolean algebras $\mathscr{B}$ and B, writing $B \subset \mathscr{L}(X)$.

We say that X is a *normed B-space* if $B \subset \mathscr{L}(X)$ and for every partition of unity $(b_\xi)_{\xi\in\Xi}$ in B the two conditions are met:

(1) If $b_\xi x = 0$ $(\xi \in \Xi)$ for some $x \in X$ then $x = 0$;

(2) If $b_\xi x = b_\xi x_\xi$ $(\xi \in \Xi)$ for $x \in X$ and some family $(x_\xi)_{\xi\in\Xi}$ in X then $\|x\| \le \sup\{\|b_\xi x_\xi\| : \xi \in \Xi\}$.

Conditions (1) and (2) amount to the respective conditions (1′) and (2′):

(1′) To each $x \in X$ there corresponds the greatest projection $b \in B$ such that $bx = 0$;

(2′) If x, (x_ξ), and (b_ξ) are the same as in (2) then $\|x\| = \sup\{\|b_\xi x_\xi\| : \xi \in \Xi\}$.

From (2′) it follows in particular that

$$\left\| \sum_{k=1}^{n} b_k x \right\| = \max_{k=1,\dots,n} \|b_k x\|$$

for $x \in X$ and disjoint projections $b_1, \dots, b_n$ in B.

Given a partition of unity (b_ξ), we refer to $x \in X$ such that $(\forall \xi)\, b_\xi x = b_\xi x_\xi$ as a *mixing* of (x_ξ) by (b_ξ). If (1) holds then there is a unique mixing x of (x_ξ) by (b_ξ). In these circumstances we naturally call x *the* mixing of (x_ξ) by (b_ξ). Condition (2) may be paraphrased as follows: The unit ball U_X of X is closed under mixing.

A normed B-space X is *B-cyclic* if we may find in X a mixing of each norm bounded family by any partition of unity in B. Considering what was said above, we note that X is a B-cyclic normed space if and only if, to a partition of unity $(b_\xi) \subset B$ and a family $(x_\xi) \subset U_X$, there is a unique element $x \in U_X$ such that $b_\xi x = b_\xi x_\xi$ for all ξ.

An isometry $\imath$ between normed B-spaces is a *B-isometry* if $\imath$ is linear and commutes with every projection in B. We say that Y is a *B-cyclic completion* of a B-space X if Y is B-cyclic and there is a B-isometry $\imath : X \to Y$ such that every B-cyclic subspace of Y including $\imath(X)$ coincides with Y. It is easy to show that each Banach B-space possesses a B-cyclic completion unique up to B-isometry.

We take a Banach space $(\mathscr{X}, \rho)$ inside $\mathbf{V}^{(B)}$. We then let Λ be the bounded part of the Kantorovich space $\mathscr{C}{\downarrow}$, i.e. the least order ideal of $\mathscr{C}{\downarrow}$ containing the unity. The bounded part of the space $\mathscr{X}\downarrow$, i.e. the set $\{x \in \mathscr{X}\downarrow : \rho\downarrow(x) \in \Lambda\}$, is called the *bounded descent* of $\mathscr{X}$ and is denoted sometimes by $\mathscr{X}\downarrow^\infty$. The bounded descent of a Banach space is a Banach space with mixed norm $|||x||| := \|p(x)\|_\infty$, where $\|z\|_\infty := \inf\{0 < \alpha \in \mathbb{R} : |z| \leq \alpha\mathbf{1}\}$ for $z \in \Lambda$.

1.8.7. Theorem. *For a Banach space X the following are equivalent:*

(1) *X is a decomposable space with mixed norm whose norm lattice is a Kantorovich space of bounded elements;*

(2) *X is a Banach B-space;*

(3) *The B-cyclic hull of X is B-isometric with the bounded descent of some Banach space inside $\mathbf{V}^{(B)}$.*

1.8.8. We assume that X is a normed B-space and Y is a B-cyclic Banach space. Let $\mathscr{X}$ and $\mathscr{Y}$ stand for the Boolean valued representations of X and Y. The space $\mathscr{L}_B(X,Y)$ is B-isometric to the bounded descent of the space $\mathscr{L}(\mathscr{X},\mathscr{Y})$ of bounded linear operators from $\mathscr{X}$ to $\mathscr{Y}$ inside $\mathbf{V}^{(B)}$. Moreover, to $T \in \mathscr{L}_B(X,Y)$ there corresponds the member $\mathscr{T} := T\uparrow$ of $\mathbf{V}^{(B)}$ determined from the formulas $[\![\mathscr{T} : \mathscr{X} \to \mathscr{Y}]\!] = \mathbf{1}$ and $[\![\mathscr{T}\imath x = \imath Tx]\!] = \mathbf{1}$ for all $x \in X$, where $\imath$ stands simultaneously for the embedding of X to $\mathscr{X}\downarrow$ and the embedding of Y to $\mathscr{Y}\downarrow$.

1.8.9. We call $X^\# := \mathscr{L}_B(X,\Lambda)$ the B-*dual* of X. Let $\mathscr{X}^*$ be the dual of $\mathscr{X}$. Denote by $\simeq$ and $\simeq_B$ the relations of isometric isomorphy and isometric B-isomorphy between Banach spaces. Suppose also that X, Y, $\mathscr{X}$, and $\mathscr{Y}$ are the same as in 1.8.8.

The following hold:

(1) $X^\# \simeq_B Y \leftrightarrow [\![\mathscr{X}^* \simeq \mathscr{Y}]\!] = \mathbf{1}$.

(2) *If $\overline{X}$ is a B-cyclic completion of X then $X^\# = \overline{X}^\#$.*

1.8.10. We suppose that A is a Stone algebra (= a commutative AW^*-algebra) and B is a complete Boolean algebra of projections of A. Consider a unital A-module X.

The mapping $\langle\cdot\,|\,\cdot\rangle : X \times X \to A$ is an A-*valued inner product* provided that for all $x, y, z \in X$ and $a \in A$ the following are satisfied:

(1) $\langle x\,|\,x\rangle \geq \mathbf{0}$; $\langle x\,|\,x\rangle = \mathbf{0} \leftrightarrow x = \mathbf{0}$;

(2) $\langle x\,|\,y\rangle = \langle y,\ x\rangle^*$;

(3) $\langle ax\,|\,y\rangle = a\langle x\,|\,y\rangle$;

(4) $\langle x + y\,|\,z\rangle = \langle x\,|\,z\rangle + \langle y\,|\,z\rangle$.

Using an A-valued inner product, we may define the norm in X as

$$(5)\ \| x \| := \sqrt{\|\langle x|x\rangle\|} \quad (x \in X),$$

and the vector norm as

$$(6)\ |x| := \sqrt{\langle x|x\rangle} \quad (x \in X).$$

In this event $\| x \| = \| |x| \|$ for all $x \in X$. Therefore, (5) defines a mixed norm on X.

It is possible to show that a pair $(X, \| \cdot \|)$ is a B-cyclic Banach space if and only if $(X, |\cdot|)$ is a Banach–Kantorovich space [60].

A *Kaplansky–Hilbert module* or an *AW^*-module* over A is a unitary A-module with A-valued inner product satisfying either of these two equivalent conditions.

1.8.11. Theorem. *The bounded descent of an arbitrary Hilbert space inside $\mathbf{V}^{(B)}$ is a Kaplansky–Hilbert module over the Stone algebra Λ. Conversely, if X is a Kaplansky–Hilbert module over Λ, then there is a Hilbert space $\mathscr{X}$ inside $\mathbf{V}^{(B)}$ whose bounded descent is unitarily equivalent to X. This Hilbert space $\mathscr{X}$ is unique up to unitary equivalence inside $\mathbf{V}^{(B)}$.*

1.8.12. As usual, we call $\mathscr{X} \in \mathbf{V}^{(B)}$ the *Boolean valued realization* of the initial Kaplansky–Hilbert module X. We assume that $\mathscr{L}^B(\mathscr{X}, \mathscr{Y})$ is the space of bounded linear operators from $\mathscr{X}$ to $\mathscr{Y}$ inside $\mathbf{V}^{(B)}$. Let $\mathrm{Hom}(X, Y)$ stand for the space of bounded Λ-linear operators from X to Y, where X and Y are some Kaplansky–Hilbert modules over the Stone algebra Λ. Obviously, $\mathrm{Hom}(X, Y) = \mathscr{L}_B(X, Y)$.

Theorem. *Let $\mathscr{X}$ and $\mathscr{Y}$ be Hilbert spaces inside $\mathbf{V}^{(B)}$. Denote by X and Y the bounded descents of $\mathscr{X}$ and $\mathscr{Y}$. If $\Phi : X \to Y$ is a bounded Λ-linear operator then $\varphi := \Phi\uparrow$ is a bounded linear operator from $\mathscr{X}$ to $\mathscr{Y}$ inside $\mathbf{V}^{(B)}$. Moreover, $[\![\|\varphi\| \le c^{\wedge}]\!] = 1$ for some $c \in \mathbb{R}$. The mapping $\Phi \mapsto \varphi$ is a B-linear isometry between the B-cyclic Banach spaces* $\mathrm{Hom}(X, Y)$ *and* $\mathscr{L}^B(\mathscr{X}, \mathscr{Y})\downarrow^{\infty}$.

1.8.13. Comments.

(1) The concept of lattice normed space appeared for the first time in the article [38] by L. V. Kantorovich. The axiom of decomposability 1.8.1 (4) looked bizarre and was often omitted in the subsequent publications of the other authors as definitely immaterial. The principal importance of this axiom was revealed only within Boolean valued analysis (see [53]). L. V. Kantorovich in the above-mentioned article also discovered some instances of dominated operators; cf. [39]. An elaborate theory of dominated operators was propounded only in the recent decades (cf. [53, 58, 63]).

(2) Spaces with mixed norm in the sense of this section were studied in [56, 58]. These articles contain applications of the concept of mixed norm to geometry of Banach spaces and operator theory. The bounded descent was first studied by G. Takeuti in connection with von Neumann algebras and C^*-algebras inside Boolean valued models [106, 107].

(3) The modern structural theory of AW^*-algebras and AW^*-modules originates with the research by I. Kaplansky [42–44]. These objects reveal themselves in algebraization of the theory of von Neumann operator algebras. As regards Boolean valued realization of AW^*-algebras and AW^*-modules, see the articles by M. Ozawa [89–92].

1.9. Nonstandard Hulls

The notion of nonstandard hull occupies a central place in the geometric theory of Banach spaces.

1.9.1. Let $(E, \|\cdot\|)$ be an internal normed space. An element $x \in E$ is called *limited* (*infinitesimal*) if $|x|$ is limited (infinitesimal). Denote by $\mathrm{fin}(E)$ and $\mu(E)$ the external sets of all finite and, respectively, infinitesimal elements of E.

We denote by $\mathrm{fin}(E)$ and $\mu(E)$ the external sets of limited and infinitesimal elements of a normed space E. Then $\mathrm{fin}(E)$ is an external vector space over the field ${}^\circ\mathbb{R}$, and $\mu(E)$ is a subspace of $\mathrm{fin}(E)$. The quotient space $\mathrm{fin}(E)/\mu(E)$ is denoted by the symbol $\widehat{E}$. We furnish $\widehat{E}$ with a norm by the formula

$$\|\pi x\| = \mathrm{st}(\|x\|) \in {}^\circ\mathbb{R} \quad (x \in \mathrm{fin}(E)),$$

where $\pi : \mathrm{fin}(E) \to \widehat{E}$ is the quotient mapping.

Furthermore, $(\widehat{E}, \|\cdot\|)$ is an external normed space, called the *nonstandard hull* of E. If the internal dimension of E is limited then $\widehat{E}$ is called a *hyperfinite-dimensional space.* If the space $(E, \|\cdot\|)$ is standard then ${}^\circ E$ with the induced norm from E is an external normed space, and the restriction of π to ${}^\circ E$ is an isometric embedding of ${}^\circ E$ into $\widehat{E}$. The inclusion ${}^\circ E \subset \widehat{E}$ is always presumed.

1.9.2. *Theorem.* *The nonstandard hull $\widehat{E}$ of E is a Banach space for every internal (not necessarily complete) normed space E.*

$\triangleleft$ Let $B_X(a, r)$ be the closed ball in X with center a and radius r. Consider a nested sequence of balls $B_{\widehat{E}}(\ddot{x}_n, r_n)$ in X such that $(x_n)_{n\in{}^\circ\mathbb{N}} \subset E$, $\ddot{x}_n = \pi x_n$, $(r_n)_{n\in{}^\circ\mathbb{N}} \subset {}^\circ\mathbb{R}$, and $\lim_{n\to\infty} r_n = 0$.

We may assume that r_n decreases. Then the sequence of internal closed balls $B_E(x_n, r_n + r_n/2^{n+1}) \subset E$ decreases too. By the idealization principle, there is an element $x \in E$ belonging to each of these balls. The element $\ddot{x} = \pi x$ is a common point of the balls $B_{\widehat{E}}(\ddot{x}_n, r_n)$. $\triangleright$

1.9.3. We suppose that E is an internal normed lattice. Then we may define an order relation in $\widehat{E}$ so that the quotient homomorphism π be positive. More precisely, if $\ddot{x} := \pi x$ and $\ddot{y} := \pi y$ then we assume by definition

$$\ddot{x} \le \ddot{y} \leftrightarrow (\exists z \in \mu(E))(x \le y + z).$$

Theorem. *The nonstandard hull $\widehat{E}$ of E is a Banach lattice with sequentially o-continuous norm. Moreover, every increasing and norm bounded sequence in $\widehat{E}$ is order bounded.*

At the same time, it is worth noting that the nonstandard hull of an internal norm lattice is not necessarily a Kantorovich space (not even a K_σ-space; for example, $\widehat{c}_0$ where c_0 is the lattice of vanishing sequences).

1.9.4. Theorem. *For an internal normed lattice E, the following are equivalent:*

(1) *$\widehat{E}$ is a Kantorovich space;*

(2) *$\widehat{E}$ is a K_σ-space;*

(3) *the norm of $\widehat{E}$ is o-continuous;*

(4) *there is no closed sublattice in $\widehat{E}$ isometric and order isomorphic to c_0.*

1.9.5. A normed lattice is said to be *rich in finite-dimensional sublattices*, if to every finite collection $x_1, \dots, x_n \in {}^\circ E, n \in {}^\circ\mathbb{N}$, and for arbitrary $0 < \varepsilon \in {}^\circ\mathbb{R}$ there are a finite-dimensional sublattice $E_0 \subset {}^\circ E$ and elements $y_1, \dots, y_n \in E_0$ such that the inequality $\|x_k - y_k\| < \varepsilon$ holds for all $k := 1, \dots, n$.

A standard Banach lattice E is rich in finite-dimensional sublattices if and only if ${}^\circ E$ is contained in a hyperfinite-dimensional subspace of the nonstandard hull $\widehat{E}$ of E.

1.9.6. We suppose now that E and F are internal normed spaces and $T : E \to F$ is an internal bounded linear operator. The set

$$c(T) := \{C \in \mathbb{R} : (\forall x \in E)\|Tx\| \le C\|x\|\}$$

is internal and bounded from above. Hence, $\|T\| := \inf c(T)$ exists.

If $\|T\|$ is limited then it follows from the inequality $\|Tx\| \le \|T\|\,\|x\|$ $(x \in E)$ that $T(\mathrm{fin}(E)) \subset \mathrm{fin}(F)$ and $T(\mu(E)) \subset \mu(F)$. Thus, the external operator $\widehat{T} : \widehat{E} \to \widehat{F}$ is soundly defined by the formula

$$\widehat{T}\pi x = \pi T x \quad (x \in E).$$

The operator $\widehat{T}$ is linear (over ${}^\circ\mathbb{R}$) and bounded; moreover, $\|\widehat{T}\| = \mathrm{st}(\|T\|)$. It is natural to call $\widehat{T}$ the *nonstandard hull* of T.

If E and F are normed lattices and T is a positive operator then $\widehat{T}$ is a sequentially o-continuous positive operator.

1.9.7. It is easy that, for bounded operators S and T, the equality $(S \circ T)^\wedge = \widehat{S} \circ \widehat{T}$ holds; and, in addition, $\widehat{I_E} = I_{\widehat{E}}$, with I_X the identity operator on X.

Thus, the operation of passing to the nonstandard hull is a covariant functor (in suitable categories of normed spaces). Many questions arise about the general properties of this functor: How does the nonstandard hull functor interact with other functors of the theory of Banach spaces (lattices)? How do the well-known properties in the geometric theory of Banach spaces (the Radon–Nikodým property, Kreĭn–Milman property, etc.) transform under this functor? What is the structure of nonstandard hulls of concrete spaces? Analogous questions can be formulated for operators and so on. The relevant ideas and methods are set forth in the surveys [24, 27, 29]. Here we will briefly outline the three important directions of research and formulate a few simple propositions of an illustrative nature.

1.9.8. The question of analytical description for nonstandard hulls is studied in detail for the classical Banach spaces; see [29].

Theorem. *The following are true:*

(1) *If E is an internal AL_p-space, where $p \geq 1$ is a limited element of $\mathbb{R}$; then $\widehat{E}$ is an AL_r-space for $r = \mathrm{st}(p)$;*

(2) *If E is an internal AL_p-space, with $p \geq 1$ an illimited element of $\mathbb{R}$, or if E is an internal AM-space; then $\widehat{E}$ is an AM-space;*

(3) *If Q is an internal compact space and $C(Q)$ is the internal space of continuous functions from Q to $\mathbb{R}$, then $\widehat{C(Q)}$ is linearly isometric to $C(\widehat{Q})$, where $\widehat{Q}$ is an external completion of Q in some uniformity.*

Only general results of this type can be obtained in axiomatic external set theory. Nevertheless, while working in the frame of the classical stance of nonstandard analysis (for instance, in a finite fragment of the von Neumann universe), a more detailed description is possible for nonstandard hulls. If, for instance, a nonstandard structure is ω_0-saturated (a restriction from below) and possesses the ω_0-isomorphism property (a restriction from above), then the nonstandard hull of the Banach lattice L_p ([0,1]) is isometrically isomorphic with the l_p-sum of k copies of the space $L_p([0,1]^k)$, where $k = 2^{\omega_0}$.

1.9.9. We now turn to the local geometry of a normed space. Some properties of such a space are "local" in the sense that they are defined by the structure and location of finite-dimensional subspaces of the space under study. In this regard, nonstandard hulls have much more preferable structure. For instance, it often happens so that a condition, satisfied "approximately" on finite-dimensional subspaces, is satisfied "exactly" in the nonstandard hull of the ambient space.

Let E and F be Banach lattices. The lattice E is said to be *finitely representable in F (as a Banach sublattice)* if for every finite-dimensional sublattice $E_0 \subset E$ and every number $\varepsilon > 0$ there is a linear and lattice isomorphism $T : E_0 \to F$ satisfying $\|x\| \le \|Tx\| \le (1+\varepsilon)\|x\|$ $(x \in E_0)$.

Theorem. *Let E be a standard Banach lattice rich in finite-dimensional sublattices (1.9.5), and let F be an internal Banach lattice. Then ${}^\circ E$ is finitely representable in F if and only if ${}^\circ E$ is linearly isometric and lattice isomorphic to a sublattice of $\widehat{F}$.*

1.9.10. We now turn to some model-theoretic properties of Banach spaces.

We start with introducing some first-order language $\mathbb{L}_B$. The signature of the language is $\{=, +, p, Q\} \cup \mathbb{Q}$, where $\mathbb{Q}$ is the rationals.

Each Banach space E may be considered as a model of $\mathbb{L}_B$ by interpreting $=$ and $+$ as equality and addition, P as $\{x \in E : \|x\| \le 1\}$, Q as $\{x \in E : \|x\| \ge 1\}$, and finally, each $r \in \mathbb{Q}$ as multiplication by r.

A formula φ of $\mathbb{L}_B$ of the shape $(Sx_1)\dots(Sx_n)(\varphi_1 \wedge \dots \wedge \varphi_n)$, where S is a restricted quantifier and φ_k is a conjunction of formulas of the shape $u = v, p(u), Q(u)$, is called a *restricted positive formula.*

If φ is such a formula and m is a natural ($\neq 0$), then φ^m is the new formula constructed as follows: in the subformulas $\varphi_1, \dots, \varphi_n$ the expression $u = v$ is replaced by $p(m(u-v))$; $p(u)$ by $p((1-1/m)u)$, and $Q(u)$ by $Q((1+1/m)u)$.

If φ^m is valid in E for all $m \in \mathbb{N}$ then we call φ *approximately valid* in E. Banach spaces E and F are called *approximately equivalent* if the same restricted positive formulas are approximately valid in them.

Theorem. *The following are true:*

(1) *Banach spaces are approximately equivalent if and only if their nonstandard hulls are isometric.*

(2) *Let μ and ν be σ-finite measures, and $1 \le p < \infty$. The spaces $L_p(\mu)$ and $L_p(\nu)$ are approximately equivalent if and only if the measures μ and ν have the same finite number of atoms or both possess infinitely many atoms.*

1.9.11. Comments.

(1) The nonstandard hull of a Banach space was invented by W. A. J. Luxemburg [80]. The ultraproducts of Banach spaces, introduced by D. Dacunha-Castelle and J. L. Krivine [12], are very similar to nonstandard hulls. Consult [24, 27, 29] about the role of these notions in the theory of Banach spaces, the most important results, and further references.

(2) The first-order language of 1.9.10 was first used by C. W. Henson [25], and later by J. Stern (see [104, 105]). The notion of finite representability had come into the theory of Banach spaces long before the set-theoretic technique. It was introduced by A. Dvoretsky (the term is due to R. C. James).

(3) About 1.9.4, 1.9.5, and 1.9.9 see [10, 29]. The results of 1.9.8 are established in [26] and [28]; and the results of 1.9.10, in [26].

1.10. The Loeb Measure

The Loeb measure is one of the most important constructions of nonstandard analysis which gave rise to applications in many sections of functional analysis, probability and stochastic modeling; see [3, 11]. We now present a few results about the structure of the Loeb measure.

1.10.1. Let $(X, \mathscr{A}, \nu)$ be an internal measure space with countably additive positive measure; more exactly, assume that $\mathscr{A}$ is an internal algebra of subsets of an internal set X and $\nu : \mathscr{A} \to \mathbb{R}$ is an internal finitely additive positive function on $\mathscr{A}$. We consider the external function ${}^\circ\nu : A \mapsto {}^\circ(\nu(A)) \in {}^\circ\mathbb{R} \cup \{+\infty\}$ $(A \in \mathscr{A})$, where ${}^\circ(\nu(A))$ is the standard part of $\nu(A)$ if $\nu(A)$ is limited and ${}^\circ(\nu(A)) = +\infty$ in the opposite case. Clearly, ${}^\circ\nu$ is finitely additive.

1.10.2. ***Theorem.*** *A finitely additive measure ${}^\circ\nu : \mathscr{A} \to {}^\circ\mathbb{R} \cup \{+\infty\}$ admits a unique countably additive extension λ to the σ-algebra $\sigma(\mathscr{A})$ generated by $\mathscr{A}$. Moreover,*

$$\lambda(B) = \inf\{{}^\circ\nu(A) : B \subset A,\ A \in \mathscr{A}\} \quad (B \in \sigma(\mathscr{A})).$$

If $\lambda(B) < +\infty$ then

$$\lambda(B) = \sup\{{}^\circ\nu(A) : A \subset B,\ A \in \mathscr{A}\} \quad (B \in \sigma(\mathscr{A}))$$

and for all $B \in \sigma(\mathscr{A})$ there is $A \in \mathscr{A}$ satisfying $\lambda(A \triangle B) = 0$.

To an arbitrary $B \in \sigma(\mathscr{A})$ either there is $A \in \mathscr{A}$ such that $A \subseteq B$ and ${}^\circ\nu(A) = +\infty$, or there is a sequence $(A_n)_{n\in\mathbb{N}}$ of sets in $\mathscr{A}$ such that $B \subseteq \bigcup_{n\in\mathbb{N}} A_n$ and ${}^\circ\nu(A_n) < +\infty$ for all $n \in \mathbb{N}$.

1.10.3. Let $S(\mathscr{A})$ stand for a completion of $\sigma(\mathscr{A})$ with respect to the measure λ, and let ν_L stand for the extension of λ to $S(\mathscr{A})$. We may show that in case $\nu_L(X) < +\infty$, the containment $B \in S(\mathscr{A})$ holds if and only if

$$\sup\{{}^\circ\nu(A) : A \subseteq B,\ A \in \mathscr{A}\} = \inf\{{}^\circ\nu(A) : B \subseteq A,\ A \in \mathscr{A}\} = \nu_L(B).$$

The triple $(X, S(\mathscr{A}), \nu_L)$, presenting a measure space with the σ-additive measure ν_L, is called the *Loeb space*; and ν_L, the *Loeb measure.*

1.10.4. A function $f : X \to {}^\circ\mathbb{R} \cup \{\pm\infty\}$ is called *Loeb-measurable*, provided that f is measurable with respect to the σ-algebra $S(\mathscr{A})$. An internal function $F : X \to \mathbb{R}$ is called $\mathscr{A}$-measurable if $\{x \in X : F(x) \leq t\} \in \mathscr{A}$ for all $t \in \mathbb{R}$. An internal function F is called *simple* if the range $\operatorname{rng}(F)$ of F is a hyperfinite set. Obviously, a simple internal function F is $\mathscr{A}$-measurable if and only if $F^{-1}(\{t\}) \in \mathscr{A}$ for all $t \in \mathbb{R}$. In this event to F there corresponds the internal integral

$$\int_X F\,d\nu = \sum_{t\in\operatorname{rng}(F)} F(t)\nu(F^{-1}(\{t\})).$$

If $A \in \mathscr{A}$ then, as usual, $\int_A F\,d\nu = \int_X F \cdot \chi_A\,d\nu$, where χ_A is the characteristic function of a set A.

We put $A_N := \{x \in X : |F(x)| \geq N\}$. A simple internal $\mathscr{A}$-measurable function $F : X \to \mathbb{R}$ is called *$\mathscr{S}$-integrable* if $\int_{A_N} F\,d\nu \approx 0$, for all infinitely large $N \in \mathbb{N}$. The next two theorems deal with the Loeb space of limited measure: $\nu_L(X) < +\infty$.

1.10.5. *Theorem.* *For each simple internal $\mathscr{A}$-measurable function $F : X \to \mathbb{R}$ the following are equivalent:*

(1) *F is $\mathscr{S}$-measurable;*

(2) *${}^\circ\int_X |F|\,d\nu < +\infty$, and $\nu(A) \approx 0$ implies $\int_A |F|\,d\nu \approx 0$ for all $A \in \mathscr{A}$;*

(3) *$\int_X {}^\circ|F|\,d\nu_L = {}^\circ\int_X |F|\,d\nu$.*

An internal $\mathscr{A}$-measurable function $F : X \to \mathbb{R}$ is called a *lifting* of a function $f : X \to {}^\circ\mathbb{R} \cup \{\pm\infty\}$ provided that $f(x) = {}^\circ F(x)$ for ν_L-almost all x.

1.10.6. *Theorem.* *The following are valid:*

(1) *A function $f : X \to {}^\circ\mathbb{R} \cup \{\pm\infty\}$ is measurable if and only if f has a lifting.*

(2) *A function $f : X \to {}^\circ\mathbb{R} \cup \{\pm\infty\}$ is integrable if and only if f has an $\mathscr{S}$-integrable lifting $F : X \to \mathbb{R}$.*

(3) *If $F : X \to \mathbb{R}$ is an $\mathscr{S}$-integrable lifting of a function $f : X \to {}^\circ\mathbb{R} \cup \{\pm\infty\}$ then $\int_X f\,d\nu_L = {}^\circ\int_X F\,d\nu$.*

1.10.7. We suppose that X is a hyperfinite set, $\mathscr{A} = \mathscr{P}(X)$, and $\nu(A) = \Delta|A|$ for all $A \in \mathscr{A}$, where Δ is the value of ν at singletons in X and $|A|$ stands for the size of A. The corresponding Loeb space is denoted $(X, S_\Delta, \nu_\Delta)$, and the measure ν_Δ is called the *uniform Loeb measure.* If $\Delta = |X|^{-1}$ then the Loeb space is called *canonical* and denoted by (X, S, ν_L) or (X, S^X, ν_L^X). In the case of uniform Loeb measures, every internal function $F : X \to \mathbb{R}$ is simple and $\mathscr{A}$-measurable; moreover, $\int_A F\,d\nu = \Delta \sum_{x \in A} F(x)$ for all $A \in \mathscr{A}$.

The Loeb measure ν_Δ is finite provided that $\Delta \cdot |X|$ is finite. In the case of a finite Loeb measure, if $F : X \to \mathbb{R}$ is an $\mathscr{S}$-integrable lifting of a function $f : X \to {}^\circ\mathbb{R} \cup \{\pm\infty\}$, then by Theorem 1.10.6

$$\int_X f\,d\nu_\Delta = {}^\circ\Big(\Delta \sum_{x \in X} F(x)\Big).$$

1.10.8. Theorem. *Let $(X, \mathscr{A}, \mu)$ be a standard measure space with σ-finite measure. Then there are an internal hyperfinite set $\mathscr{X} \subset X$ and a positive real $\Delta \in \mathbb{R}$ such that*

$$\int_X f\,d\mu = {}^\circ\Big(\Delta \sum_{\xi \in \mathscr{X}} f(\xi)\Big)$$

for every integrable standard function $f : X \to \mathbb{R}$.

In other words, under the hypotheses of the theorem there are a natural $N \in \mathbb{N} \setminus {}^\circ\mathbb{N}$, a collection of elements $\mathscr{X} := \{x_1, \dots, x_N\} \subset X$, and a real $\Delta \in \mathbb{R}$ satisfying

$$\int_X f\,d\mu = {}^\circ\Big(\Delta \sum_{k=1}^{N} f(x_k)\Big).$$

Theorem 1.10.8 remains valid in the case of a σ-finite measure [19].

1.10.9. We now apply the construction of the Loeb measure to a measurable family of measures. Let $(X, \mathscr{A})$ be a measurable space and let $(Y, \mathscr{B}, \nu)$ stand for a measure space; i.e., we assume as usual that X and Y are nonempty sets, $\mathscr{A}$ and $\mathscr{B}$ are some σ-algebras of subsets of X and Y respectively, and ν is a measure on $\mathscr{B}$.

A *random measure* we call a function $\lambda : \mathscr{A} \times Y \to \mathbb{R}$ that satisfies the conditions:

(1) the function $\lambda_A := \lambda(A, \cdot) : Y \to \mathbb{R}$ is $\mathscr{B}$-measurable for all $A \in \mathscr{A}$;

(2) there is a subset $\overline{Y} \subset Y$ of full ν-measure such that the function $\lambda_y := \lambda(\cdot, y) : \mathscr{A} \to \mathbb{R}$ is a measure on $\mathscr{A}$ for all $y \in \overline{Y}$.

We will write $\lambda : \mathscr{A} \times Y_{\mathscr{B}} \to \mathbb{R}$, emphasizing that Y is viewed as furnished with the σ-algebra $\mathscr{B}$.

In the sequel, the objects $(X, \mathscr{A})$, $(Y, \mathscr{B}, \nu)$ and λ are internal, with λ bounded by a standard constant. We will briefly denote the Loeb space $L(X, S(\mathscr{B}), \nu_L)$ by

$L(\mathscr{B}) := L(\mathscr{B}, \nu)$. Given $y \in \overline{Y}$ and considering λ_y, we construct the Loeb measure $(\lambda_y)_L : L(\mathscr{A}, \lambda_y) \to {}^\circ\mathbb{R}$. Let $\sigma(\mathscr{A})$ stand for the least external σ-algebra including the algebra $\mathscr{A}$. By the construction of the Loeb measure, we have $\sigma(\mathscr{A}) \subset L(\mathscr{A}, \lambda_y)$ for all $y \in \overline{Y}$.

We define the function $\lambda^L : \sigma(\mathscr{A}) \times Y \to {}^\circ\mathbb{R}$ as follows: Given $y \in \overline{Y}$ and $A \in \sigma(\mathscr{A})$, we put $\lambda^L(A, y) := (\lambda_y)_L(A)$, defining λ^L on $Y \setminus \overline{Y}$ arbitrarily.

1.10.10. *Theorem.* *A function $\lambda^L : \sigma(\mathscr{A}) \times Y_{L(\mathscr{B})} \to {}^\circ\mathbb{R}$ is an external random measure.*

$\triangleleft$ Note first that $\lambda^L_y = (\lambda_y)_L$ and $\nu_L(Y \backslash \overline{Y}) = 0$, which implies that λ^L_y is a measure for ν_L-almost all $y \in Y$. Denote by $\mathfrak{M}$ the set of $A \in \sigma(\mathscr{A})$ such that $\lambda^L_A = \lambda^L(A, \cdot)$ is a $L(\mathscr{B})$-measurable function. If $A \in \mathscr{A}$ then $\lambda^L_A(y) = \lambda^L_y(A) = {}^\circ\lambda_y(A) = {}^\circ\lambda_A(y)$ for all $y \in \overline{Y}$. Consequently, λ_A is a lifting of λ^L_A. Since λ_A is a $\mathscr{B}$-measurable function, by Theorem 1.10.6 of lifting λ^L_A is $L(\mathscr{B})$-measurable, i.e. $\mathscr{A} \subset \mathfrak{M}$.

Let now $(A_n)_{n \in {}^\circ\mathbb{N}}$ be a nested sequence of sets in $\mathfrak{M}$, with $A = \lim_{n \to \infty} A_n$. Then $A \in \sigma(\mathscr{A})$.

Since $\lambda^L_y = \lim_{n \to \infty} \lambda^L_y(A_n)$ for all $y \in \overline{Y}$; therefore, λ^L_A is an $L(\mathscr{B})$-measurable function as the limit of the sequence of $L(\mathscr{B})$-measurable functions $(\lambda^L_{A_n})_{n \in {}^\circ\mathbb{N}}$. Consequently, $\mathfrak{M}$ is a monotonic class and so $\sigma(\mathscr{A}) \subset \mathfrak{M}$. It remains to observe that $\mathfrak{M} \subset \sigma(\mathscr{A})$ by construction. $\triangleright$

We consider the family of measures $\lambda^L_{(\cdot)}$ only on $\sigma(\mathscr{A})$, since it may fail to be a random measure on a wider σ-algebra.

1.10.11. The above properties of the Loeb measure may be helpful in "discretizing" operators, i.e., in constructing their hyperfinite approximants.

We consider a standard family of measure spaces $(X, \mathscr{A}, \lambda_y)_{y \in \mathscr{Y}}$, each with a σ-finite measure $\lambda_y = \lambda(\cdot, y)$ for some standard function $\lambda : \mathscr{A} \times Y \to \mathbb{R}$. Introduce the notations: $\mathscr{F}(Y) := \mathbb{R}^Y$ and $\mathscr{L}_1(X) := \{f : X \to \mathbb{R} : f$ is λ_y-integrable for all $y \in Y\}$. Given a finite collection $\mathscr{X} = \{x_1, \dots, x_N\}$ of members of X, we let the symbol π_X stand for the "projection" from $\mathscr{L}_1(X)$ to $\mathbb{R}^N$ which sends each function $f \in \mathscr{L}_1(\mathscr{X})$ to the vector $(f(x_1), \dots, f(x_N))$. By analogy, granted a finite collection $\mathscr{Y} = \{y_1, \dots, y_M\}$ of members of Y, we define $\pi_Y : \mathscr{F}(Y) \to \mathbb{R}^M$ by the rule $\pi_Y(F) = (F(y_1), \dots, F(y_M))$.

We denote by T the pseudointegral operator that acts from $\mathscr{L}_1(X)$ to $\mathscr{F}(Y)$ as follows:

$$(Tf)(y) = \int\limits_X f\, d\lambda_y \quad (f \in \mathscr{L}_1(X)).$$

1.10.12. *Theorem.* *In the spaces X and Y there are finite collections of elements $\mathscr{X} := \{x_1, \dots, x_N\}$ and $\mathscr{Y} := \{y_1, \dots, y_M\}$, together with an $N \times M$*

matrix Λ, *such that* $\pi_Y(Tf) \approx \Lambda\pi_X(f)$ *for every standard function* $f \in \mathscr{L}_1(X)$; *i.e.*,

$$\int f\,d\lambda_{y_l} \approx \sum_{k=1}^{N} f(x_k)\Lambda_{kl} \quad (l = 1, \dots, M).$$

In other words, the following diagram commutes to within infinitesimals:

$$\begin{array}{ccc} \mathscr{L}_1(X) & \xrightarrow{T} & \mathscr{F}(Y) \\ {\scriptstyle \pi_X}\downarrow & & \downarrow{\scriptstyle \pi_Y} \\ \mathbb{R}^N & \xrightarrow{\Lambda} & \mathbb{R}^M \end{array}$$

1.10.13. *Theorem.* *There are a collection* $\mathscr{Y} = \{y_1, \dots, y_M\}$ *of members of* Y *and a matrix* $\Lambda := (\Lambda_{kl})$ *such that* $\Lambda_{kl} = \Delta \cdot K(x_k, y_j)$ *and* $\pi_Y(Tf) \approx \Lambda\pi_X(f)$.

1.10.14. Comments.

(1) The content of 1.10.1–1.10.7 is well known; cf. [3, 11]. Our presentation follows [21]. Theorem 1.10.8 belongs to E. I. Gordon [19]. In a series of articles, E. I. Gordon has elaborated a technique for hyperfinite approximation of integral operators [17, 21] as well as some nonstandard methods for discretization in harmonic analysis [20, 21].

(2) The construction of the Loeb measure generalizes straightforwardly to the case of vector measures ranging in a Banach space. However, the problem becomes more involved in the case of measures taking values in a "norm-free" vector lattice even on replacing norm completeness with Dedekind completeness.

(3) Theorems 1.10.12 and 1.10.13 belong to V. G. Troitskiĭ [110].

1.11. Boolean Valued Modeling in a Nonstandard Universe

Boolean valued analysis introduces a new important class of mathematical objects, the structures with the cyclic property, that is, closed under mixing; see 1.2.6 (2). These objects are the descents of analogous formations in the Boolean valued universe $\mathbf{V}^{(B)}$; cf. 1.2.8.

On the other hand, the methodology of infinitesimal analysis rests on a special machinery for studying filters, *monadology*.

Indeed, let $\mathscr{F}$ be a standard filter; ${}^{\circ}\mathscr{F}$, its standard core; and let ${}^{a}\mathscr{F} := \mathscr{F} \setminus {}^{\circ}\mathscr{F}$ be the external set of *astray* or *remote* elements of $\mathscr{F}$. Note that

$$\mu(\mathscr{F}) := \bigcap {}^{\circ}\mathscr{F} = \bigcup {}^{a}\mathscr{F}$$

is the *monad* of $\mathscr{F}$. Also, $\mathscr{F} = {}^*\operatorname{fil}(\{\mu(\mathscr{F})\})$; i.e., $\mathscr{F}$ is the standardization of the collection $\operatorname{fil}(\mu(\mathscr{F}))$ of supersets of $\mu(\mathscr{F})$.

The notion of monad is central to external set theory. In this connection the development of combined methods (in particular, application of infinitesimals and the technique of descending and ascending simultaneously in Kantorovich space theory) requires adaptation of the notion of monad for filters and their implementations in a Boolean valued universe.

In this section we pursue an approach in which the ordinary monadology is applied to descents of objects. An alternative way of applying the standard monadology inside $\mathbf{V}^{(B)}$ while ascending and descending will be treated in the next section.

1.11.1. We start with recalling some constructions from the theory of filters in $\mathbf{V}^{(B)}$. Let $\mathscr{G}$ be a filterbase on X, with $X \in \mathscr{P}(\mathbf{V}^{(B)})$. We put

$$\mathscr{G}' := \{F \in \mathscr{P}(X\uparrow)\downarrow : (\exists G \in \mathscr{G})\, [\![F \supset G\uparrow]\!] = \mathbf{1}\};$$
$$\mathscr{G}'' := \{G\uparrow : G \in \mathscr{G}\}.$$

Then $\mathscr{G}'\uparrow$ and $\mathscr{G}''\uparrow$ are bases of the same filter $\mathscr{G}^{\uparrow}$ on $X\uparrow$ inside $\mathbf{V}^{(B)}$. The filter $\mathscr{G}^{\uparrow}$ is called the *ascent* of $\mathscr{G}$. If $\operatorname{mix}(\mathscr{G})$ is the set of all mixings of nonempty families of elements of $\mathscr{G}$ and $\mathscr{G}$ consists of cyclic sets; then $\operatorname{mix}(\mathscr{G})$ is a filterbase on X and $\mathscr{G}^{\uparrow} = \operatorname{mix}(\mathscr{G})^{\uparrow}$.

If $\mathscr{F}$ is a filter on X inside $\mathbf{V}^{(B)}$ then we put $\mathscr{F}^{\downarrow} := \operatorname{fil}(\{F\downarrow : F \in \mathscr{F}\downarrow\})$. The filter $\mathscr{F}^{\downarrow}$ is called the *descent* of $\mathscr{F}$. A filterbase $\mathscr{G}$ on $X\downarrow$ is called *extensional* if there is a filter $\mathscr{F}$ on X such that $\operatorname{fil}(\mathscr{G}) = \mathscr{F}$.

Finally, the descent of an ultrafilter on X is called a *proultrafilter* on $X\downarrow$. A filter having a base of cyclic sets is called *cyclic*. Proultrafilters are maximal cyclic filters.

1.11.2. We fix a standard complete Boolean algebra B and the corresponding Boolean valued universe $\mathbf{V}^{(B)}$ thought of as composed of internal sets. If A is an external set then the *cyclic hull* $\operatorname{mix}(A)$ is introduced as follows: We say that an element $x \in \mathbf{V}^{(B)}$ belongs to $\operatorname{mix}(A)$ if there are an internal family $(a_\xi)_{\xi\in\Xi}$ of elements of A and an internal partition $(b_\xi)_{\xi\in\Xi}$ of unity in B such that x is the mixing of $(a_\xi)_{\xi\in\Xi}$ by $(b_\xi)_{\xi\in\Xi}$; i.e., $b_\xi x = b_\xi a_\xi$ for $\xi \in \Xi$ or, equivalently, $x = \operatorname{mix}_{\xi\in\Xi}(b_\xi a_\xi)$.

1.11.3. *Theorem.* *Given a filter $\mathscr{F}$ on $X\downarrow$, let*

$$\mathscr{F}\uparrow\downarrow := \operatorname{fil}(\{F\uparrow\downarrow : F \in \mathscr{F}\}).$$

Then $\operatorname{mix}(\mu(\mathscr{F})) = \mu(\mathscr{F}\uparrow\downarrow)$ and $\mathscr{F}\uparrow\downarrow$ is the greatest cyclic filter coarser than $\mathscr{F}$.

In connection with this theorem, the monad of $\mathscr{F}$ is called *cyclic* if $\mu(\mathscr{F}) = \mathrm{mix}(\mu(\mathscr{F}))$. Unfortunately, the cyclicity of a monad is not completely responsible for extensionality of a filter. To obviate this shortcoming, we introduce the *cyclic monad hull* $\mu_c(U)$ of an external set U. Namely, we put

$$x \in \mu_c(U) \leftrightarrow (\forall^{\mathrm{st}} V = V{\uparrow}{\downarrow}) V \supset U \rightarrow x \in \mu(V).$$

In particular, if $B = \{0, 1\}$, then $\mu_c(U)$ coincides with the monad of the standardization of the external filter of supersets of U, i.e. with the so-called (*discrete*) *monad hull* $\mu_d(U)$ (the word "monadic" is also in common parlance).

1.11.4. *The cyclic monad hull of a set is the cyclic hull of its monad hull*

$$\mu_c(U) = \mathrm{mix}(\mu_d(U)).$$

1.11.5. A special role is played by the *essential points* of $X{\downarrow}$ constituting the external set ${}^{\mathrm{e}}X$. By definition, ${}^{\mathrm{e}}X$ consists of the elements of proultrafilter monads on $X{\downarrow}$.

Criterion for Essentiality. *A point is essential if and only if it can be separated by a standard set from every standard cyclic set not containing the point.*

1.11.6. If there is an essential point in the monad of an ultrafilter $\mathscr{F}$ then $\mu(\mathscr{F}) \subset {}^{\mathrm{e}}X$; moreover, $\mathscr{F}{\uparrow}{\downarrow}$ is a proultrafilter.

The next statement follows from the constructions and considerations presented above.

Criterion for a Filter to Be Extensional. *A filter is extensional if and only if its monad is the cyclic monad hull of the set of its own essential points.*

A standard set is cyclic if and only if it is the cyclic monad hull of its own essential points.

1.11.7. Nonstandard Criterion for the Mixing of Filters. *Let* $(\mathscr{F}_\xi)_{\xi \in \Xi}$ *be a standard family of extensional filters, and let* $(b_\xi)_{\xi \in \Xi}$ *be a standard partition of unity. The filter* $\mathscr{F}$ *is the mixing of* $(\mathscr{F}_\xi)_{\xi \in \Xi}$ *by* $(b_\xi)_{\xi \in \Xi}$ *if and only if*

$$(\forall^{\mathrm{St}} \xi \in \Xi) b_\xi \mu(\mathscr{F}) = b_\xi \mu(\mathscr{F}_\xi).$$

A peculiarity of the above approach reveals itself in applications to the descents of topological spaces through a special new role of essential points. In this connection, we note some properties of the latter.

1.11.8. *The following are true:*

(1) *The image of an essential point under an extensional mapping is an essential point of the image;*

(2) *Let E be a standard set, and let X be a standard element of $\mathbf{V}^{(B)}$. Consider the product $X^{E^\wedge}$ inside $\mathbf{V}^{(B)}$, where $E^\wedge$ is the standard name of E in $\mathbf{V}^{(B)}$. If x is an essential point of $X^{E^\wedge}{\downarrow}$ then for every standard $e \in E$ the point $x{\downarrow}(e)$ is essential in $X{\downarrow}$;*

(3) *Let $\mathscr{F}$ be a cyclic filter in $X{\downarrow}$, and let ${}^e\mu(\mathscr{F}) := \mu(\mathscr{F}) \cap {}^eX$ be the set of essential points of its monad. Then ${}^e\mu(\mathscr{F}) = {}^e\mu(\mathscr{F}^{\uparrow\downarrow})$.*

Let $(X, \mathscr{U})$ be a uniform space inside $\mathbf{V}^{(B)}$. The uniform space $(X{\downarrow}, \mathscr{U}^{\downarrow})$ is called *procompact* or *cyclically compact* if $(X, \mathscr{U})$ is compact inside $\mathbf{V}^{(B)}$. A similar sense resides in the notion of *pro-total-boundedness* and so on.

1.11.9. Nonstandard Criterion for Procompactness. *Every essential point of $X{\downarrow}$ is nearstandard, i.e., infinitesimally close to a standard point, if and only if $X{\downarrow}$ is procompact.*

It is easy from Theorem 1.11.9 that the Boolean valued criterion of procompactness differs from its usual analog: "A compact space is a space formed by nearstandard points." Existence of a great number of procompact but not compact spaces provides the variety of examples of inessential points.

We note here that a combined application of 1.11.7 and 1.11.8 (2), of course, allows us to produce a nonstandard proof for a natural analog of Tychonoff's Theorem for a product of procompact spaces, the "descent of Tychonoff's Theorem in $\mathbf{V}^{(B)}$.

1.11.10. Nonstandard Criterion for Proprecompactness. *A standard space is the descent of a totally bounded uniform space if and only if its every essential point is prenearstandard, i.e. belongs to the monad of a Cauchy filter.*

We will apply this approach to describing o-convergence in a Kantorovich space Y. To save space, we restrict exposition to the filters containing order intervals or, equivalently, filters with bounded monad. Moreover, with the same end in mind, we assume Y to be a universally complete Kantorovich space. By Gordon's Theorem, we may view the space Y as the descent $\mathscr{R}{\downarrow}$ of the reals $\mathscr{R}$ inside the Boolean valued universe $\mathbf{V}^{(B)}$ over the base B of Y.

We denote by $\mathscr{E}$ the filter of order units in Y, i.e. the set $\mathscr{E} := \{\varepsilon \in Y_+ : [\![\varepsilon = 0]\!] = \mathbf{0}\}$. We write $x \approx y$ whenever elements $x, y \in Y$ are infinitely close with respect to the descent of the natural topology of $\mathscr{R}$ inside $\mathbf{V}^{(B)}$, i.e., $x \approx y \leftrightarrow (\forall^{\mathrm{st}}\varepsilon \in \mathscr{E})\,(|x - y| < \varepsilon)$. Given $a, b \in Y$, we write $a < b$ if $[\![a < b]\!] = \mathbf{1}$; in other words, $a > b \leftrightarrow a - b \in \mathscr{E}$. Thus, there is some deviation from the understanding of the theory of ordered vector spaces. Clearly, this is done in order to adhere to the principles of introducing notations while descending and ascending.

Let ${}^{\approx}Y$ be the *nearstandard part* of Y. Given $y \in {}^{\approx}Y$, we denote by ${}^{\circ}y$ (or by $\mathrm{st}(y)$) the *standard part* of y, i.e. the unique standard element infinitely close to y.

1.11.11. *Theorem.* *For a standard filter $\mathscr{F}$ in Y and a standard $z \in Y$, the following are true:*

$$\text{(1)}\ \inf_{F\in\mathscr{F}} \sup F \le z \leftrightarrow (\forall y \in {}^{\cdot}\mu(\mathscr{F}\uparrow\downarrow))\, {}^{\circ}y \le z \leftrightarrow (\forall y \in {}^{e}\mu(\mathscr{F}\uparrow\downarrow))\, {}^{\circ}y \le z;$$

$$\text{(2)}\ \sup_{F\in\mathscr{F}} \inf F \ge z \leftrightarrow (\forall y \in {}^{\cdot}\mu(\mathscr{F}\uparrow\downarrow))\, {}^{\circ}y \ge z \leftrightarrow (\forall y \in {}^{e}\mu(\mathscr{F}\uparrow\downarrow))\, {}^{\circ}y \ge z;$$

$$\text{(3)}\ \inf_{F\in\mathscr{F}} \sup F \ge z \leftrightarrow (\exists y \in {}^{\cdot}\mu(\mathscr{F}\uparrow\downarrow))\, {}^{\circ}y \ge z \leftrightarrow (\exists y \in {}^{e}\mu(\mathscr{F}\uparrow\downarrow))\, {}^{\circ}y \ge z;$$

$$\text{(4)}\ \sup_{F\in\mathscr{F}} \inf F \le z \leftrightarrow (\exists y \in {}^{\cdot}\mu(\mathscr{F}\uparrow\downarrow))\, {}^{\circ}y \le z \leftrightarrow (\exists y \in {}^{e}\mu(\mathscr{F}\uparrow\downarrow))^{\circ}y \le z;$$

$$\text{(5)}\ \mathscr{F} \overset{(o)}{\to} z \leftrightarrow (\forall y \in {}^{e}\mu(\mathscr{F}\uparrow\downarrow))y \approx z \leftrightarrow (\forall y \in \mu(\mathscr{F}^{\uparrow\downarrow}))y \approx z.$$

Here ${}^{\cdot}\mu(\mathscr{F}\uparrow\downarrow) := \mu(\mathscr{F}\uparrow\downarrow) \cap {}^{\approx}Y$, and, as usual, ${}^{e}\mu(\mathscr{F}\uparrow\downarrow)$ is the set of essential points of the monad $\mu(\mathscr{F}\uparrow\downarrow)$, i.e. ${}^{e}\mu(\mathscr{F}\uparrow\downarrow) = \mu(\mathscr{F}\uparrow\downarrow) \cap {}^{e}\mathscr{R}$.

◁ By way of illustration, we will prove (3).

Suppose first that in the greater set ${}^{\cdot}\mu(\mathscr{F}\uparrow\downarrow)$ there is an element y satisfying ${}^{\circ}y \ge z$. For every standard $F \in \mathscr{F}$ we have $y \in F\uparrow\downarrow$. Hence, if $\varepsilon \in {}^{\circ}\mathscr{E}$ then $y > z - \varepsilon$ and $\sup F = \sup F\uparrow\downarrow > z - \varepsilon$.

By Leibniz's Principle we infer $(\forall^{\mathrm{st}} F \in \mathscr{F})\, (\forall^{\mathrm{st}} \varepsilon > 0)\, (\sup F \ge z)$, i.e. $(\forall F \in \mathscr{F})\, (\sup F \ge z)$ and $\inf_{F\in\mathscr{F}} \sup F \ge z$.

To prove the rest of the claim, begin with noting that by the properties of the upper limit in $\mathbb{R}$ and by the transfer principle of Boolean valued analysis we have

$$[\![(\exists \mathscr{G})\, (\mathscr{G} \text{ is an ultrafilter on } \mathscr{R} \wedge \mathscr{G} \supset \mathscr{F}^{\uparrow} \wedge \inf_{G\in\mathscr{G}} \sup G \ge z)]\!] = \mathbf{1}.$$

According to the maximum principle, there is a proultrafilter $\mathscr{G}$ such that $\mathscr{G} \supset \mathscr{F}^{\uparrow\downarrow}$ and $\inf_{G\in\mathscr{G}} \sup G \ge z$. Using the transfer and idealization principles, we successively find

$$(\forall^{\mathrm{st}} G \in \mathscr{G})\ \sup G \ge z \leftrightarrow (\forall^{\mathrm{st}} G \in \mathscr{G})\, [\![\sup(G\uparrow) = z]\!] = \mathbf{1}$$

$$\leftrightarrow\ (\forall^{\mathrm{st}} G \in \mathscr{G})\, [\![(\forall \varepsilon > 0)\, (\exists g \in G\uparrow)\, g > z - \varepsilon]\!] = \mathbf{1}$$

$$\leftrightarrow (\forall^{\mathrm{st}} G \in \mathscr{G})\, (\forall \varepsilon > 0)\, (\exists g \in G\uparrow\downarrow) g > z - \varepsilon$$

$$\leftrightarrow\ (\forall^{\mathrm{st}} G \in \mathscr{G})\, (\forall^{\mathrm{st}} \varepsilon > 0)\, (\exists g \in G\uparrow\downarrow) g > z - \varepsilon$$

$$\leftrightarrow (\forall^{\mathrm{st\,fin}}\mathscr{G}_0 \supset \mathscr{G})\,(\forall^{\mathrm{st\,fin}}\mathscr{E}_0 \subset \mathscr{E})\,(\exists g)(\forall G \in \mathscr{G}_0)\,(\forall \varepsilon \in \mathscr{E}_0)\,(g \in G{\uparrow\downarrow} \wedge g > z - \varepsilon)$$

$$\leftrightarrow (\exists g)\,(\forall^{\mathrm{st}} G \in \mathscr{G})\,(\forall^{\mathrm{st}} \varepsilon > 0)\,(g \in G{\uparrow\downarrow} \wedge g > z - \varepsilon)$$

$$\leftrightarrow (\exists g \in \mu(\mathscr{G}^{\uparrow\downarrow}))\,{}^{\circ}g \geq z \leftrightarrow (\exists g \in \mu(\mathscr{G}))^{\circ}g = z.$$

The observation

$$\mu(\mathscr{G}) \subset {}^{e}\mu(\mathscr{F}^{\uparrow\downarrow}) = {}^{e}\mu(\mathscr{F}{\uparrow\downarrow}) \subset {}^{\circ}\mu(\mathscr{F}{\uparrow\downarrow})$$

completes the proof. ▷

1.11.12. Comments.

(1) Monadology as a philosophical doctrine is a creation of G. W. Leibniz [78]. The general theory of the monads of filters was proposed by W. A. J. Luxemburg [80]. Cyclic topologies are widely used in Boolean valued analysis. The theory of cyclic compactness and the principles of descending and ascending filters are given in [51, 53, 73, 74]. Setting forth cyclic monadology, we proceed along the lines of [73, 74].

(2) Considering ultraproducts inside a Boolean valued universe causes no difficulty in principle and was carried out in several papers. We do not discuss here how Robinson's standardization is introduced in $\mathbf{V}^{(B)}$; as a matter of fact, an axiomatic approach is also possible. A decisive element is the appearance of excrescences on the Kantorovich spaces under study which sprout in general out of the uncustomary way of standardizing the source space (the effect of essential points). Our presentation follows [59].

1.12. Infinitesimal Modeling in a Boolean Valued Universe

In this section we assume given a complete Boolean algebra B and a separated universe $\mathbf{V}^{(B)}$.

Applying the methods of infinitesimal analysis, we adhere to the classical approach of A. Robinson inside $\mathbf{V}^{(B)}$. In other words, in a particular situation the classical and internal universes and the corresponding $*$-map (Robinson's standardization) are understood to be members of $\mathbf{V}^{(B)}$. Moreover, the nonstandard world is supposed to be properly saturated.

1.12.1. The descent of the $*$-map is referred to as *descent standardization.* Alongside the term "descent standardization" we also use the expressions like "B-standardization," "prostandardization," etc. Furthermore, we denote the Robinson standardization of a B-set A by the symbol ${}^{*}A$.

By analogy, the *descent standardization* of a set A with B-structure, i.e. a subset of $\mathbf{V}^{(B)}$, is defined as $({}^*(A\uparrow))\downarrow$ and is denoted by the symbol ${}_*A$ (it is meant here that $A\uparrow$ is an element of the standard universe located inside $\mathbf{V}^{(B)}$).

Thus, ${}^*a \in {}_*A \leftrightarrow a \in A\uparrow\downarrow$. The *descent standardization* ${}_*\Phi$ of an *extensional correspondence* Φ is also defined in a natural way.

While considering descent standardizations of the standard names of elements of the von Neumann universe $\mathbf{V}$, for convenience we will use the abbreviations ${}^*x := {}^*(x^\wedge)$ and ${}_*x := ({}^*x)\downarrow$ for $x \in \mathbf{V}$. The rules of placing and omitting asterisks (by default) in descent standardization are also assumed as free as those for the Robinson $*$-map.

1.12.2. Transfer Principle. *Let $\varphi = \varphi(x, y)$ be a formula of ZFC without any free variables other than x and y. Then we have:*

$$(\exists x \in {}_*F)\,[\![\varphi(x, {}^*z)]\!] = \mathbf{1} \leftrightarrow (\exists x \in F\downarrow)\,[\![\varphi(x, z)]\!] = \mathbf{1};$$

$$(\forall x \in {}_*F)\,[\![\varphi(x, {}^*z)]\!] = \mathbf{1} \leftrightarrow (\forall x \in F\downarrow)\,[\![\varphi(x, z)]\!] = \mathbf{1}$$

for a nonempty element F in $\mathbf{V}^{(B)}$ and for every z.

1.12.3. Idealization Principle. *Let $X\uparrow$ and Y be classical elements of $\mathbf{V}^{(B)}$, and let $\varphi = \varphi(x, y, z)$ be a formula of ZFC. Then*

$$(\forall^{\mathrm{fin}} A \subset X)\,(\exists y \in {}_*Y)\,(\forall x \in A)\,[\![\varphi({}^*x, y, z)]\!] = \mathbf{1}$$

$$\leftrightarrow (\exists y \in {}_*Y)\,(\forall x \in X)\,[\![\varphi({}^*x, y, z)]\!] = \mathbf{1}$$

for an internal element z in $\mathbf{V}^{(B)}$.

Given a filter $\mathscr{F}$ of sets with B-structure, we define its *descent monad* $m(\mathscr{F})$ as follows:

$$m(\mathscr{F}) := \bigcap_{F \in \mathscr{F}} {}_*F.$$

1.12.4. Theorem. *Let $\mathscr{E}$ be a set of filters, and let $\mathscr{E}^\uparrow := \{\mathscr{F}^\uparrow : \mathscr{F} \in \mathscr{E}\}$ be its ascent to $\mathbf{V}^{(B)}$. The following are equivalent:*

(1) *the set of cyclic hulls of $\mathscr{E}$, i.e. $\mathscr{E}\uparrow\downarrow := \{\mathscr{F}\uparrow\downarrow : \mathscr{F} \in \mathscr{E}\}$, is bounded above;*

(2) *the set $\mathscr{E}^\uparrow$ is bounded above inside $\mathbf{V}^{(B)}$;*

(3) $\bigcap\{m(\mathscr{F}) : \mathscr{F} \in \mathscr{E}\} \neq \varnothing$.

Moreover, if the conditions (1)–(3) *are met then*

$$m(\sup \mathscr{E}{\uparrow}{\downarrow}) = \bigcap\{m(\mathscr{F}) : \mathscr{F} \in \mathscr{E}\};$$

$$\sup \mathscr{E}^{\uparrow} = (\sup \mathscr{E})^{\uparrow}.$$

It is worth noting that for an infinite set of descent monads, its union, and even the cyclic hull of this union, is not a descent monad in general. The situation here is the same as for ordinary monads.

1.12.5. Nonstandard Criteria for a Proultrafilter. *The following are equivalent:*

(1) $\mathscr{U}$ *is a proultrafilter;*

(2) $\mathscr{U}$ *is an extensional filter with inclusion-minimal descent monad;*

(3) *the representation* $\mathscr{U} = (x)^{\downarrow} := \operatorname{fil}(\{U{\uparrow}{\downarrow} : x \in {}_*A\})$ *holds for each point* x *of the descent monad* $m(\mathscr{U})$;

(4) $\mathscr{U}$ *is an extensional filter whose descent monad is easily caught by a cyclic set; i.e. either* $m(\mathscr{U}) \subset {}_*U$ *or* $m(\mathscr{U}) \subset {}_*(X \setminus U)$ *for every* $U = U{\uparrow}{\downarrow}$;

(5) $\mathscr{U}$ *is a cyclic filter satisfying the condition: for every cyclic* U, *if* ${}_*U \cap m(\mathscr{A}) \neq \varnothing$ *then* $U \in \mathscr{U}$.

1.12.6. ***Nonstandard Criterion for a Mixing of Filters.*** *Let* $(\mathscr{F}_\xi)_{\xi\in\Xi}$ *be a family of filters, let* $(b_\xi)_{\xi\in\Xi}$ *be a partition of unity, and let* $\mathscr{F} = \operatorname{mix}_{\xi\in\Xi}(b_\xi \mathscr{F}_\xi^{\uparrow})$ *be the mixing of* $\mathscr{F}_\xi^{\uparrow}$ *by* b_ξ. *Then*

$$m(\mathscr{F}^{\downarrow}) = \operatorname*{mix}_{\xi\in\Xi}(b_\xi m(\mathscr{F}_\xi)).$$

It is useful to compare 1.12.6 with 1.11.7.

A point y of the set ${}_*X$ is called *descent-nearstandard* or simply *nearstandard* if there is no danger of misunderstanding whenever ${}^*x \approx y$ for some $x \in X{\downarrow}$; i.e., $(x, y) \in m(\mathscr{U}^{\downarrow})$, with $\mathscr{U}$ the uniformity on X.

1.12.7. ***Nonstandard Criterion for Procompactness.*** *A set* $A{\uparrow}{\downarrow}$ *is procompact if and only if every point of* ${}_*A$ *is descent-nearstandard.*

It is reasonable to compare 1.12.7 with 1.11.9.

1.12.8. Finally, we will formulate a few general tools for descent standardization. We start with the following observation.

Let $\varphi = \varphi(x)$ be a formula of ZFC. The truth value of φ is constant on the descent monad of every proultrafilter $\mathscr{A}$; i.e.,

$$(\forall x, y \in m(\mathscr{A}))\, [\![\varphi(x)]\!] = [\![\varphi(y)]\!].$$

We are in a position to state particular instances of the transfer principle which are useful in working with descent standardization.

1.12.9. Theorem. *Let $\varphi = \varphi(x, y, z)$ be a formula of ZFC, and let $\mathscr{F}$ and $\mathscr{G}$ be filters of sets with B-structure.*

The following quantification rules are valid (for internal y, z in $\mathbf{V}^{(B)}$):

(1) $(\exists x \in m(\mathscr{F}))\, [\![\varphi(x, y, z)]\!] = \mathbf{1}$
$\leftrightarrow (\forall F \in \mathscr{F})\, (\exists x \in {}^*F)\, [\![\varphi(x, y, z)]\!] = \mathbf{1};$

(2) $(\forall x \in m(\mathscr{F}))\, [\![\varphi(x, y, z)]\!] = \mathbf{1}$
$\leftrightarrow (\exists F \in \mathscr{F}^{\uparrow\downarrow})(\forall x \in {}_*F)\, [\![\varphi(x, y, z)]\!] = \mathbf{1};$

(3) $(\forall x \in m(\mathscr{F}))\, (\exists y \in m(\mathscr{G}))[\![\varphi(x, y, z)]\!] = \mathbf{1}$
$\leftrightarrow (\forall G \in \mathscr{G})\, (\exists F \in \mathscr{F}^{\uparrow\downarrow})\, (\forall x \in {}^*F)\, (\exists y \in {}^*G)$
$[\![\varphi(x, y, z)]\!] = \mathbf{1};$

(4) $(\exists x \in m(\mathscr{F}))\, (\forall y \in m(\mathscr{G}))\, [\![\varphi(x, y, z)]\!] = \mathbf{1}$
$\leftrightarrow (\exists G \in \mathscr{G}^{\uparrow\downarrow})\, (\forall F \in \mathscr{F})\, (\exists x \in {}^*F)\, (\forall y \in {}^*G)$
$[\![\varphi(x, y, z)]\!] = \mathbf{1}.$

Moreover, given standardized free variables, we have:

(1′) $(\exists x \in m(\mathscr{F}))[\![\varphi(x, {}^*y, {}^*z)]\!] = \mathbf{1}$
$\leftrightarrow (\forall F \in \mathscr{F})(\exists x \in F{\uparrow}{\downarrow})[\![\varphi(x, y, z)]\!] = \mathbf{1};$

(2′) $(\forall x \in m(\mathscr{F}))[\![\varphi(x, {}^*y, {}^*z)]\!] = \mathbf{1}$
$\leftrightarrow (\exists F \in \mathscr{F}^{\uparrow\downarrow})(\forall x \in F)[\![\varphi(x, y, z)]\!] = \mathbf{1};$

(3′) $(\forall x \in m(\mathscr{F}))(\exists y \in m(\mathscr{G}))[\![\varphi(x, y, {}^*z)]\!] = \mathbf{1}$
$\leftrightarrow (\forall G \in \mathscr{G})(\exists F \in \mathscr{F}^{\uparrow\downarrow})(\forall x \in F)(\exists y \in G{\uparrow}{\downarrow})$
$[\![\varphi(x, y, z)]\!] = \mathbf{1};$

(4′) $(\exists x \in m(\mathscr{F}))(\forall y \in m(\mathscr{G}))[\![\varphi(x, y, {}^*z)]\!] = \mathbf{1}$
$\leftrightarrow (\exists G \in \mathscr{G}^{\uparrow\downarrow})(\forall F \in \mathscr{F})(\exists x \in F{\uparrow}{\downarrow})(\forall y \in G)$
$[\![\varphi(x, y, z)]\!] = \mathbf{1}.$

◁ All claims are verified by straightforward calculation. ▷

1.13. Extension and Decomposition of Positive Operators

We will demonstrate in this section that many questions of the theory of order bounded and dominated operators can be reduced to the case of functionals by using Boolean valued analysis.

1.13.1. The fact that E is a vector lattice may be rewritten as a restricted formula, say, $\varphi(E, \mathbb{R})$. Hence, recalling the restricted transfer principle, we come to the equality $[\![\varphi(E^\wedge, \mathbb{R}^\wedge)]\!] = \mathbf{1}$; i.e., $E^\wedge$ is a vector lattice over the ordered field $\mathbb{R}^\wedge$ inside $\mathbf{V}^{(B)}$.

Let $E^{\wedge\sim}$ be the space of regular $\mathbb{R}^\wedge$-linear functionals from $E^\wedge$ to $\mathscr{R}$. It is easy that $E^{\wedge\sim} := L^\sim(E^\wedge, \mathscr{R})$ is a Kantorovich space inside $\mathbf{V}^{(B)}$. Since $E^{\wedge\sim}$ is a Kantorovich space, the descent $E^{\wedge\sim}{\downarrow}$ of $E^{\wedge\sim}$ is a Kantorovich space too.

We consider the universally complete Kantorovich space $F := \mathscr{R}{\downarrow}$ (see 1.5.4). We recall that for every operator $T \in L^\sim(E, F)$ the ascent $T{\uparrow}$ is defined by the equality $[\![Tx = T{\uparrow}(x^\wedge)]\!] = \mathbf{1}$ for all $x \in E$. We note that if $\tau \in E^{\wedge\sim}$, then $[\![\tau : E^\wedge \to \mathscr{R}]\!] = \mathbf{1}$; hence, the operator $\tau{\downarrow} : E \to F$ is available. Moreover, $\tau{\downarrow}{\uparrow} = \tau$. On the other hand, $T{\uparrow}{\downarrow} = T$.

1.13.2. *Theorem.* *For every $T \in L^\sim(E, F)$ the ascent $T{\uparrow}$ is a regular $\mathbb{R}^\wedge$-functional on $E^\wedge$ inside $\mathbf{V}^{(B)}$; i.e., $[\![T{\uparrow} \in E^{\wedge\sim}]\!] = \mathbf{1}$. The mapping $T \mapsto T{\uparrow}$ is a linear and lattice isomorphism between the Kantorovich spaces $L^\sim(E, F)$ and $E^{\wedge\sim}{\downarrow}$.*

1.13.3. We now formulate a few corollaries to 1.13.2. First of all, we introduce necessary definitions. An operator $S \in L^\sim(E, F)$ is called a *fragment* of an operator $0 \leq T \in L^\sim(E, F)$ if $S \wedge (T - S) = 0$. We say that T is an *F-discrete* operator whenever $[0, T] = [0, I_F] \circ T$; i.e., for every $0 \leq S \leq T$ there is an operator $0 \leq \alpha \leq I_F$ satisfying $S = \alpha \circ T$. Let $L_a^\sim(E, F)$ be the band of the space $L^\sim(E, F)$ generated by F-discrete operators, and write $L_d^\sim(E, F) := L_a^\sim(E, F)^\perp$. The bands $(E^{\wedge\sim})_a$ and $(E^{\wedge\sim})_d$ are introduced similarly. The elements of $L_d^\sim(E, F)$ are usually referred to as *F-spread* or *F-diffuse* operators. $\mathbb{R}$-discrete or $\mathbb{R}$-diffuse operators are called for the sake of brevity *discrete* or *diffuse* functionals.

Consider $S, T \in L^\sim(E, F)$ and put $\tau := T{\uparrow}$, $\sigma := S{\uparrow}$. The following are true:

(1) $T \geq 0 \leftrightarrow [\![\tau \geq 0]\!] = \mathbf{1}$;

(2) $[\![$ *S is a fragment of T* $]\!] \leftrightarrow [\![$ *σ is a fragment of τ* $]\!] = \mathbf{1}$;

(3) $[\![$ *T is F-discrete* $]\!] \leftrightarrow [\![$ *τ is discrete* $]\!] = \mathbf{1}$;

(4) $T \in L_a^\sim(E, F) \leftrightarrow [\![\tau \in (E^{\wedge\sim})_a]\!] = \mathbf{1}$;

(5) $T \in L_d^{\sim}(E,F) \leftrightarrow [\![\tau \in (E^{\wedge\sim})_d]\!] = \mathbb{1}$.

We need one more fact that follows from 1.13.2 by direct computation of Boolean truth values:

(6) $[\![T$ *is a lattice homomorphism*$]\!]$
$\leftrightarrow [\![\tau$ *is a lattice homomorphism*$]\!] = \mathbb{1}$.

1.13.4. Theorem. *Let E be a vector lattice, let F be a Kantorovich space, and suppose that $T \in L^{\sim}(E,F)$. The following are equivalent:*

(1) *T is an F-discrete element of the Kantorovich space $L^{\sim}(E,F)$;*

(2) *T is a lattice homomorphism;*

(3) *T preserves disjointness, i.e. if $x, y \in E$ and $x \perp y$ then $Tx \perp Ty$.*

$\lhd$ Appeal to 1.13.2, 1.13.3 and use the well-known result on characterization of discrete functionals, Theorem 1.13.4 for $F = \mathbb{R}$. $\rhd$

1.13.5. It is easy to verify that if a functional $f \in E^{\sim}$ preserves disjointness, then $|f|$ has the same property (see [31]). By 1.13.4 (1) the functionals f^+ and f^- are multiples of $|f|$; and since $f^+ \perp f^-$, either $f^+ = 0$ or $f^- = 0$. This means that either $f \geq 0$ or $f \leq 0$. In particular, for the functional $\tau := T\!\uparrow$ we have $[\![\tau \geq 0]\!] \vee [\![\tau \leq 0]\!] = \mathbb{1}$. If $\pi := \chi[\![\tau \geq 0]\!]$ then $\pi^{\perp} \leq \chi[\![\tau \leq 0]\!]$ and the inequalities $\pi\tau \geq 0$ and $\pi^{\perp}\tau \leq 0$ are true. Descending leads to the following conclusion:

For a regular disjointness preserving operator $T \in L^{\sim}(E,F)$ there exists a projection $\pi \in \mathfrak{P}(F)$ such that $\pi T = T^+$ and $\pi^{\perp} T = T^-$. In particular, for all $0 \leq x$, $y \in E$ we have $(Tx)^+ \perp (Ty)^-$.

1.13.6. A subspace $E_0 \subset E$ is called *massive*, or *coinitial*, or even *cofinal* whenever for every $x \in E$ there are some $\underline{x}$ and $\overline{x}$ in E_0 satisfying $\underline{x} \leq x \leq \overline{x}$. Suppose that $T_0 \in L(E_0, E)$ and write $\tau_0 := T_0\!\uparrow$. Obviously, the following take place:

(1) $[\![E_0$ is massive in $E]\!] \leftrightarrow [\![E_0^{\wedge}$ is massive in $E^{\wedge}]\!] = \mathbb{1}$;

(2) $[\![T$ is an extension of $T_0]\!] \leftrightarrow [\![\tau$ is an extension of $\tau_0]\!] = \mathbb{1}$.

The Kreĭn–Rutman Theorem states that a positive functional defined on a massive subspace admits a positive extension to the whole space. This theorem remains valid with the word "positive" replaced by "discrete." Putting these facts into $\mathbf{V}^{(B)}$ and by transfer and (1), (2) and 1.13.3 (3), we obtain the following:

1.13.7. Kantorovich Theorem. *Let F be an arbitrary Kantorovich space. If E_0 is a massive subspace of E, then every positive operator $T_0 : E_0 \to F$ admits a positive extension $T \in L^{\sim}(E,F)$.*

1.13.8. Theorem. *Assume that F is a Kantorovich space and E_0 is a massive subspace of E. Each F-discrete operator $T_0 : E_0 \to F$ admits an F-discrete extension $T : E \to F$. In particular, if E_0 is a sublattice then to each lattice homomorphism $T_0 : E_0 \to F$ there is a lattice homomorphism $T : E \to F$ extending T_0.*

1.13.9. In the case when E_0 is a massive sublattice of E, Theorem 1.13.7 may be strengthened essentially. We first denote by $\varepsilon^+(S_0) \subset L^\sim(E, F)$ the set of extensions of a positive operator $S_0 : E_0 \to F$ to E.

(1) Theorem. *Let E and F be vector lattices, with F Dedekind complete. Assume that E_0 is a massive sublattice of E and $S_0 : E_0 \to F$ is a positive operator. Then the set of extreme points of the convex set $\varepsilon^+(S_0) \subset L^\sim(E, F)$ is nonempty.*

(2) Theorem. *Let (Y, F) be a Banach–Kantorovich space. Assume that $T_0 : E_0 \to Y$ is a dominated operator and S is an arbitrary extreme point of $\varepsilon^+(|T_0|)$. Then there is a unique dominated operator $T: E \to Y$ such that T is an extension of T_0 and $|T| = S$.*

$\lhd$ The above method of Boolean valued realization reduces the situation to the case in which $F = \mathscr{R}{\downarrow}$. We may thus assume that Y is a Banach space while $|T_0|$ and S are positive functionals. Existence of an extreme extension $S \in \varepsilon^+(|T_0|)$ follows from (1). Define the seminorm $p(e) := S(|e|)$ for $e \in E$. Then E_0 becomes dense in E with respect to the locally convex topology determined from p. This ensues from the following criterion in [79]: S is an extreme point of $\varepsilon^+(S_0)$ with $S_0 \in L_+(E_0, F)$ if and only if $\inf\{S(|e - e_0|) : e_0 \in E_0\} = 0$ for every $e \in E$. The operator T_0 is continuous and may be extended by continuity to some operator T defined on the whole of E; moreover, T is dominated and, clearly, $|T| = S$. For details, see [48]. $\rhd$

1.13.10. Theorem. *For a positive operator $T : E \to F$, the following are equivalent:*

(1) *T is F-diffuse;*

(2) *for all $0 \le x \in E$, $0 \le \varepsilon \in F$, and $b \in B$ such that $b\varepsilon \ne 0$ there are a nonzero projection $\rho \le b$ and disjoint positive operators $T_1, \dots, T_n$ such that*

$$T = T_1 + \dots + T_n, \quad |\rho T_k x| \le \varepsilon \quad (k := 1, \dots, n);$$

(3) *for all $0 \le x \in E$, $0 \le \varepsilon \in F$, and $b \in B$ such that $b\varepsilon \ne 0$ there is a countable partition of unity (b_n) such that for every $n \in \mathbb{N}$*

the operator T decomposes into the sum of disjoint positive operators $T_{1,n}, \dots, T_{k_n,n}$, satisfying the inequalities $b_n|T_{k,n}x| \le \varepsilon$ ($k := 1, \dots, k_n$).

$\lhd$ The proof is obtained by interpreting inside $\mathbf{V}^{(B)}$ the following scalar fact: A positive functional f is diffuse if for all $x \ge 0$ and $0 < \varepsilon \in \mathbb{R}$ there are disjoint positive functionals $f_1, \dots, f_n$ such that $f = f_1 + \dots + f_n$ and $|f_k(x)| < \varepsilon$ ($k := 1, \dots, n$) (see [46]). $\rhd$

1.13.11. We say that $\pi \in \mathfrak{P}(F)$ is a (γ, E)-homogeneous projection if to each nonzero projection $\rho \le \pi$ and every set $\mathscr{H}$ of disjoint lattice homomorphisms from E to ρF satisfying $(\operatorname{im} S)^{\perp\perp} = \rho F$ for all $S \in \mathscr{H}$, we have $\operatorname{card}(\mathscr{H}) \ge \gamma$. Introduce the notation $\operatorname{Orth}(T, F) := \operatorname{Orth}(G, F)$, where G is the order ideal in F generated by $T(E)$.

Theorem. *Let E and F be vector lattices, with F Dedekind complete. Then there is a set of cardinals Γ, to each cardinal $\gamma \in \Gamma$ there are a projection $\pi_\gamma \in \mathfrak{P}(F)$ and a family of disjoint lattice homomorphisms $(\Phi_{\gamma,\alpha})_{\alpha<\gamma}$ from E to F such that the following hold:*

(1) *$(\pi_\gamma)_{\gamma\in\Gamma}$ is a partition of unity in the Boolean algebra $\mathfrak{P}(F)$; moreover, $\pi_\gamma \ne 0$ for all $\gamma \in \Gamma$;*

(2) *$\operatorname{im} \Phi_{\gamma,\alpha} = \pi_\gamma(F)$ ($\gamma \in \Gamma, \alpha < \gamma$);*

(3) *π_γ is a (γ, E)-homogeneous projection;*

(4) *each operator $T \in L^\sim(E, F)$ admits a unique representation of the shape*

$$T = T_0 + o\text{-}\sum_{\gamma\in\Gamma} o\text{-}\sum_{\alpha<\gamma} \sigma_{\gamma,\alpha} \circ \Phi_{\gamma,\alpha},$$

with $T_0 \in L_d^\sim(E, F)$ and $\sigma_{\gamma,\alpha} \in \operatorname{Orth}(\Phi_{\gamma,\alpha}, \pi_\gamma(F))$.

$\lhd$ The operator T_0 is uniquely determined, whereas the family (T_ξ) is unique up to rearrangement and "mixing." To prove the theorem, we use the fact that by transfer each Kantorovich space inside $\mathbf{V}^{(B)}$ (in particular, our $E^{\wedge\sim}$) splits into the direct sum of the band of diffuse elements and the band spanned by discrete elements; the latter is the direct sum of one-dimensional bands, i.e. bands spanned by discrete elements. We then appeal to 1.13.3 (3–5). $\rhd$

1.13.12. Comments.

(1) The material of this section can be viewed as an illustration to the following heuristic principle formulated by L. V. Kantorovich in the article [38] where he had introduced Kantorovich spaces: "Introduction of these spaces allows us to study linear operations of one general class (operations with values in such spaces) as linear functionals."

(2) The elementary Theorem 1.13.2 serves as a technical basis for enhancing the Kantorovich heuristic principle and making it a rigorous research tool for the problems we have presented. Other versions and abstractions are collected in [53].

(3) The equivalence (1)$\leftrightarrow$(2) in Theorem 1.13.4 was obtained by S. S. Kutateladze standardly in 1976. The scalar case ($F = \mathbb{R}$) is well known. As regards 4.1.5, see [4].

(4) A standard proof of Theorem 1.13.7 is presented in many monographs (see, for instance, [2, 53, 61]). The theorem is also valid if E is an ordered vector space. Extension of a positive operator with additional properties is rather topical. We only note here that this theme is close to research into the extreme structure of convex sets; e.g., see [61].

(5) Theorem 1.13.9 (1) is a particular case of one theorem by S. S. Kutateladze which was established in [71]; also see [61]. Theorem 1.13.9 (2) was obtained in [48]. Theorem 1.13.11 seems new. Analogs of Theorems 1.13.9 (2), 1.13.10, and 1.13.11 for vector measures are given in [63, 64, 66].

1.14. Fragments of Positive Operators

In this section we address the problem of presenting the fragments of a positive operator. This problem can be scrutinized in detail by consistent usage of nonstandard analysis. As in the previous section, let E stand for a vector lattice and F, for a Kantorovich space.

1.14.1. A set $\mathscr{P}$ of band projections in the Kantorovich space $L^{\sim}(E,F)$ *generates the fragments of an operator* T, $0 \leq T \in L^{\sim}(E,F)$, provided that $Tx^{+} = \sup\{pTx : p \in \mathscr{P}\}$ for all $x \in E$. In the event this happens for all $0 \leq T \in L^{\sim}(E,F)$, we call the set $\mathscr{P}$ *generating.*

We put $F := \mathscr{R}{\downarrow}$ and let p be a band projection in $L^{\sim}(E,F)$. Then

(1) *there is a unique element* $p{\uparrow} \in \mathbf{V}^{(B)}$ *such that* $[\![p{\uparrow}$ *is a band projection in* $E^{\wedge\sim}]\!] = \mathbf{1}$ *and* $(pT){\uparrow} = p{\uparrow}\, T{\uparrow}$ *for all* $T \in L^{\sim}(E,F)$.

We now consider a set $\mathscr{P}$ of band projections in $L^{\sim}(E,F)$ and a positive operator $T \in L^{\sim}(E,F)$. Put $\tau := T{\uparrow}$ and $\mathscr{P}{\uparrow} := \{p{\uparrow} : p \in \mathscr{P}\}{\uparrow}$. Then $[\![\mathscr{P}{\uparrow}$ is a set of band projections in $E^{\wedge\sim}]\!] = \mathbf{1}$ and the following are true:

(2) $[\![\mathscr{P}$ *generates the fragments of* $T]\!]$
$\leftrightarrow$ $[\![\mathscr{P}{\uparrow}$ *generates the fragments of* $\tau]\!] = \mathbf{1}$;

(3) $[\![\mathscr{P}$ *is a generating set* $]\!]$
$\leftrightarrow$ $[\![\mathscr{P}{\uparrow}$ *is a generating set* $]\!] = \mathbf{1}$.

1.14.2. Given a set A in a Kantorovich space, we denote by $A^{\vee}$ the result of adjoining to A suprema of its every nonempty finite subset. The symbol $A^{\uparrow}$ stands for the result of adjoining to A suprema of nonempty increasing nets of elements of A. The symbols $A^{\uparrow\downarrow}$ and $A^{\uparrow\downarrow\uparrow}$ are understood naturally. The sign $\approx$ in a Kantorovich space F has the ordinary meaning: $x \approx y$ for $x, y \in F$ symbolizes that $(\forall^{\mathrm{St}} e \in \mathscr{E})|x - y| \leq e$, where $\mathscr{E}$ is the filter of order units in F. It is clear that if $F := \mathbb{R}$ then $x - y$ is an infinitesimal real.

Our results on positive operators will be obtained along the same lines as in Section 1.13 by using Boolean valued models. With this in mind, we must start with the case of functionals.

We will use the notation $\mathscr{P}(f) := \{pf : p \in \mathscr{P}\}$. Henceforth, in Subsections 1.14.3–1.14.5 we let E stand for a vector lattice over a dense subfield of $\mathbb{R}$ while $\mathscr{P}$ is a set of band projections in $E^{\sim}$.

1.14.3. *Theorem.* *The following are equivalent:*

(1) $\mathscr{P}(f)^{\vee(\uparrow\downarrow\uparrow)} = \mathfrak{E}(f)$;

(2) $\mathscr{P}$ *generates the fragments of* f;

(3) $(\forall x \in {}^{\circ}E)(\exists p \in \mathscr{P}) pf(x) \approx f(x^{+})$;

(4) *a functional* g *in* $[0, f]$ *is a fragment of* f *if and only if*

$$\inf_{p \in \mathscr{P}} (p^{\perp} g(x) + p(f - g)(x)) = 0$$

for every $0 \leq x \in E$;

(5) $(\forall g \in {}^{\circ}\mathfrak{E}(f))(\forall x \in {}^{\circ}E_{+})(\exists p \in \mathscr{P})|pf - g|(x) \approx 0$;

(6) $\inf\{|pf - g|(x) : p \in \mathscr{P}\} = 0$ *for all fragments* $g \in \mathfrak{E}(f)$ *and positive elements* $x \geq 0$;

(7) *for* $x \in E_{+}$ *and* $g \in \mathscr{E}(f)$ *there is an element* $p \in \mathscr{P}(f)^{\vee(\uparrow\downarrow\uparrow)}$, *satisfying* $|pf - g|(x) = 0$.

$\triangleleft$ The implications (1) $\rightarrow$ (2) $\rightarrow$ (3) are obvious.

(3) $\rightarrow$ (4): We will work within the *standard entourage*; i.e., we presume that all free variables are standard. Note first that validity of the sought equality for all functionals g and f satisfying $0 \leq g \leq f$ amounts to existence of $p \in \mathscr{P}$, given a standard $x \geq 0$, such that $p^{\perp} g(x) \approx 0$ and $p(f - g)(x) \approx 0$. (As usual, $p^{\perp}$ is the *complementary band projection* to p.) Thus, ${}^{\circ}p(g \wedge (f - g))(x) \leq {}^{\circ}p(f - g)(x) = 0$ and ${}^{\circ}p^{\perp}((f - g) \wedge g)(x) \leq {}^{\circ}p^{\perp} g(x) = 0$, i.e. $g \wedge (f - g) = 0$.

Prove now that, on assuming (3), the sought equality ensues from the conventional criterion for disjointness:

$$\inf\{g(x_1) + (f - g)(x_2) : x_1 \geq 0,\, x_2 \geq 0,\, x_1 + x_2 = x\} = 0.$$

Given a standard x, find internal positive x_1 and x_2 such that $x = x_1 + x_2$ and, moreover, $g(x_1) \approx 0$ and $f(x_2) \approx g(x_2)$. By (3), it follows from the Kreĭn–Milman Theorem that the fragment g belongs to the weak closure of $\mathscr{P}(f)$. In particular, there is an element $p \in \mathscr{P}$ satisfying $g(x_1) \approx pf(x_1)$ and $g(x_2) \approx pf(x_2)$. Thus, $p^{\perp}g(x_2) \approx 0$, because $p^{\perp}g \leq p^{\perp}f$. Finally, $p^{\perp}g(x) \approx 0$. Hence,

$$\begin{aligned} p(f-g)(x) &= pf(x_2) + pf(x_1) - pg(x) \\ &\approx g(x_2) + g(x_1) - pg(x) \approx p^{\perp}g(x) \approx 0. \end{aligned}$$

This ensures the needed equality.

(4) $\to$ (5): Using the equality $|pf - g|(x) = p^{\perp}g(x) + p(f-g)(x)$, we may find $p \in \mathscr{P}$ so that $p^{\perp}g(x) \approx 0$ and $p(f-g)(x) \approx 0$. This justifies the claim.

The equivalence (5) $\leftrightarrow$ (6) is clear.

The implications (5) $\to$ (7) $\to$ (1) can be proved as in [1, 47, 67]. ▷

1.14.4. Theorem. *For positive functionals f and g and for a generating set of band projections $\mathscr{P}$, the following are equivalent:*

(1) $g \in \{f\}^{\perp\perp}$;

(2) *If x is a limited element of E, i.e. $x \in {}^{\mathrm{fin}}E := \{x \in E : (\exists \overline{x} \in {}^{\circ}E)|x| \leq \overline{x}\}$, then $pg(x) \approx 0$ whenever $pf(x) \approx 0$ for $p \in \mathscr{P}$;*

(3) $(\forall x \in E_+)(\forall \varepsilon > 0)(\exists \delta > 0)(\forall p \in \mathscr{P})pf(x) \leq \delta \to pg(x) \leq \varepsilon$.

1.14.5. Theorem. *Let f and g be positive functionals on E, and let x be a positive element of E. The following representations of the band projection π_f onto the band $\{f\}^{\perp\perp}$ are valid:*

(1) $\pi_f g(x) \doteq \inf{}^*\{{}^{\circ}pg(x) : p^{\perp}f(x) \approx 0,\, p \in \mathscr{P}\}$ *(the symbol $\doteq$ means that the formula is exact, i.e., the equality is attained);*

(2) $\pi_f g(x) = \sup_{\varepsilon > 0} \inf\{pg(x) : p^{\perp}f(x) \leq \varepsilon,\, p \in \mathscr{P}\}$;

(3) $\pi_f g(x) \doteq \inf{}^*\{{}^{\circ}g(y) : f(x-y) \approx 0,\, 0 \leq y \leq x\}$;

(4) $(\forall \varepsilon > 0)\,(\exists \delta > 0)\,(\forall p \in \mathscr{P})\,pf(x) < \delta \to \pi_f g(x) \leq p^{\perp}g(x) + \varepsilon$;

$(\forall \varepsilon > 0)\,(\forall \delta > 0)\,(\exists p \in \mathscr{P})\,pf(x) < \delta \wedge p^{\perp}g(x) \leq \pi_f g(x) + \varepsilon$;

(5) $(\forall \varepsilon > 0)\,(\exists \delta > 0)\,(\forall 0 \leq y \leq x)\,f(x-y) \leq \delta \to \pi_f g(x) \leq g(y) + \varepsilon$;

$$(\forall \varepsilon > 0)\,(\forall \delta > 0)\,(\exists 0 \leq y \leq x)\, f(x-y) \leq \delta \wedge g(y) \leq \pi_f g(x) + \varepsilon.$$

Transferring 1.14.3–1.14.5 to $\mathbf{V}^{(B)}$ and using 1.14.1, we infer the next Propositions 1.14.6–1.14.9.

1.14.6. *For a set of band projections $\mathscr{P}$ in $L^{\sim}(E,F)$ and $0 \leq S \in L^{\sim}(E,F)$ the following are equivalent:*

(1) $\mathscr{P}(S)^{\vee(\uparrow\downarrow\uparrow)} = \mathfrak{E}(S)$;

(2) *$\mathscr{P}$ generates the fragments of S;*

(3) *an operator $T \in [0,S]$ is a fragment of S if and only if*

$$\inf_{p \in \mathscr{P}} (p^{\perp}Tx + p(S-T)x) = 0$$

for all $0 \leq x \in E$;

(4) $(\forall x \in {}^{\circ}E)\,(\exists p \in \mathscr{P}\uparrow\downarrow)\, pSx \approx Sx^{+}$.

1.14.7. *For positive operators S and T and a generating set $\mathscr{P}$ of band projections in $L^{\sim}(E,F)$, the following are equivalent:*

(1) $T \in \{S\}^{\perp\perp}$;

(2) $(\forall x \in {}^{\mathrm{fin}}E)\,(\forall p \in \mathscr{P})\,(\forall \pi \in B)\, \pi pSx \approx 0 \rightarrow \pi pTx \approx 0$;

(3) $(\forall x \in {}^{\mathrm{fin}}E)\,(\forall \pi \in B)\, \pi Sx \approx 0 \rightarrow \pi Tx \approx 0$;

(4) $(\forall x \geq 0)\,(\forall \varepsilon \in \mathscr{E})\,(\exists \delta \in \mathscr{E})\,(\forall p \in \mathscr{P})\,(\forall \pi \in B)$
$\pi pSx \leq \delta \rightarrow \pi pTx \leq \varepsilon$;

(5) $(\forall x \geq 0)\,(\forall \varepsilon \in \mathscr{E})\,(\exists \delta \in \mathscr{E})\,(\forall \pi \in B)$
$\pi Sx \leq \delta \rightarrow \pi Tx \leq \varepsilon$.

1.14.8. Theorem. *Let E be a vector lattice, and let F be a Kantorovich space having the filter of order units $\mathscr{E}$ and the base B. Suppose that S and T are positive operators in $L^{\sim}(E,F)$ and R is the band projection of T to the band $\{S\}^{\perp\perp}$. For a positive $x \in E$, the following are valid:*

(1) $Rx = \sup_{\varepsilon \in \mathscr{E}} \inf\{\pi Ty + \pi^{\perp}Sx : 0 \leq y \leq x,$
$\pi \in B,\, \pi S(x-y) \leq \varepsilon\}$;

(2) $Rx = \sup_{\varepsilon \in \mathscr{E}} \inf\{(\pi p)^{\perp}Tx : \pi pSx \leq \varepsilon,\, p \in \mathscr{P}, \pi \in B\}$,
where $\mathscr{P}$ is a generating set of band projections in F.

1.14.9. Given an element $0 \le e \in E$, we define the operator $\pi_e S$ by the formulas:

$$(\pi_e S)x := \sup_{n\in\mathbb{N}} S(x \wedge ne) \quad (x \in E_+),$$

$$(\pi_e S)x := (\pi_e S)x^+ - (\pi_e S)x^- \quad (x \in E).$$

It is easy to see that $\pi_e S \in L^\sim(E, F)$. Moreover, $\pi_e S$ is a fragment of S and the mapping $S \mapsto \pi_e S$ $(S \ge 0)$ can be naturally extended to $L^\sim(E, F)$ to become a band projection.

The set of band projections $\mathscr{P} := \{\pi_e : 0 \le e \in E\}$ is generating. Hence, 1.14.6 implies the formula

$$\mathfrak{E}(S) = \{(\rho \circ \pi_e)S : \rho \in \mathfrak{P}(F), 0 \le e \in E\}^{\wedge(\uparrow\downarrow\uparrow)}.$$

1.14.10. Comments.

(1) The formulas for band projections in 1.14.8 (1, 2) have been constructed gradually. A glimpse into this history may be caught on using [4, 93]. A general approach of [75] is a basis for our exposition. This approach allows us to derive various formulas for band projections by specifying generating sets.

(2) A formula like 1.14.9 was established for the first time by B. de Pagter (see [93]) under two essential restrictions: F admits a point separating set of o-continuous functionals, and E is Dedekind complete. The first restriction was removed in [67]; the second, in [1, 47]. All these cases correspond to different generating sets of band projections.

(3) The main idea of [75] is as follows: The fragments of a positive operator T are the extreme points of the order interval $[0, T]$. The latter coincides with the subdifferential at zero (the supporting set) ∂p of the sublinear operator $px = Tx_+$. Thus, study of the fragments of a positive operator reduces to description for the extreme structure of some subdifferential. This description for a general sublinear operator was first obtained by S. S. Kutateladze (see details in [61]). Note that the approach of [70, 71] solves in particular the problem of extreme extension of a positive operator (for references on this subject; see [7, 61]).

1.15. Order Continuous Operators

The methods of the two previous sections are not applicable directly to order continuous operators since we lose order continuity in ascending an operator (see 1.13.2). Here we pursue another approach that rests on D. Maharam's ideas.

1.15.1. A positive operator $T : E \to F$ satisfies the *Maharam condition* if $T[0, x] = [0, Tx]$ for every $0 \le x \in E$; i.e., if to all $0 \le x \in E$ and $0 \le z \le Tx$ there is some $0 \leq y \in E$ such that $Ty = z$ and $0 \leq y \leq x$. An order continuous positive operator, enjoying this condition, is customarily called a *Maharam operator.*

Throughout this section E and F are Kantorovich spaces, and for simplicity we assume F universally complete. By the symbol E_T we denote the *carrier* of the operator T, i.e. the set $\{x \in E : T(|x|) = 0\}^{\perp}$. We put $F_T := (\operatorname{im} T)^{\perp\perp}$, and let $\mathscr{D}_m(T)$ stand for the greatest order dense ideal in a universal completion of E onto which we may extend the operator T preserving its order continuity. If $E_T = E$ and $T \geq 0$ then T is called *essentially positive.*

1.15.2. Theorem. *Let E be a Kantorovich space, $F := \mathscr{R}{\downarrow}$, and let $T : E \to F$ be a Maharam operator such that $E = E_T = \mathscr{D}_m(T)$ and $F = F_T$. Then there are $\mathscr{E} \in \mathbf{V}^{(B)}$ and $\tau \in \mathbf{V}^{(B)}$ satisfying the following:*

(1) $\mathbf{V}^{(B)} \models [\![\mathscr{E}$ *is a Kantorovich space and* $\tau : \mathscr{E} \to \mathscr{R}$ *is an essentially order continuous functional* $]\!]$;

(2) *$\mathscr{E}{\downarrow}$ is also a Kantorovich space, $\tau{\downarrow} : \mathscr{E}{\downarrow} \to \mathscr{R}{\downarrow}$ is a Maharam operator; moreover, $\mathscr{E}{\downarrow} = \mathscr{D}_m(\tau{\downarrow})$;*

(3) *there exists a linear and lattice isomorphism h from E onto $\mathscr{E}{\downarrow}$ such that $T = \tau{\downarrow} \circ h$.*

1.15.3. The decomposition of 1.13.10 can be elaborated for a Maharam operator. Let e be an order unit in E. Then $[\![e$ is an order unit in $\mathscr{E}]\!] = \mathbf{1}$. The functional τ is representable as $\tau = \tau_0 + \sum_{k=1}^{\infty} \tau_k$, where τ_0 is a diffuse functional and τ_k are order continuous lattice homomorphisms. All these functionals are uniquely defined by measures on the base of unit elements. Furthermore, to τ_0 there corresponds an atomless measure, while to τ_k there corresponds a two-valued measure.

To transfer this situation to $\mathbf{V}^{(B)}$, we first recall that the Maharam condition for a positive vector measure $\mu : \mathfrak{C}(e) \to F$ has precisely the same meaning as in 1.15.1; i.e., $\mu[0, a] = [0, \mu(a)]$ $(a \in \mathfrak{C}(e))$. If μ is an isomorphism of Boolean algebras then μ^* stands for the resulting isomorphism of the corresponding universally complete Kantorovich spaces.

1.15.4. Theorem. *Let E be a Kantorovich space with unity e, and let $T : E \to F$ be an essentially positive Maharam operator. Then there are sequences $(e_k)_{k=0}^{\infty}$, $(c_k)_{k=1}^{\infty}$, $(\mu_k)_{k=0}^{\infty}$, and $(\alpha_k)_{k=1}^{\infty}$ such that*

(1) *(e_k) is a partition of unity in the Boolean algebra $\mathfrak{C}(e)$ and (c_k) is a sequence of fragments of the element $c := Te$;*

(2) *$\mu : \mathfrak{C}(e_0) \to F$ is a strictly positive order continuous measure satisfying the Maharam condition;*

(3) $\mu_k : \mathscr{E}(e_k) \to \mathscr{E}(c_k)$ *is a Boolean isomorphism and* α_k *is a positive invertible orthomorphism in* $\{c_k\}^{\perp\perp}$;

(4) *the representation holds:*

$$Tx = \int_{-\infty}^{\infty} \lambda \, d\mu_0(e_\lambda^{x_0}) + \sum_{k=1}^{\infty} \alpha_k \mu_k^*(x_k),$$

with x_k *the band projection of* x *to* $\{e_k\}^{\perp\perp}$.

The dual analogs of 1.13.4 and 1.13.5 are valid for Maharam operators.

1.15.5. *Theorem.* *Let* $T : E \to F$ *be a positive order continuous operator. The following are equivalent:*

(1) T *satisfies the Maharam condition;*

(2) *for every operator* $0 \le S \le T$ *there is an orthomorphism* $\alpha : E \to E$, $0 \le \alpha \le I_E$, *such that* $Sx = S\alpha x$ $(x \in E)$;

(3) *if* $Tx = f_1 + f_2$ *for some* $0 \le x \in E$ *and* $0 \le f_1, f_2 \in F$, *and if* $f_1 \perp f_2$, *then there are* $0 \le x_1, x_2 \in E$, *such that* $x = x_1 + x_2$, $x_1 \perp x_2$ *and* $Tx_k = f_k$ $(k = 1, 2)$.

$\lhd$ Without loss of generality we may assume T essentially positive. If (1) is true then $T = \tau{\downarrow} \circ h$ (see 1.15.2). Since τ is $\mathscr{R}$-linear, T is $\mathscr{R}{\downarrow}$-linear. If $0 \le S \le T$ then S is also $\mathscr{R}{\downarrow}$-linear and, hence, a Maharam operator. By 1.15.2, $S = \sigma{\downarrow} \circ h$ where $[\![\sigma \in \mathscr{E}^{\sim}]\!] = [\![0 \le \sigma \le \tau]\!] = \mathbb{1}$. The claim (2) for the functionals τ and σ follows from the Radon–Nikodým Theorem. Taking descents, we obtain (2) for the operators T and S. The remaining implications are straightforward. $\rhd$

1.15.6. *Let* $S : E \to F$ *be a regular operator such that* $T := |S|$ *is a Maharam operator. Then there is a band projection* $\pi \in \mathfrak{P}(E)$ *such that* $S^+ = S \circ \pi$ *and* $S^- = S \circ \pi^\perp$.

$\lhd$ Again we may assume that $T = \tau{\downarrow}$, where τ is an essentially positive o-continuous functional inside $\mathbf{V}^{(B)}$. As in 1.15.5 we infer that there exists a regular functional $\sigma \in \mathscr{E}$ satisfying $\tau = |\sigma|$.

Let p be the band projection in $\mathscr{E}$ to the carrier (= the band of essential positivity) of σ^+. Order continuous functionals are disjoint if and only if their carriers are disjoint. Hence, $\sigma^+ = \sigma \circ p$ and $\sigma^- = \sigma \circ p^\perp$. Writing $\pi := p{\downarrow}$ and descending, we complete the proof. $\rhd$

1.15.7. Thus, the general properties of Maharam operators can be deduced from the corresponding facts about functionals with the help of Theorem 1.15.2. Nevertheless, these methods may be also useful in studying arbitrary regular operators.

We fix a positive operator Φ that acts from a vector lattice X to F. By Theorem 1.13.2, there is a positive $\mathbb{R}^\wedge$-linear functional $\varphi : X^\wedge \to \mathcal{R}$ such that the identity $[\![\Phi(x) = \varphi(x^\wedge)]\!] = \mathbb{1}$ holds for all elements $x \in X$.

We equip $X^\wedge$ with the seminorm $\rho(x) := \varphi(|x|)$. Let $\mathcal{X}$ be a completion of the quotient lattice $X^\wedge/\rho^{-1}(0)$ with respect to the quotient norm. Then $\mathcal{X}$ becomes a Banach lattice and there is a unique positive ($\mathcal{R}$-linear) functional $\overline{\varphi} : \mathcal{X} \to \mathcal{R}$ such that $\varphi = \overline{\varphi} \circ \iota$, where $\iota : X^\wedge \to \mathcal{X}$ is the quotient homomorphism. Moreover, $\overline{\varphi}$ is order continuous and essentially positive.

By ascending and descending, we come to the following

1.15.8. *Theorem.* *There are a Kantorovich space $\overline{X}$ and an essentially positive Maharam operator $\overline{\Phi} : \overline{X} \to F$ satisfying the conditions:*

(1) *there are lattice homomorphisms $\imath : X \to \overline{X}$ and $\jmath : \mathcal{Z}(X) \to \mathcal{Z}(\overline{X})$, with $\mathcal{Z}(X)$ the ideal of X spanned by the identity operator, such that $\jmath$ is also a ring homomorphism and $\alpha\Phi x = \overline{\Phi}(\jmath(\alpha)\imath(x))$ for all elements $x \in X$ and $\alpha \in \mathcal{Z}(F)$; in particular, $\Phi(x) = \overline{\Phi}(\imath(x))$;*

(2) *$\imath(X)$ is a massive sublattice in $\overline{X}$ and $\jmath(\mathcal{Z}(F))$ is an o-closed sublattice and subring of $\mathcal{Z}(\overline{X})$;*

(3) *$\overline{X} = b(X \otimes \mathcal{Z}(F))^{\downarrow\uparrow}$, with $b : X \otimes \mathcal{Z}(F) \to \overline{X}$ the linear operator defined by the relation $b(x \otimes \alpha) := \jmath(\alpha)\imath(x)$ for $x \in X$ and $\alpha \in \mathcal{Z}(F)$.*

The pair $(\overline{X}, \overline{\Phi})$ is defined uniquely up to isomorphism. Moreover, if an essentially positive Maharam operator $\overline{\Phi}_1 : \overline{X}_1 \to F$ and a lattice homomorphism $\imath_1 : \overline{X}_1 \to F$ satisfy the condition $\Phi = \overline{\Phi}_1 \circ \imath_1$, then there is an isomorphism h from $\overline{X}$ onto an o-closed sublattice in $\overline{X}$ such that $\overline{\Phi} = \overline{\Phi}_1 \circ h$ and $h \circ \imath = \imath_1$.

We denote by $m\overline{X}$ a universal completion of a Kantorovich space $\overline{X}$. Fixing an order unity in $m\overline{X}$, we define the corresponding unique structure of an f-algebra. Let $L_1(\Phi)$ be the greatest order dense ideal in $m\overline{X}$, onto which $\overline{\Phi}$ can be extended with preserving o-continuity. The following result is a variant of the Radon–Nikodým Theorem for positive operators.

1.15.9. *Theorem.* *For every operator $T \in \{\Phi\}^{\perp\perp}$ there is a unique element $z \in m\overline{X}$ satisfying*

$$Tx = \overline{\Phi}(z \cdot \imath(x)) \quad (x \in X).$$

The correspondence $T \mapsto z$ establishes a linear and lattice isomorphism between the band $\{\Phi\}^{\perp\perp}$ and the order dense ideal in $m\overline{X}$ defined by the formula $\{z \in m\overline{X} : z \cdot \imath(X) \subset L_1(\Phi)\}$.

1.15.10. Comments.

(1) In a long sequence of papers which was published in the 1950s, D. Maharam proposed an original approach to studying positive operators (see the survey [86]). The concept of Maharam operator and the idea of extending a positive operator to a Maharam operator (see 1.15.8) stem from these papers. It is worth emphasizing that D. Maharam's approach is notable within Boolean valued analysis for the clarity and simplicity of the idea, because a considerable part of the theory reduces to manipulating numerical measures and integrals in a suitable Boolean valued model.

(2) Several results of D. Maharam were translated to vector lattices by W. A. J. Luxemburg and A. R. Schep (see [81]). Theorem 1.15.2 is due to A. G. Kusraev [50].

(3) The equivalence (1)$\leftrightarrow$(2) in 1.15.5 is a restricted version of the Radon–Nikodým Theorem for a Maharam operator. The complete form of this theorem was proved in [81] by standard methods; and in [50, 53], with the help of 1.15.2. Proposition 1.15.6 is an operator variant of the Hahn Decomposition Theorem for measures (see [81, 85]). D. Maharam established Theorem 1.15.4 for an operator between spaces of measurable functions by her original method.

(4) The problem of extending a positive operator to a Maharam operator was thoroughly studied (see [1, 53]). The details of 1.15.8 and 1.15.9 may also be found in these papers. The structure of such an extension is rather complicated, but sometimes admits a functional description.

1.16. Cyclically Compact Operators

The Boolean valued interpretation of compactness gives rise to the new notions of *cyclically compact sets* and *operators* which deserves an independent study. A part of the corresponding theory is presented in this section.

1.16.1. Let B be a complete Boolean algebra and A, a nonempty set. We recall that $B(\mathrm{A})$ denotes the set of all partitions of unity in B with the fixed index set A. The sets $B(\mathrm{A})$ and $\mathrm{A}^\wedge\!\downarrow$ are bijective, and so they are frequently identified. If A is an ordered set then we may furnish $B(\mathrm{A})$ with some order by the rule

$$\nu \leq \mu \leftrightarrow (\forall \alpha, \beta \in \mathrm{A})\ \big(\nu(\alpha) \wedge \mu(\beta) \neq 0 \rightarrow \alpha \leq \beta\big) \quad \big(\nu, \mu \in B(\mathrm{A})\big).$$

Moreover, if A is a directed set then so is $B(\mathrm{A})$.

We consider a normed B-space X and a net $(x_\alpha)_{\alpha\in\mathrm{A}}$ in it. Given $\nu \in B(\mathrm{A})$, we put $x_\nu := \mathrm{mix}_{\alpha\in\mathrm{A}}(\nu(\alpha)x_\alpha)$. If all mixings exist then we come to a new net $(x_\nu)_{\nu\in B(\mathrm{A})}$ in X. Every subnet of the net $(x_\nu)_{\nu\in B(\mathrm{A})}$ is called a *cyclic subnet* of the original net $(x_\alpha)_{\alpha\in\mathrm{A}}$. If $s : \mathrm{A} \to X$ and $\varkappa : \mathrm{A}' \to B(\mathrm{A})$ then the mapping $s \bullet \varkappa : \mathrm{A}' \to X$ is defined by $s \bullet \varkappa(\alpha) := x_\nu$ where $\nu = \varkappa(\alpha)$. A *cyclic subsequence* of a sequence $(x_k)_{k\in\mathbb{N}} \subset X$ is a sequence of the form $(x_{\nu_k})_{k\in\mathbb{N}}$ where $(\nu_k)_{k\in\mathbb{N}}$ is a sequence in $B(\mathbb{N})$ with $\nu_k \leq \nu_{k+1}$ for all $k \in \mathbb{N}$.

According to 1.8.7 there is no loss of generality in assuming that X is a decomposable subspace of the Banach–Kantorovich space $\mathscr{X}{\downarrow}$, where $\mathscr{X}$ is a Banach space inside $\mathbf{V}^{(B)}$ and every projection $b \in B$ coincides with the restriction of $\chi(b)$ onto X. More precisely, we will assume that X is the *bounded descent* of $\mathscr{X}$; i.e., $X = \{x \in \mathscr{X}{\downarrow} : |x| \in \Lambda\}$, where Λ is the Stone algebra $\mathscr{S}(B)$ identified with the bounded part of the complex algebra $\mathscr{C}{\downarrow}$.

A subset $C \in X$ is said to be *cyclically compact* if C is cyclic and every sequence in C has a cyclic subsequence that converges (in norm) to some element of C. A subset in X is called *relatively cyclically compact* if it is contained in a cyclically compact set. Comparing these definitions with 1.11.8 and appealing to 1.11.9, it is easy to prove that a subset C of X is cyclically compact (relatively cyclically compact) if and only if $C{\uparrow}$ is a compact (relatively compact) subset of $\mathscr{X}$.

The following fact results from transferring the Hausdorff Criterion for compactness to a Boolean valued model.

1.16.2. Theorem. *A cyclic subset C of a B-space X is relatively cyclically compact if and only if to all $\varepsilon > 0$ there are a countable partition of unity (π_n) in the Boolean algebra B and a sequence (θ_n) of finite sets $\theta_n \subset C$ such that $\pi_n(\mathrm{mix}(\theta_n))$ is a ε-net for $\pi_n(C)$ for all $n \in \mathbb{N}$. The latter means that if $\theta_n := \{x_{n,1}, \dots, x_{n,l(n)}\}$, then to each $x \in \pi_n(C)$ there is a partition of unity $\{\rho_{n,1}, \dots, \rho_{n,l(n)}\}$ in B such that*

$$\left\| x - \sum_{k=1}^{l(n)} \pi_n \rho_{n,k} x_{n,k} \right\| \leq \varepsilon.$$

1.16.3. We consider the second operator dual (or second B-dual) of a space X defined as $X^{\#\#} := (X^{\#})^{\#} := \mathscr{L}_B(X^{\#}, \Lambda)$. Given $x \in X$ and $f \in X^{\#}$, we put $x^{\#\#} := \iota(x)$ where $\iota(x) : f \mapsto f(x)$. Clearly, $\iota(x) \in L(X^{\#}, \Lambda)$. Moreover,

$$\begin{aligned} |x^{\#\#}| = |\iota(x)| &= \sup\{|\iota(x)(f)| : |f| \leq 1\} \\ &= \sup\{|f(x)| : (\forall x \in X)\, |f(x)| \leq |x|\} \\ &= \sup\{|f(x)| : f \in \partial(|\cdot|)\} = |x|. \end{aligned}$$

Therefore, $\iota(x) \in X^{\#\#}$ for all $x \in X$. It is obvious that the operator $\iota : X \to X^{\#\#}$, acting by the rule $\iota : x \mapsto \iota(x)$, is a linear isometry. The operator ι is referred to as the *canonical embedding* of X into the second B-dual of X. As in the case of Banach spaces, it is convenient to treat x and $x^{\#\#} := \iota x$ as the same element and consider X as a subspace of $X^{\#\#}$. A B-normed space X is said to be *B-reflexive* if X and $X^{\#\#}$ coincide under the indicated embedding ι.

Theorem. *A normed B-space is B-reflexive if and only if its unit ball is cyclically $\sigma_\infty(X, X^\#)$-compact.*

$\lhd$ The Kakutani Criterion claims that a normed space is reflexive if and only if its unit ball is weakly compact. $\rhd$

1.16.4. Let X and Y be normed B-spaces.

An operator $T \in \mathscr{L}_B(X, Y)$ is called *cyclically compact* (in symbols, $T \in \mathscr{K}_B(X, Y)$) if the image $T(C)$ of every bounded subset $C \subset X$ is relatively cyclically compact in Y. It is easy to see that $\mathscr{K}_B(X, Y)$ is a decomposable subspace of the Banach–Kantorovich space $\mathscr{L}_B(X, Y)$.

Let $\mathscr{X}$ and $\mathscr{Y}$ be Boolean valued representations of X and Y. We recall that the immersion mapping $T \mapsto T^\sim$ over operators is a linear isometric embedding of the lattice normed spaces $\mathscr{L}_B(X, Y)$ into $\mathscr{L}^B(\mathscr{X}, \mathscr{Y})\!\downarrow$. Moreover, a bounded operator T from X to Y is cyclically compact if and only if $[\![T^\sim$ is a compact operator from $\mathscr{X}$ to $\mathscr{Y}]\!] = \mathbb{1}$.

1.16.5. Theorem. *Let X and Y be some Kaplansky–Hilbert modules over Λ, and let T in $\mathscr{K}_B(X, Y)$ be a cyclically compact operator from the Kaplansky–Hilbert module X to the Kaplansky–Hilbert module Y. There are orthonormal families $(e_k)_{k\in\mathbb{N}}$ in X, $(f_k)_{k\in\mathbb{N}}$ in Y, and a family $(\mu_k)_{k\in\mathbb{N}}$ in Λ such that the following hold:*

(1) *$\mu_{k+1} \le \mu_k$ $(k \in \mathbb{N})$ and $o\text{-}\lim_{k\to\infty} \mu_k = 0$;*

(2) *there is a band projection π_∞ in Λ such that $\pi_\infty \mu_k$ is a weak order unity in $\pi_\infty \Lambda$ for all $k \in \mathbb{N}$;*

(3) *there is a partition $(\pi_k)_{k=0}^\infty$ of the band projection $\pi_\infty^\perp$ such that $\pi_0 \mu_1 = 0$, $\pi_k \le [\mu_k]$, and $\pi_k \mu_{k+1} = 0$, $k \in \mathbb{N}$;*

(4) *the following representation holds:*

$$T = \pi_\infty \sum_{k=1}^{\infty} \mu_k e_k' \otimes f_k + \sum_{n=1}^{\infty} \pi_n \sum_{k=1}^{n} \mu_k e_k' \otimes f_k.$$

$\lhd$ By virtue of 1.8.11 we may assume that X and Y coincide with the bounded descents of Hilbert spaces $\mathscr{X}$ and $\mathscr{Y}$, respectively. The operator $T\!\uparrow : \mathscr{X} \to \mathscr{Y}$ is compact and we may apply inside $\mathbf{V}^{(B)}$ the theorem of ZFC on the general form of a compact operator in Hilbert space. $\rhd$

1.16.6. A variant of the Fredholm Alternative holds for cyclically compact operators. We will call it the *Fredholm B-Alternative.*

We now consider a B-cyclic Banach space X and a bounded B-linear operator T in X. In this case X and $X^{\#}$ are modules over the Stone algebra $\Lambda := \mathscr{S}(B)$ and T is Λ-linear. A subset $\mathscr{E} \subset X$ is said to be *locally linearly independent* if whenever $e_1, \dots, e_n \in \mathscr{E}$, $\lambda_1, \dots, \lambda_n \in \mathbb{C}$, and $\pi \in B$ with $\pi(\lambda_1 e_1 + \dots + \lambda_n e_n) = 0$ we have $\pi \lambda_k e_k = 0$ for all $k := 1, \dots, n$.

We say that the *Fredholm B-Alternative* is valid for an operator T if there exists a countable partition of unity (b_n) in B such that the following are fulfilled:

(1) The homogeneous equation $b_0 \circ Tx = 0$ has a sole solution, zero. The homogeneous conjugate equation $b_0 \circ T^{\#}y^{\#} = 0$ has a sole solution, zero. The equation $b_0 \circ Tx = b_0 y$ is solvable and has a unique solution given an arbitrary $y \in X$. The conjugate equation $b_0 \circ T^{\#}y^{\#} = b_0 x^{\#}$ is solvable and has a unique solution given an arbitrary $x^{\#} \in X^{\#}$.

(2) For every $n \in \mathbb{N}$ the homogeneous equation $b_n \circ Tx = 0$ has n locally linearly independent solutions $x_{1,n}, \dots, x_{n,n}$ and the homogeneous conjugate equation $b_n \circ T^{\#}y^{\#} = 0$ has n locally linearly independent (hence, nonzero) solutions $y^{\#}_{1,n}, \dots, y^{\#}_{n,n}$.

(3) The equation $Tx = y$ is solvable if and only if $b_n \circ y^{\#}_{k,n}(y) = 0$ $(n \in \mathbb{N},\ k \leq n)$. The conjugate equation $T^{\#}y^{\#} = x^{\#}$ is solvable if and only if $b_n \circ x^{\#}(x_{k,n}) = 0$ $(n \in \mathbb{N},\ k \leq n)$.

(4) The general solution x of the equation $Tx = y$ has the form

$$x = bo\text{-}\sum_{n=1}^{\infty} b_n \left(x_n + \sum_{k=1}^{n} \lambda_{k,n} x_{k,n} \right),$$

where x_n is a particular solution of the equation $b_n \circ Tx = b_n y$ and $(\lambda_{k,n})_{n \in \mathbb{N}, k \leq n}$ are arbitrary elements in Λ.

The general solution $y^{\#}$ of the conjugate equation $T^{\#}y^{\#} = x^{\#}$ has the form

$$y^{\#} = bo\text{-}\sum_{n=1}^{\infty} b_n \left(y^{\#}_n + \sum_{k=1}^{n} \lambda_{k,n} y^{\#}_{k,n} \right),$$

where $y^{\#}_n$ is a particular solution of the equation $b_n \circ T^{\#}y^{\#} = b_n x^{\#}$ and $\lambda_{k,n}$ are arbitrary elements in Λ for $n \in \mathbb{N}$ and $k \leq n$.

1.16.7. *Theorem.* *If S is a cyclically compact operator in a B-cyclic space X then the Fredholm B-Alternative is valid for the operator $T := I_X - S$.*

1.16.8. Comments.

(1) Cyclically compact sets and operators in lattice normed spaces were introduced in [51] and [53], respectively. A standard proof of Theorem 1.16.7 can be extracted from [53] wherein a more general approach is developed. Certain variants of Theorems 1.16.5 and 1.16.7 for operators in Banach–Kantorovich spaces can be also found in [53].

References

1. Akilov G. P., Kolesnikov E. V., and Kusraev A. G., "On order continuous extension of a positive operator," Sibirsk. Mat. Zh., **29**, No. 5, 24–35 (1988).
2. Akilov G. P. and Kutateladze S. S., Ordered Vector Spaces [in Russian], Nauka, Novosibirsk (1978).
3. Albeverio S., Fenstad J. E., et al., Nonstandard Methods in Stochastic Analysis and Mathematical Physics, Academic Press, Orlando etc. (1990).
4. Aliprantis C. D. and Burkinshaw O., Positive Operators, Academic Press, New York (1985).
5. Ballard D. and Hrbáček K., "Standard foundations of nonstandard analysis," J. Symbolic Logic, **57**, No. 2, 741–748 (1992).
6. Bell J. L., Boolean-Valued Models and Independence Proofs in Set Theory, Clarendon Press, New York etc. (1985).
7. Bukhvalov A. V., "Order bounded operators in vector lattices and spaces of measurable functions," in: Mathematical Analysis [in Russian], VINITI, Moscow, 1988, pp. 3–63 (Itogi Nauki i Tekhniki, **26**).
8. Kutateladze S. S. (ed.), Vector Lattices and Integral Operators, Kluwer Academic Publishers, Dordrecht (1996).
9. Cohen P. J., Set Theory and the Continuum Hypothesis, Benjamin, New York etc. (1966).
10. Cozart D. and Moore L. C. Jr., "The nonstandard hull of a normed Riesz space," Duke Math. J., **41**, 263–275 (1974).
11. Cutland N., "Nonstandard measure theory and its applications," Proc. London Math. Soc., **15**, No. 6, 530–589 (1983).
12. Dacunha-Castelle D. and Krivine J. L., "Applications des ultraproducts a l'etude des espaces et des algebres de Banach," Studia Math., **41**, 315–334 (1972).
13. Gordon H., "Decomposition of linear functionals on Riesz spaces," Duke Math. J., **27**, 323–332 (1960).
14. Gordon E. I., "Real numbers in Boolean-valued models of set theory and K-spaces," Dokl. Akad. Nauk SSSR, **237**, No. 4, 773–775 (1977).

15. Gordon E. I., "K-spaces in Boolean-valued models of set theory," Dokl. Akad. Nauk SSSR, **258**, No. 4, 777–780 (1981).

16. Gordon E. I., "To the theorems of relation preservation in K-spaces," Sibirsk. Mat. Zh., **23**, No. 5, 55–65 (1982).

17. Gordon E. I., "Nonstandard finite-dimensional analogs of operators in $L_2(\mathbb{R}^n)$," Sibirsk. Mat. Zh., **29**, No. 2, 45–59 (1988).

18. Gordon E. I., "Relative standard elements in E. Nelson's internal set theory," Sibirsk. Mat. Zh., **30**, No. 1, 89–95 (1989).

19. Gordon E. I., "On Loeb measures," Izv. Vyssh. Uchebn. Zaved. Mat., No. 2, 25–33 (1991).

20. Gordon E. I., "Nonstandard analysis and compact Abelian groups," Sibirsk. Mat. Zh., **32**, No. 2, 26–40 (1991).

21. Gordon E. I., Nonstandard Methods in Commutative Harmonic Analysis, Amer. Math. Soc., Providence, RI (1997).

22. Gumilëv N. S., Verses and Poems [in Russian], Sov. Pisatel′, Leningrad (1988).

23. Halmos P. R., Lectures on Boolean Algebras, Van Nostrand, Toronto, New York, and London (1963).

24. Heinrich S., "Ultraproducts in Banach space theory," J. Reine Angem. Math., **313**, 72–104 (1980).

25. Henson C. W., "When do the Banach spaces have isometrically isomorphic nonstandard hulls?" Israel J. Math., **22**, 57–67 (1975).

26. Henson C. W., "Nonstandard hulls of Banach spaces," Israel J. Math., **25**, 108–114 (1976).

27. Henson C. W., "Infinitesimals in functional analysis," in: Nonstandard Analysis and Its Applications, Cambridge Univ. Press, Cambridge etc., 1988, pp. 140–181.

28. Henson C. W. and Moore L. C. Jr., "Nonstandard hulls of the classical Banach spaces," Duke Math. J., **41**, No. 2, 277–284 (1974).

29. Henson C. W. and Moore L. C. Jr., "Nonstandard analysis and the theory of Banach spaces," in: Nonstandard Analysis. Recent Development, Springer-Verlag, Berlin etc., 1983, pp. 27–112. (Lecture Notes in Math., **983**.)

30. Hrbáček K., "Axiomatic foundations for nonstandard analysis," Fund. Math., **98**, No. 1, 1–24 (1978).

31. Jameson G. J. O., Ordered Linear Spaces, Springer-Verlag, Berlin etc. (1970). (Lecture Notes in Math., **141**.)

32. Jech T. J., The Axiom of Choice, North-Holland, Amsterdam etc. (1973).

33. Jech T. J., "Abstract theory of abelian operator algebras: an application of forcing," Trans. Amer. Math. Soc., **289**, No. 1, 133–162 (1985).

34. Jech T. J., "First order theory of complete Stonean algebras (Boolean-valued real and complex numbers)," Canad. Math. Bull., **30**, No. 4, 385–392 (1987).

35. Jech T. J., "Boolean-linear spaces," Adv. in Math., **81**, No. 2, 117–197 (1990).

36. Jech T. J., Set Theory and the Method of Forcing, Springer-Verlag, Berlin etc. (1997).

37. Kanovei V. and Reeken M., "Internal approach to external sets and universes," Studia Logica, part 1: **55**, 227–235 (1995); part II: **55**, 347–376 (1995); part III: **56**, 293–322 (1996).

38. Kantorovich L. V., "On semiordered linear spaces and their applications to the theory of linear operations," Dokl. Akad. Nauk SSSR, **4**, No. 1–2, 11–14 (1935).

39. Kantorovich L. V., "To the general theory of operations in semiordered spaces," Dokl. Akad. Nauk SSSR, **1**, No. 7, 271–274 (1936).

40. Kantorovich L. V. and Akilov G. P., Functional Analysis, Pergamon Press, Oxford and New York (1982).

41. Kantorovich L. V., Vulikh B. Z., and Pinsker A. G., Functional Analysis in Semiordered Spaces [in Russian], Gostekhizdat, Moscow and Leningrad (1950).

42. Kaplansky I., "Projections in Banach algebras," Ann. of Math., **53**, No. 2, 235–249 (1951).

43. Kaplansky I., "Algebras of type I," Ann. of Math., **56**, 460–472 (1952).

44. Kaplansky I., "Modules over operator algebras," Amer. J. Math., **75**, No. 4, 839–858 (1953).

45. Kawai T., "Axiom systems of nonstandard set theory," in: Logic Symposia, Hakone 1979, 1980, Springer-Verlag, Berlin etc., 1981, pp. 57–65.

46. Kawai T., "Nonstandard analysis by axiomatic method," in: Southeast Asian Conference on Logic, North Holland, Amsterdam etc., 1983, pp. 55–76.

47. Kolesnikov E. V., "Decomposition of a positive operator," Sibirsk. Mat. Zh., **30**, No. 5, 70–73 (1989).

48. Kolesnikov E. V., Kusraev A. G., and Malyugin S. A., On Dominated Operators [in Russian] [Preprint, No. 26], Sobolev Inst. Mat., Novosibirsk (1988).

49. Krivine J. L., "Langages a valeurs reelles et applications," Fund. Math., **81**, No. 3, 213–253 (1974).

50. Kusraev A. G., "General desintegration formulas," Dokl. Akad. Nauk SSSR, **265**, No. 6, 1312–1316 (1982).

51. Kusraev A. G., "Boolean valued analysis of duality between universally complete modules," Dokl. Akad. Nauk SSSR, **267**, No. 5, 1049–1052 (1982).

52. Kusraev A. G., "On some categories and functors of Boolean valued analysis," Dokl. Akad. Nauk SSSR, **271**, No. 1, 281–284 (1983).

53. Kusraev A. G., Vector Duality and Its Applications [in Russian], Nauka, Novosibirsk (1985).

54. Kusraev A. G., "On Banach–Kantorovich spaces," Sibirsk. Mat. Zh., **26**, No. 2, 119–126 (1985).

55. Kusraev A. G., "Numeric systems in Boolean-valued models of set theory," in: Proceedings of the VIII All-Union Conference on Mathematical Logic (Moscow), Moscow, 1986, p. 99.

56. Kusraev A. G., "Linear operators in lattice normed spaces," in: Studies on Geometry in the Large and Mathematical Analysis. Vol. 9 [in Russian], Trudy Inst. Mat. (Novosibirsk), Novosibirsk, 1987, pp. 84–123.

57. Kusraev A. G., "On functional representation of type I AW^*-algebras," Sibirsk. Mat. Zh., **32**, No. 3, 78–88 (1991).

58. Kusraev A. G., "Dominated operators," in: Linear Operators Compatible with Order [in Russian], Sobolev Institute Press, Novosibirsk, 1995, pp. 212–292.

59. Kusraev A. G. and Kutateladze S. S., "Combining nonstandard methods," Sibirsk. Mat. Zh., **31**, No. 5, 111–119 (1990).

60. Kusraev A. G. and Kutateladze S. S., Boolean Valued Analysis, Kluwer Academic Publishers, Dordrecht (1999).

61. Kusraev A. G. and Kutateladze S. S., Subdifferentials: Theory and Applications, Kluwer Academic Publishers, Dordrecht (1995).

62. Kusraev A. G. and Kutateladze S. S., Nonstandard Methods of Analysis, Kluwer Academic Publishers, Dordrecht (1994).

63. Kusraev A. G. and Malyugin S. A., Some Questions of the Theory of Vector Measures [in Russian], Inst. Mat. (Novosibirsk), Novosibirsk (1988).

64. Kusraev A. G. and Malyugin S. A., "On atomic decomposition of vector measures," Sibirsk. Mat. Zh., **30**, No. 5, 101–110 (1989).

65. Kusraev A. G. and Malyugin S. A., "The product and projective limit of vector measures," in: Contemporary Problems of Geometry and Analysis [in Russian], Nauka, Novosibirsk, 1989, pp. 132–152.

66. Kusraev A, G. and Malyugin S. A., "On extension of finitely additive measures," Mat. Zametki, **48**, No. 1, 56–60 (1990).

67. Kusraev A. G. and Strizhevskiĭ V. Z., "Lattice normed spaces and dominated operators," in: Studies on Geometry and Functional Analysis. Vol. 7 [in Russian], Trudy Inst. Mat. (Novosibirsk), Novosibirsk, 1987, pp. 132–158.

68. Kutateladze S. S., "Supporting sets of sublinear operators," Dokl. Akad. Nauk SSSR, **230**, No. 5, 1029–1032 (1976).

69. Kutateladze S. S., "Subdifferentials of convex operators," Sibirsk. Mat. Zh., **18**, No. 5, 1057–1064 (1977).

70. Kutateladze S. S., "Extreme points of subdifferentials," Dokl. Akad. Nauk SSSR, **242**, No. 5, 1001–1003 (1978).

71. Kutateladze S. S., "The Kreĭn–Milman theorem and its converse," Sibirsk. Mat. Zh., **21**, No. 1, 130–138 (1980).

72. Kutateladze S. S., "Descents and ascents," Dokl. Akad. Nauk SSSR, **272**, No. 2, 521–524 (1983).

73. Kutateladze S. S., "Cyclic monads and their applications," Sibirsk. Mat. Zh., **27**, No. 1, 100–110 (1986).

74. Kutateladze S. S., "Monads of ultrafilters and extensional filters," Sibirsk. Mat. Zh., **30**, No. 1, 129–133 (1989).

75. Kutateladze S. S., "Fragments of a positive operator," Sibirsk. Mat. Zh., **30**, No. 5, 111–119 (1989).

76. Kutateladze S. S., "Credenda of nonstandard analysis," Siberian Adv. Math., **1**, No. 1, 109–137 (1991).

77. Kutateladze S. S., Fundamentals of Functional Analysis, Kluwer Academic Publishers, Dordrecht (1996).

78. Leibniz G. W., Discourse on Metaphysics and Other Essays, Hackett Publising Co., Indianapolis (1991).

79. Lipecki Z., Plachky D., and Thompsen W., "Extensions of positive operators and extreme points. I," Colloq. Math., **42**, 279–284 (1979).

80. Luxemburg W. A. J., "A general theory of monads," in: Applications of Model Theory to Algebra, Analysis and Probability, Holt, Rinehart and Minston, New York, 1966, pp. 18–86.

81. Luxemburg W. A. J. and Schep A. R., "A Radon–Nikodým type theorem for positive operators and a dual," Indag. Math., **40**, 357–375 (1978).

82. Luxemburg W. A. J. and Zaanen A. C., Riesz Spaces. Vol. 1, North-Holland, Amsterdam and London (1971).

83. Lyubetskiĭ V. A., "Truth-values and bundles. On some problems of nonstandard analysis," Uspekhi Mat. Nauk, **55**, No. 6, 99–153 (1979).

84. Lyubetskiĭ V. A. and Gordon E. I., "Boolean completion of uniformities," in: Studies on Nonclassical Logics and Formal Systems [in Russian], Nauka, Moscow, 1983, pp. 82–153.

85. Maharam D., "On kernel representation of linear operators," Trans. Amer. Math. Soc., **79**, No. 1, 229–255 (1955).

86. Maharam D., "On positive operators," Contemporary Math., **26**, 263–277 (1984).

87. Nelson E., "Internal set theory. A new approach to nonstandard analysis," Bull. Amer. Math. Soc., **83**, No. 6, 1165–1198 (1977).

88. Nelson E., "The syntax of nonstandard analysis," Ann. Pure Appl. Logic, **38**, No. 2, 123–134 (1988).

89. Ozawa M., "Boolean valued interpretation of Hilbert space theory," J. Math. Soc. Japan, **35**, No. 4, 609–627 (1983).
90. Ozawa M., "A classification of type I AW^*-algebras and Boolean valued analysis," J. Math. Soc. Japan, **36**, No. 4, 589–608 (1984).
91. Ozawa M., "A transfer principle from von Neumann algebras to AW^*-algebras," J. London Math. Soc. (2), **32**, No. 1, 141–148 (1985).
92. Ozawa M., "Nonuniqueness of the cardinality attached to homogeneous AW^*-algebras," Proc. Amer. Math. Soc., **93**, 681–684 (1985).
93. Pagter B. de, "The components of a positive operator," Indag. Math., **45**, No. 2, 229–241 (1983).
94. Péraire Y., "Une nouvelle théorie des infinitesimaux," C. R. Acad. Sci. Paris Ser. I, **301**, No. 5, 157–159 (1985).
95. Péraire Y., "A general theory of infinitesimals," Sibirsk. Mat. Zh., **31**, No. 3, 107–124 (1990).
96. Péraire Y., "Théorie relative des ensembles internes," Osaka J. Math., **29**, No. 2, 267–297 (1992).
97. Péraire Y., "Some extensions of the principles of idealization, transfer and choice in the relative internal set theory," Arch. Math. Logic, **34**, 269–277 (1995).
98. Schaefer H. H., Banach Lattices and Positive Operators, Springer-Verlag, Berlin etc. (1974).
99. Schwarz H.-U., Banach Lattices and Operators, Teubner, Leipzig (1984).
100. Sikorski R., Boolean Algebras, Springer-Verlag, Berlin etc. (1964).
101. Sobolev V. I., "On a poset valued measure of a set, measurable functions, and some abstract integrals," Dokl. Akad. Nauk SSSR, **91**, No. 1, 23–26 (1953).
102. Solovay R. M., "A model of set theory in which every set of reals is Lebesgue measurable," Ann. of Math. (2), **92**, No. 2, 1–56 (1970).
103. Solovay R. and Tennenbaum S., "Iterated Cohen extensions and Souslin's problem," Ann. Math., **94**, No. 2, 201–245 (1972).
104. Stern J., "Some applications of model theory in Banach space theory," Ann. Math. Logic, **9**, No. 1, 49–121 (1976).
105. Stern J., "The problem of envelopes for Banach spaces," Israel J. Math., **24**, No. 1, 1–15 (1976).
106. Takeuti G., Two Applications of Logic to Mathematics, Princeton, Iwanami and Princeton Univ. Press, Tokyo (1978).
107. Takeuti G., "A transfer principle in harmonic analysis," J. Symbolic Logic, **44**, No. 3, 417–440 (1979).
108. Takeuti G., "Boolean valued analysis," in: Applications of Sheaves (Proc. Res. Sympos. Appl. Sheaf Theory to Logic, Algebra and Anal., Univ. Durham, Durham, 1977), Springer-Verlag, Berlin etc., 1979.

109. Takeuti G. and Zaring W. M., Axiomatic Set Theory, Springer-Verlag, New York (1973).
110. Troitskiĭ V. G., "Nonstandard discretization and the Loeb extension of a family of measures," Siberian Math. J., **34**, No. 3, 566–573 (1993).
111. Vladimirov D. A., Boolean Algebras [in Russian], Nauka, Moscow (1969).
112. Vopěnka P., "General theory of ∇-models," Comment. Math. Univ. Carolin., **7**, No. 1, 147–170 (1967).
113. Vopěnka P., Mathematics in the Alternative Set Theory, Teubner, Leipzig (1979).
114. Vopěnka P. and Hajek P., The Theory of Semisets, North-Holland, Amsterdam (1972).
115. Vulikh B. Z., Introduction to the Theory of Partially Ordered Spaces, Noordhoff, Groningen (1967).
116. Wang H. and McNaughton R., Les Systemes Axiomatiques de la Theorie des Ensemblas, Paris and Louvain (1962).
117. Zaanen A. C., Riesz Spaces. Vol. 2, North-Holland, Amsterdam etc. (1983).

CHAPTER 2

Functional Representation of a Boolean Valued Universe

BY

A. E. Gutman and G. A. Losenkov

S.S. Kutateladze (ed.), Nonstandard Analysis and Vector Lattices, 81–104.

The methods of Boolean valued analysis rest on multivalued nonstandard models of set theory. More exactly, the truth value of an assertion in such a model acts into some complete Boolean algebra.

At present, Boolean valued analysis is a rather powerful theory rich in deep results and various applications mostly to set theory. As regards functional analysis, the methods of Boolean valued analysis found successful applications in such domains as the theory of vector lattices and lattice normed spaces, the theory of positive and dominated operators, the theory of von Neumann algebras, convex analysis, and the theory of vector measures.

Contemporary methods of Boolean valued analysis, due to their nature, involve a rather bulky logical technique. From a pragmatic viewpoint, this technique might distract the user-analyst from a concrete aim: to apply the results of Boolean valued analysis for solving analytical problems.

Various function spaces reside in functional analysis, and so the intention is natural of replacing an abstract Boolean valued system by some function analog, a model whose elements are functions and in which the basic logical operations are calculated "pointwise." An example of such a model is given by the class $\mathbf{V}^Q$ of all functions defined on a fixed nonempty set Q and acting into the class $\mathbf{V}$ of all sets. Truth values in the model $\mathbf{V}^Q$ are various subsets of Q and, in addition, the truth value $[\![\varphi(u_1,\dots,u_n)]\!]$ of an assertion $\varphi(t_1,\dots,t_n)$ at functions $u_1,\dots,u_n \in \mathbf{V}^Q$ is calculated as follows:

$$[\![\varphi(u_1,\dots,u_n)]\!] = \{q \in Q : \varphi(u_1(q),\dots,u_n(q))\}.$$

In the present article, a solution is proposed to the above problem. To this end, we introduce and study a new notion of continuous polyverse which is a continuous bundle of models of set theory. It is shown that the class of continuous sections of a continuous polyverse is a Boolean valued system satisfying all basic principles of Boolean valued analysis and, conversely, every Boolean valued algebraic system can be represented as the class of sections of a suitable continuous polyverse.

2.1. Preliminaries

2.1.1. Let X and Y be topological spaces. A mapping $f : X \to Y$ is called *open* if f satisfies one (and hence all) of the following equivalent conditions:

(1) for every open subset $A \subset X$, the image $f(A)$ is open in Y;

(2) for every point $x \in X$ and every neighborhood $A \subset X$ about x, the image $f(A)$ is a neighborhood about $f(x)$ in Y;

(3) $f^{-1}(\mathrm{cl}\, B) \subset \mathrm{cl}\, f^{-1}(B)$ for every subset $B \subset Y$. Observe that the equality $f^{-1}(\mathrm{cl}\, B) = \mathrm{cl}\, f^{-1}(B)$ holds for all subsets $B \subset Y$ if and only if f is a continuous and open mapping.

A mapping $f : X \to Y$ is called *closed* if f satisfies one (and hence all) of the following equivalent conditions:

(1) for every closed subset $A \subset X$, the image $f(A)$ is closed in Y;

(2) $\mathrm{cl}\, f(A) \subset f(\mathrm{cl}\, A)$ for every subset $A \subset X$. The equality $\mathrm{cl}\, f(A) = f(\mathrm{cl}\, A)$ holds for every subset $A \subset X$ if and only if $f : X \to Y$ is a continuous and closed mapping.

2.1.2. Given a class X, the symbol $\mathscr{P}(X)$ denotes the class of all subsets of X. Let X be a class. A subclass $\tau \subset \mathscr{P}(X)$ is called a *topology* on X whenever

(1) $\cup\tau = X$;

(2) $U \cap V \in \tau$ for all $U, V \in \tau$;

(3) $\cup\mathscr{U} \in \tau$ for every subset $\mathscr{U} \subset \tau$. As usual, a class X endowed with a topology is called a *topological space.*

All basic topological concepts (such as neighborhood about a point, closed set, interior, closure, continuous function, Hausdorff space, etc.) can be introduced by analogy to the case of a topology on a set. However, observe that not all classical approaches to the definition of these concepts remain formally valid in the case of a class-topology. For instance, considering the two definitions of a closed set

(a) as a subset of X whose complement belongs to τ,

(b) as a subset of X whose complement, together with each point of it, contains an element of τ,

we should choose the second.

Defining the closure of a set A as the smallest closed subset of X that contains A, we take a risk: some sets may turn out to have no closure. However, the problem disappears if the topology τ is Hausdorff. (Indeed, in the case of a Hausdorff topology, every convergent filter has a unique limit and, hence, the totality of all limits of convergent filters over a given set makes a set rather than a proper class.)

The symbol $\mathrm{Clop}(X)$ denotes the class of all clopen subsets of X (i.e., subsets that are closed and open simultaneously). Henceforth the notation $U \sqsubset X$ means that $U \in \mathrm{Clop}(X)$. The class $\{A \sqsubset X : x \in A\}$ is denoted by $\mathrm{Clop}(x)$.

A topology is called *extremally disconnected* if the closure of every open set is again open.

Most of the necessary information about topological spaces can be found, for instance, in [1, 2].

2.1.3. Let B be a complete Boolean algebra. A triple $(\mathfrak{U}, [\![\cdot = \cdot]\!], [\![\cdot \in \cdot]\!])$ is called a *Boolean valued algebraic system* over B (or a *B-valued algebraic system*) if the classes $[\![\cdot = \cdot]\!]$ and $[\![\cdot \in \cdot]\!]$ are class-functions from $\mathfrak{U} \times \mathfrak{U}$ to B satisfying the following conditions:

(1) $[\![u = u]\!] = \mathbf{1}$;

(2) $[\![u = v]\!] = [\![v = u]\!]$;

(3) $[\![u = v]\!] \wedge [\![v = w]\!] \leqslant [\![u = w]\!]$;

(4) $[\![u = v]\!] \wedge [\![v \in w]\!] \leqslant [\![u \in w]\!]$;

(5) $[\![u = v]\!] \wedge [\![w \in v]\!] \leqslant [\![w \in u]\!]$ for all $u, v, w \in \mathfrak{U}$.

The class-functions $[\![\cdot = \cdot]\!]$ and $[\![\cdot \in \cdot]\!]$ are called the Boolean valued (B-valued) *truth values* of equality and membership.

Instead of $(\mathfrak{U}, [\![\cdot = \cdot]\!], [\![\cdot \in \cdot]\!])$, we usually write simply $\mathfrak{U}$, furnishing the symbols of truth values with the index: $[\![\cdot = \cdot]\!]_{\mathfrak{U}}$ and $[\![\cdot \in \cdot]\!]_{\mathfrak{U}}$ if need be.

A Boolean valued system $\mathfrak{U}$ is called *separated* whenever, for all $u, v \in \mathfrak{U}$, the equality $[\![u = v]\!] = \mathbf{1}$ implies $u = v$.

2.1.4. Consider Boolean valued algebraic systems $\mathfrak{U}$ and $\mathfrak{V}$ over complete Boolean algebras B and C and assume that there is a Boolean isomorphism $\jmath : B \to C$. By an *isomorphism between the Boolean valued algebraic systems* $\mathfrak{U}$ *and* $\mathfrak{V}$ *(associated with the isomorphism* $\jmath$*)* we mean a bijective class-function $\imath : \mathfrak{U} \to \mathfrak{V}$ that satisfies the following relations:

$$\jmath([\![u_1 = u_2]\!]_{\mathfrak{U}}) = [\![\imath(u_1) = \imath(u_2)]\!]_{\mathfrak{V}},$$
$$\jmath([\![u_1 \in u_2]\!]_{\mathfrak{U}}) = [\![\imath(u_1) \in \imath(u_2)]\!]_{\mathfrak{V}}$$

for all $u_1, u_2 \in \mathfrak{U}$. Boolean valued systems are said to be *isomorphic* if there is an isomorphism between them. In case $\mathfrak{U}$ and $\mathfrak{V}$ are Boolean valued algebraic systems over the same algebra B, each isomorphism $\imath : \mathfrak{U} \to \mathfrak{V}$ is assumed by default to be associated with the identity isomorphism: $[\![u_1 = u_2]\!]_{\mathfrak{U}} = [\![\imath(u_1) = \imath(u_2)]\!]_{\mathfrak{V}}$, $[\![u_1 \in u_2]\!]_{\mathfrak{U}} = [\![\imath(u_1) \in \imath(u_2)]\!]_{\mathfrak{V}}$. For emphasizing this convention, whenever necessary, we call such an isomorphism *B-isomorphism* and refer to the corresponding systems as *B-isomorphic*.

2.1.5. In what follows, using an expression like $\varphi(t_1, \dots, t_n)$, we assume that φ is a set-theoretic formula with all free variables listed in $(t_1, \dots, t_n)$.

An arbitrary tuple $(u_1, \dots, u_n)$ of elements of a system $\mathfrak{U}$ is called a *valuation* of the list of variables $(t_1, \dots, t_n)$. By recursion on the length of a formula, the (Boolean) *truth value* $[\![\varphi(u_1, \dots, u_n)]\!]$ of a formula $\varphi(t_1, \dots, t_n)$ can be defined by assignment $(u_1, \dots, u_n)$ to the variables $(t_1, \dots, t_n)$. If a formula φ is atomic, i.e., has the form $t_1 = t_2$ or $t_1 \in t_2$; then the truth value of φ by assignment (u_1, u_2) is defined to be $[\![u_1 = u_2]\!]$ or $[\![u_1 \in u_2]\!]$. Considering compound formulas, we define their truth values as follows:

$$
\begin{aligned}
[\![\varphi(u_1, \dots, u_n) \mathbin{\&} \psi(u_1, \dots, u_n)]\!] &:= [\![\varphi(u_1, \dots, u_n)]\!] \wedge [\![\psi(u_1, \dots, u_n)]\!], \\
[\![\varphi(u_1, \dots, u_n) \vee \psi(u_1, \dots, u_n)]\!] &:= [\![\varphi(u_1, \dots, u_n)]\!] \vee [\![\psi(u_1, \dots, u_n)]\!], \\
[\![\varphi(u_1, \dots, u_n) \to \psi(u_1, \dots, u_n)]\!] &:= [\![\varphi(u_1, \dots, u_n)]\!] \Rightarrow [\![\psi(u_1, \dots, u_n)]\!], \\
[\![\neg\varphi(u_1, \dots, u_n)]\!] &:= [\![\varphi(u_1, \dots, u_n)]\!]^{\perp}, \\
[\![(\forall t)\, \varphi(t, u_1, \dots, u_n)]\!] &:= \bigwedge_{u \in \mathfrak{U}} [\![\varphi(u, u_1, \dots, u_n)]\!], \\
[\![(\exists t)\, \varphi(t, u_1, \dots, u_n)]\!] &:= \bigvee_{u \in \mathfrak{U}} [\![\varphi(u, u_1, \dots, u_n)]\!],
\end{aligned}
$$

where the symbol $b^{\perp}$ denotes the complement of b in the Boolean algebra B. A formula $\varphi(t_1, \dots, t_n)$ is said to be *true* in an algebraic system $\mathfrak{U}$ by assignment $(u_1, \dots, u_n)$ if the equality $[\![\varphi(u_1, \dots, u_n)]\!] = \mathbf{1}$ holds. In this case, we write $\mathfrak{U} \models \varphi(u_1, \dots, u_n)$.

2.1.6. *Proposition.* *If a formula $\varphi(t_1, \dots, t_n)$ is provable in the first-order predicate calculus then $[\![\varphi(u_1, \dots, u_n)]\!] = \mathbf{1}$ for all $u_1, \dots, u_n \in \mathfrak{U}$.*

$\triangleleft$ It is easy to verify that all axioms of the first-order predicate calculus are true in $\mathfrak{U}$ and the rules of inference increase the truth value. The latter means that derivability (in the first-order predicate calculus) of a formula φ from formulas $\varphi_1, \dots, \varphi_n$ ensures the inequality $[\![\varphi_1 \wedge \dots \wedge \varphi_n]\!] \leqslant [\![\varphi]\!]$. $\triangleright$

In particular, the last proposition implies that, for all $\varphi(t, t_1, \dots, t_n)$ and $u, v, w_1, \dots, w_n \in \mathfrak{U}$, we have the inequality

$$
[\![u = v]\!] \wedge [\![\varphi(u, w_1, \dots, w_n)]\!] \leqslant [\![\varphi(v, w_1, \dots, w_n)]\!].
$$

2.1.7. Let $u \in \mathfrak{U}$ be such that $\mathfrak{U} \models u \neq \varnothing$. The *descent* of u is the class $\{v \in \mathfrak{U} : \mathfrak{U} \models v \in u\}$ denoted by $u{\downarrow}$.

2.1.8. Let $(u_\xi)_{\xi \in \Xi}$ be a family of elements in $\mathfrak{U}$ and let $(b_\xi)_{\xi \in \Xi}$ be a family of elements in the Boolean algebra B. An element $u \in \mathfrak{U}$ is called an *ascent of the family* $(u_\xi)_{\xi \in \Xi}$ *by (weights)* $(b_\xi)_{\xi \in \Xi}$, if $[\![v \in u]\!] = \bigvee_{\xi \in \Xi} b_\xi \wedge [\![v = u_\xi]\!]$ for all $v \in \mathfrak{U}$.

Let $\mathscr{U}$ be a subset of $\mathfrak{U}$. An element $\overline{u} \in \mathfrak{U}$ is called an *ascent of the set* $\mathscr{U}$, if $[\![v \in \overline{u}]\!] = \bigvee_{u \in \mathscr{U}} [\![v = u]\!]$ for all $v \in \mathfrak{U}$, i.e., $\overline{u}$ is an ascent of the family $(u)_{u \in \mathscr{U}}$ by unit weights.

Assume that $(b_\xi)_{\xi\in\Xi}$ is an antichain in the algebra B. An element $u \in \mathfrak{U}$ is called a *mixing* of the family $(u_\xi)_{\xi\in\Xi}$ by $(b_\xi)_{\xi\in\Xi}$, if $[\![u = u_\xi]\!] \geqslant b_\xi$ for all $\xi \in \Xi$, and $[\![u = \varnothing]\!] \geqslant (\bigvee_{\xi\in\Xi} b_\xi)^\perp$.

If the system $\mathfrak{U}$ is separated and the axiom of extensionality is true in $\mathfrak{U}$, then an ascent (mixing) of a family $(u_\xi)_{\xi\in\Xi}$ by $(b_\xi)_{\xi\in\Xi}$ is uniquely determined. In this case, whenever the ascent (mixing) exists we denote it by $\mathrm{asc}_{\xi\in\Xi}\, b_\xi u_\xi$ $(\mathrm{mix}_{\xi\in\Xi}\, b_\xi u_\xi)$. For the ascent of a set $\mathscr{U} \subset \mathfrak{U}$, we use the notation $\mathscr{U}\!\uparrow$.

2.1.9. A key role in Boolean valued analysis is played by the three basic principles: the maximum principle, the mixing principle, and the ascent principle. The reason behind this is the fact that, in algebraic systems satisfying these principles, there is a possibility of constructing new elements from those available.

In the current section, we state the above-mentioned principles and study interrelations between them, leaving aside verification of the principles for concrete algebraic systems.

Let B be a complete Boolean algebra, and let $\mathfrak{U}$ be a B-valued algebraic system.

Maximum Principle. *For every formula $\varphi(t, t_1, \dots, t_n)$ and arbitrary elements $u_1, \dots, u_n \in \mathfrak{U}$, there exists an element $u \in \mathfrak{U}$ such that $[\![(\exists t)\, \varphi(t, u_1, \dots, u_n)]\!] = [\![\varphi(u, u_1, \dots, u_n)]\!]$.*

Mixing Principle. *For every family $(u_\xi)_{\xi\in\Xi}$ of elements in $\mathfrak{U}$ and every antichain $(b_\xi)_{\xi\in\Xi}$ in the algebra B, there exists a mixing $(u_\xi)_{\xi\in\Xi}$ by $(b_\xi)_{\xi\in\Xi}$.*

Ascent Principle. *The following hold:*

(1) *For every family $(u_\xi)_{\xi\in\Xi}$ of elements in $\mathfrak{U}$ and every family $(b_\xi)_{\xi\in\Xi}$ of elements in the algebra B, there exists an ascent $(u_\xi)_{\xi\in\Xi}$ by $(b_\xi)_{\xi\in\Xi}$.*

(2) *For every element $u \in \mathfrak{U}$, there exist a family $(u_\xi)_{\xi\in\Xi}$ of elements in $\mathfrak{U}$ and a family $(b_\xi)_{\xi\in\Xi}$ of elements in the algebra B such that u is an ascent of $(u_\xi)_{\xi\in\Xi}$ by $(b_\xi)_{\xi\in\Xi}$.*

2.1.10. Theorem. *If a B-valued system $\mathfrak{U}$ satisfies the mixing principle then $\mathfrak{U}$ satisfies the maximum principle.*

$\lhd$ Considering a formula $\varphi(t, t_1, \dots, t_n)$, denote by $\vec{u}$ a tuple of arbitrary elements $u_1, \dots, u_n \in \mathfrak{U}$ and put $b = [\![(\exists t)\, \varphi(t, \vec{u})]\!]$. By the definition of truth value, $b = \bigvee_{v\in\mathfrak{U}} [\![\varphi(v, \vec{u})]\!]$. According to the exhaustion principle, there exist an antichain $(b_\xi)_{\xi\in\Xi}$ in the algebra B and a family $(v_\xi)_{\xi\in\Xi}$ of elements in $\mathfrak{U}$ such that $\bigvee_{\xi\in\Xi} b_\xi = b$ and $b_\xi \leqslant [\![\varphi(v_\xi, \vec{u})]\!]$. By the hypothesis of the theorem, there exists a mixing $v \in \mathfrak{U}$ of the family $(v_\xi)_{\xi\in\Xi}$ by $(b_\xi)_{\xi\in\Xi}$. In particular, $[\![v = v_\xi]\!] \geqslant b_\xi$. In view of Proposition 2.1.6, the following inequalities hold: $[\![\varphi(v, \vec{u})]\!] \geqslant [\![v = v_\xi]\!] \wedge [\![\varphi(v_\xi, \vec{u})]\!] \geqslant b_\xi$. Consequently, $[\![\varphi(v, \vec{u})]\!] \geqslant \bigvee_{\xi\in\Xi} b_\xi = b$. The inequality $[\![\varphi(v, \vec{u})]\!] \leqslant b$ is obvious. $\rhd$

2.1.11. Theorem. *Let a B-valued algebraic system $\mathfrak{U}$ satisfy the ascent principle and let the axiom of extensionality be true in $\mathfrak{U}$. Then the mixing principle is valid for $\mathfrak{U}$.*

$\lhd$ Let $(u_\xi)_{\xi\in\Xi}$ be a family of elements in $\mathfrak{U}$ and let $(b_\xi)_{\xi\in\Xi}$ be an antichain in the algebra B. By the hypothesis of the theorem, for every $\xi \in \Xi$, there exist a family $(u_\xi^\alpha)_{\alpha\in A(\xi)}$ of elements in $\mathfrak{U}$ and a family $(b_\xi^\alpha)_{\alpha\in A(\xi)}$ of elements in the algebra B such that

$$[\![v \in u_\xi]\!] = \bigvee_{\alpha\in A(\xi)} b_\xi^\alpha \wedge [\![v = u_\xi^\alpha]\!] \quad \text{for all } v \in \mathfrak{U}.$$

Consider the set $\Gamma = \{(\xi,\alpha) : \xi \in \Xi,\ \alpha \in A(\xi)\}$ and, for each pair $\gamma = (\xi,\alpha) \in \Gamma$, put $c_\gamma = b_\xi \wedge b_\xi^\alpha$ and $v_\gamma = u_\xi^\alpha$. Let $u \in \mathfrak{U}$ be an ascent of the family $(v_\gamma)_{\gamma\in\Gamma}$ by $(c_\gamma)_{\gamma\in\Gamma}$. Using straightforward calculation and employing definitions, we obtain:

$$\begin{aligned}[\![v \in u]\!] &= \bigvee_{\gamma\in\Gamma} c_\gamma \wedge [\![v = v_\gamma]\!] \\ &= \bigvee_{\xi\in\Xi}\ \bigvee_{\alpha\in A(\xi)} b_\xi \wedge b_\xi^\alpha \wedge [\![v = u_\xi^\alpha]\!] \\ &= \bigvee_{\xi\in\Xi} b_\xi \wedge [\![v \in u_\xi]\!].\end{aligned}$$

Show that u is a mixing of the family $(u_\xi)_{\xi\in\Xi}$ by $(b_\xi)_{\xi\in\Xi}$. We begin with establishing the inequality $[\![u = u_\xi]\!] \geqslant b_\xi$. Since the axiom of extensionality is true, it is sufficient to show that $([\![v \in u]\!] \Leftrightarrow [\![v \in u_\xi]\!]) \geqslant b_\xi$ or, which is equivalent, $b_\xi \wedge [\![v \in u]\!] = b_\xi \wedge [\![v \in u_\xi]\!]$. Since $b_\xi \wedge b_\eta = \mathbf{0}$ for $\xi \neq \eta$, we have:

$$b_\xi \wedge [\![v \in u]\!] = \bigvee_{\eta\in\Xi} b_\xi \wedge b_\eta \wedge [\![v \in u_\eta]\!] = b_\xi \wedge [\![v \in u_\xi]\!].$$

We now show that $[\![u \neq \varnothing]\!] \leqslant \bigvee_{\xi\in\Xi} b_\xi$. Indeed,

$$[\![u \neq \varnothing]\!] = [\![(\exists t)\, t \in u]\!] = \bigvee_{v\in\mathfrak{U}} [\![v \in u]\!] = \bigvee_{v\in\mathfrak{U}}\ \bigvee_{\xi\in\Xi} b_\xi \wedge [\![v \in u_\xi]\!] \leqslant \bigvee_{\xi\in\Xi} b_\xi. \ \rhd$$

2.1.12. Theorem. *If a B-valued algebraic system $\mathfrak{U}$ satisfies the maximum and ascent principles then $\mathfrak{U}$ satisfies the mixing principle.*

$\lhd$ Let $\varnothing^\wedge \in \mathfrak{U}$ be an ascent of the empty subset of $\mathfrak{U}$. It is easy to verify that $[\![\varnothing^\wedge = \varnothing]\!] = \mathbf{1}$. (Here and in the sequel, the notation $u = \varnothing$ means $(\forall t)\, t \notin u$.)

Consider a family $(u_\xi)_{\xi\in\Xi}$ of elements in $\mathfrak{U}$ and an antichain $(b_\xi)_{\xi\in\Xi}$ in the algebra B. Put $b = (\bigvee_{\xi\in\Xi} b_\xi)^\perp$. Define a family $(v_\xi)_{\xi\in\Xi'}$ and a partition of unity

$(c_\xi)_{\xi\in\Xi'}$ as follows: $\Xi' = \Xi \cup \{\Xi\}$, $v_\xi = u_\xi$, $c_\xi = b_\xi$ for $\xi \in \Xi$, and $v_\Xi = \varnothing^\wedge$, $c_\Xi = b$. Let $u \in \mathfrak{U}$ be an ascent of the family $(v_\xi)_{\xi\in\Xi'}$ by $(c_\xi)_{\xi\in\Xi'}$. It is easily seen that $[\![u \neq \varnothing]\!] = \mathbf{1}$. Indeed, $[\![v_\xi \in u]\!] \geqslant c_\xi$ for $\xi \in \Xi'$, which implies

$$[\![u \neq \varnothing]\!] = \bigvee_{v\in\mathfrak{U}} [\![v \in u]\!] \geqslant \bigvee_{\xi\in\Xi'} c_\xi = \mathbf{1}.$$

Thus, $[\![(\exists t)\, t \in u]\!] = \mathbf{1}$. According to the maximum principle, there exists an element $v \in \mathfrak{U}$ such that $[\![v \in u]\!] = \mathbf{1}$. Then, by the definition of ascent,

$$c_\xi = \mathbf{1} \wedge c_\xi = \bigvee_{\eta\in\Xi'} c_\eta \wedge [\![v = v_\eta]\!] \wedge c_\xi = [\![v = v_\xi]\!] \wedge c_\xi$$

and, hence, $[\![v = v_\xi]\!] \geqslant c_\xi$ for all $\xi \in \Xi'$. In particular, for $\xi \in \Xi$, we have $[\![v = u_\xi]\!] \geqslant b_\xi$. In addition, by Proposition 2.1.6, the following relations hold:

$$\Big(\bigvee_{\xi\in\Xi} b_\xi\Big)^{\perp} \leqslant [\![v = \varnothing^\wedge]\!] = [\![v = \varnothing^\wedge]\!] \wedge [\![\varnothing^\wedge = \varnothing]\!] \leqslant [\![v = \varnothing]\!].$$

Consequently, v is a mixing of the family $(u_\xi)_{\xi\in\Xi}$ by $(b_\xi)_{\xi\in\Xi}$. $\rhd$

2.1.13. Let B be a complete Boolean algebra and let $\mathfrak{U}$ be a B-valued algebraic system. The system $\mathfrak{U}$ is called a *Boolean valued universe over B* (a *B-valued universe*) if it satisfies the following three conditions:

(1) $\mathfrak{U}$ is separated;

(2) $\mathfrak{U}$ satisfies the ascent principle;

(3) the axioms of extensionality and regularity are true in $\mathfrak{U}$.

Theorem ([6]). *For every complete Boolean algebra B, there is a unique B-valued universe up to isomorphism.*

A detailed presentation of the theories of Boolean algebras and Boolean valued algebraic systems can be found in [3–5, 7].

2.2. The Concept of Continuous Bundle

2.2.1. Let Q be an arbitrary nonempty set and let $V^Q \subset Q \times \mathbf{V}$ be a class-correspondence. As usual, $\mathbf{V}$ denotes the class of all sets. Given $q \in Q$, denote the class

$$\{q\} \times V^Q(q) = \{(q, x) : (q, x) \in V^Q\}$$

by V^q. Obviously, $V^p \cap V^q = \varnothing$ for $p \neq q$. The correspondence V^Q is called a *bundle* on Q and the class V^q is called the *stalk* of V^Q at a point q.

Let $D \subset Q$. A function $u : D \to V^Q$ is called a *section* of the bundle V^Q on D if $u(q) \in V^q$ for all $q \in D$. The class of all sections of V^Q on D is denoted by $S(D, V^Q)$. The sections defined on Q are called *global.* If X is a subset of V^Q then the symbol $S(D, X)$ stands for the set of all sections of X on D.

A point $q \in Q$ is called the *projection of an element* $x \in V^Q$ and denoted by $\mathrm{pr}(x)$ if $x \in V^q$. The *projection of a set* $X \subset V^Q$ is defined to be $\{\mathrm{pr}(x) : x \in X\}$ and denoted by $\mathrm{pr}(X)$.

2.2.2. Assume now Q to be a topological space and suppose that some topology is given on a class $V^Q \subset Q \times \mathbf{V}$. In this case, we call V^Q a *continuous bundle* on Q.

By a *continuous section* of the bundle V^Q we mean a section that is a continuous function. Given a subset $D \subset Q$, the symbol $C(D, V^Q)$ stands for the class of all continuous sections of V^Q on D. Analogously, if X is a subset of V^Q then $C(D, X)$ stands for the totality of all continuous sections of X on D. Obviously, $C(D, X) = C(D, V^Q) \cap S(D, X)$.

Henceforth we suppose that Q is an extremally disconnected Hausdorff compact space and assume satisfied the following conditions:

(1) $(\forall q \in Q)\ (\forall x \in V^q)\ (\exists u \in C(Q, V^Q))\ u(q) = x$;

(2) $(\forall u \in C(Q, V^Q))\ (\forall A \sqsubset Q)\ u(A) \sqsubset V^Q$.

2.2.3. Proposition. *The continuous bundle V^Q possesses the following properties:*

(1) *the topology of V^Q is Hausdorff;*

(2) *for every $u \in C(Q, V^Q)$ and $q \in Q$, the family $\{u(A) : A \in \mathrm{Clop}(q)\}$ is a neighborhood base of the point $u(q)$;*

(3) *all elements of $C(Q, V^Q)$ are open and closed mappings (see 2.1.1).*

$\lhd$ Let x and y be different elements of V^Q. Put $p = \mathrm{pr}(x)$ and $q = \mathrm{pr}(y)$. In view of 2.2.2 (1), there are sections $u, v \in C(Q, V^Q)$ such that $u(p) = x$ and $v(q) = y$.

Suppose first that $p = q$. The set

$$A = \{q \in Q : u(q) \neq v(q)\} = Q \setminus u^{-1}(v(Q))$$

is clopen in view of 2.2.2 (2). Then $u(A)$ and $v(A)$ are disjoint neighborhoods about the points x and y.

Suppose now that $p \neq q$. In this case, there exist $A, B \sqsubset Q$ such that $A \cap B = \varnothing$, $p \in A$, and $q \in B$. Then $u(A)$ and $v(B)$ are disjoint neighborhoods about the points x and y.

Assertion (2) follows readily from 2.2.2 (2).

Assertion (3) is equivalent to 2.2.2 (2) due to the fact that $\mathrm{Clop}(Q)$ is a base both for the open and close topologies of Q. $\rhd$

2.2.4. Lemma. *A subset $X \subset V^Q$ is clopen if and only if $u^{-1}(X) \sqsubset Q$ for all $u \in C(Q, V^Q)$.*

$\lhd$ Only sufficiency requires some comments. Consider an arbitrary element $x \in V^Q$. Let a section $u \in C(Q, V^Q)$ and a point $q \in Q$ be such that $u(q) = x$.

Suppose first that $x \in X$. The set $A = u^{-1}(X)$ is clopen in Q and, therefore, $u(A)$ is a neighborhood about x lying in X. Since x is arbitrary, we conclude that X is open.

If $x \notin X$ then the set $A = Q \backslash u^{-1}(X)$ is clopen in Q and, hence, $u(A)$ is a neighborhood about x disjoint from X. Since x is arbitrary, we conclude that X is closed. $\rhd$

2.2.5. *Proposition.* *The topology of V^Q is extremally disconnected.*

$\lhd$ Let X be an open subset of V^Q. Since the topology of V^Q is Hausdorff, the closure $\mathrm{cl}\, X$ is a set (see 2.1.2). Furthermore, for every section $u \in C(Q, V^Q)$, the set $u^{-1}(\mathrm{cl}\, X) = \mathrm{cl}\, u^{-1}(X)$ is clopen. In view of Lemma 2.2.4, the set $\mathrm{cl}\, X$ is open. $\rhd$

2.2.6. Lemma. *For every subset $X \subset V^Q$ the following hold:*

$$X = \bigcup_{u \in C(Q, V^Q)} u(u^{-1}(X)),$$

$$\mathrm{int}\, X = \bigcup_{u \in C(Q, V^Q)} u(\mathrm{int}\, u^{-1}(X));$$

$$\mathrm{cl}\, X = \bigcup_{u \in C(Q, V^Q)} u(\mathrm{cl}\, u^{-1}(X)).$$

$\lhd$ The claim is an obvious consequence of 2.2.2 (1) and the fact that all continuous sections are open. $\rhd$

2.2.7. Lemma. *Let X and Y be subclasses of V^Q. The equality $X = Y$ holds if and only if $u^{-1}(X) = u^{-1}(Y)$ for all $u \in C(Q, V^Q)$.*

$\lhd$ Take arbitrary $q \in Q$ and $x \in V^q$ and consider a section $u \in C(Q, V^Q)$ such that $u(q) = x$. If $x \in X$ then $q \in u^{-1}(X) = u^{-1}(Y)$ and, consequently, $x = u(q) \in Y$. The reverse inclusion can be established similarly. $\rhd$

2.2.8. Proposition. *A section $u \in S(D, V^Q)$ defined on an open subset $D \subset Q$ is continuous if and only if $\operatorname{im} u$ is an open subset of V^Q.*

$\lhd$ Suppose that a section u is continuous. For every $q \in D$, choose a section $u_q \in C(Q, V^Q)$ such that $u_q(q) = u(q)$. The set $D_q = \{p \in D : u(p) = u_q(p)\} = u^{-1}(\operatorname{im} u_q)$ is open in D and, hence, it is also open in Q. Therefore, the image $u(D_q) = u_q(D_q)$ is open in view of the fact that global continuous sections are open. Obviously, $D = \bigcup_{q \in D} D_q$, since $q \in D_q$. Thus, $\operatorname{im} u = u(D) = u\big(\bigcup_{q \in D} D_q\big) = \bigcup_{q \in D} u(D_q)$ is an open set.

Suppose now that $\operatorname{im} u$ is an open set. Consider an arbitrary point $q \in D$ and choose a section $u_q \in C(Q, V^Q)$ such that $u(q) = u_q(q)$. The open set $\{p \in D : u(p) = u(p)\} = u^{-1}(\operatorname{im} u)$ is a neighborhood about q, whence it follows that u is continuous at q. $\rhd$

2.2.9. Lemma. *For every subset $X \subset V^Q$, the following hold:*

(1) $\operatorname{pr}(\operatorname{cl} X) \subset \operatorname{cl} \operatorname{pr}(X)$;

(2) $\operatorname{pr}(\operatorname{int} X) \subset \operatorname{int} \operatorname{pr}(X)$.

$\lhd$ Consider an arbitrary section $u \in C(Q, V^Q)$. In view of the properties of the closure, we have the relations $u^{-1}(\operatorname{cl} X) = \operatorname{cl} u^{-1}(X) \subset \operatorname{cl} \operatorname{pr}(X)$, whence, due to the equality $\operatorname{pr}(X) = \bigcup_{u \in C(Q, V^Q)} u^{-1}(X)$, it follows that $\operatorname{pr}(\operatorname{cl} X) \subset \operatorname{cl} \operatorname{pr}(X)$.

Relation (2) can be established similarly. $\rhd$

2.3. A Continuous Polyverse

2.3.1. Consider a nonempty set Q and a bundle $V^Q \subset Q \times \mathbf{V}$. Suppose that, for each point $q \in Q$, the class V^q is an algebraic system of signature $\{\in\}$.

Given an arbitrary formula $\varphi(t_1, \dots, t_n)$ and sections $u_1, \dots, u_n$ of the bundle V^Q, we denote by $\{\varphi(u_1, \dots, u_n)\}$ the set

$$\{q \in \operatorname{dom} u_1 \cap \dots \cap \operatorname{dom} u_n : V^q \models \varphi(u_1(q), \dots, u_n(q))\}.$$

For every element $x \in V^q$, put $x{\downarrow} = \{y \in V^q : V^q \models y \in x\}$. Obviously, if the axiom of extensionality is true in the system V^q, then $x{\downarrow} = y{\downarrow} \leftrightarrow x = y$ for all $x, y \in V^q$. If X is a subset of V^Q then the symbol $\sqcup X$ denotes the union $\bigcup_{x \in X} x{\downarrow}$.

Henceforth we assume that Q is an extremally disconnected Hausdorff compact space and V^Q is a continuous bundle on Q.

For an arbitrary section $u \in C(Q, V^Q)$, the class $\bigcup_{q \in Q} u(q){\downarrow}$ is called the *unpack* of the section u and denoted by $\llcorner u \lrcorner$.

2.3.2. A continuous bundle V^Q is called a *continuous polyverse* on Q, if the axioms of extensionality and regularity are true in each stalk V^q $(q \in Q)$ and, in addition, the following conditions hold:

(1) $(\forall q \in Q)\ (\forall x \in V^q)\ (\exists u \in C(Q, V^Q))\ u(q) = x$;

(2) $(\forall u \in C(Q, V^Q))\ (\forall A \in \mathrm{Clop}(Q))\ u(A) \in \mathrm{Clop}(V^Q)$;

(3) $(\forall u \in C(Q, V^Q))\ \llcorner u \lrcorner \in \mathrm{Clop}(V^Q)$;

(4) $(\forall X \in \mathrm{Clop}(V^Q))\ (\exists u \in C(Q, V^Q))\ \llcorner u \lrcorner = X$.

2.3.3. For arbitrary sections $u, v \in C(Q, V^Q)$, the equalities $\{u = v\} = u^{-1}(\mathrm{im}\, v)$ and $\{u \in v\} = u^{-1}(\llcorner v \lrcorner)$ imply that the sets $\{u = v\}$ and $\{u \in v\}$ are clopen, which allows us to introduce the two class-functions

$$[\![\cdot = \cdot]\!], [\![\cdot \in \cdot]\!] : C(Q, V^Q) \times C(Q, V^Q) \to \mathrm{Clop}(Q)$$

by letting $[\![u = v]\!] = \{u = v\}$ and $[\![u \in v]\!] = \{u \in v\}$.

It is easy to verify that the triple $\big(C(Q, V^Q),\ [\![\cdot = \cdot]\!],\ [\![\cdot \in \cdot]\!]\big)$ is a separated $\mathrm{Clop}(Q)$-valued algebraic system (see 2.1.3).

The definition 2.3.2 (4) of continuous polyverse implies that there exists a continuous section $\varnothing^\wedge$ satisfying the condition $\llcorner \varnothing^\wedge \lrcorner = \varnothing$. Obviously, this section is unique. It is easy that $V^q \models \varnothing^\wedge(q) = \varnothing$, $[\![\varnothing^\wedge = \varnothing]\!] = Q$, and, in addition, $[\![u = \varnothing^\wedge]\!] = [\![u = \varnothing]\!]$ for all $u \in C(Q, V^Q)$.

2.3.4. Lemma. *For every subset $X \subset V^Q$, the following hold:*

(1) *if $X \sqsubset V^Q$ then* $\mathrm{pr}(X) \sqsubset Q$;

(2) *if X is open then* $\mathrm{pr}(\mathrm{cl}\, X) = \mathrm{cl}\, \mathrm{pr}(X)$.

$\lhd$ (1): If $X \sqsubset V^Q$ then there is a section $u \in C(Q, V^Q)$ such that $\sqcup \mathrm{im}\, u = \llcorner u \lrcorner = X$. Obviously, $\mathrm{pr}\big(\sqcup \mathrm{im}\, u\big) = [\![u \neq \varnothing]\!]$, whence $\mathrm{pr}(X)$ is clopen.

(2): Let X be an open subset of V^Q. Then the closure $\mathrm{cl}\, X$ is clopen, the same is true of its projection $\mathrm{pr}(\mathrm{cl}\, X)$. The obvious inclusion $\mathrm{pr}(X) \subset \mathrm{pr}(\mathrm{cl}\, X)$ implies $\mathrm{cl}\, \mathrm{pr}(X) \subset \mathrm{pr}(\mathrm{cl}\, X)$. The reverse inclusion is established in 2.2.9. $\rhd$

2.3.5. The *support* $\mathrm{supp}\, u$ of a section $u \in S(D, V^Q)$ on $D \subset Q$ is defined to be the set $\{q \in D : V^q \models u(q) \neq \varnothing\}$. Obviously, $\mathrm{supp}\, u = \{u \neq \varnothing\} = \{u \neq \varnothing^\wedge\}$. So, if $u \in C(Q, V^Q)$ then $\mathrm{supp}\, u$ is a clopen set.

Let u be a continuous section of V^Q and let D be a subset of $\mathrm{supp}\, u$. The symbol $C(D, u)$ denotes the class

$$\big\{v \in C(D, V^Q) : (\forall q \in D)\ V^q \models v(q) \in u(q)\big\}.$$

Obviously, $C(D, u) = C(D, \llcorner u \lrcorner)$.

By the *descent* of a section u we mean the class $C(\mathrm{supp}\, u, u)$ and denote the latter by $u{\downarrow}$. It is easily seen that $u{\downarrow} = C(\mathrm{supp}\, u, \llcorner u \lrcorner)$. Obviously, in case $[\![u \neq \varnothing]\!] = Q$, the descent of u is the descent of the section u regarded as an element of a Boolean valued algebraic system (see 2.1.7).

2.3.6. Proposition. *For arbitrary $X \sqsubset V^Q$ and $u \in C(Q, V^Q)$, the following are equivalent:*

(1) $\llcorner u \lrcorner = X$;

(2) $u(q){\downarrow} = X \cap V^q$ *for all* $q \in Q$;

(3) $\operatorname{supp} u = \operatorname{pr}(X)$ *and* $u{\downarrow} = C(\operatorname{pr}(X), X)$;

(4) $[\![v \in u]\!] = v^{-1}(X)$ *for all* $v \in C(Q, V^Q)$.

$\lhd$ (1)$\to$(3): It suffices to observe that $\operatorname{supp} u = [\![u \neq \varnothing]\!] = \operatorname{pr}(\llcorner u \lrcorner)$ and employ the equality $u{\downarrow} = C(\operatorname{supp} u, \llcorner u \lrcorner)$.

(3)$\to$(2): Put $A = \operatorname{supp} u$. It is clear that $X \cap V^q = \varnothing = u(q){\downarrow}$ for all $q \in Q \backslash A$. Given an arbitrary point $q \in A$, there are $x \in u(q){\downarrow}$ and $v_q \in C(Q, V^Q)$ such that $v_q(q) = x$. Put $B_q = [\![v_q \in u]\!]$. The family $(B_q)_{q \in A}$ is an open cover of the compact set A; therefore, we can refine a subcover $(B_q)_{q \in F}$, where $F \subset A$ is finite. By the exhaustion principle, there is an antichain $(C_q)_{q \in F}$ such that $C_q \subset B_q$ for $q \in F$ and $\bigcup_{q \in F} C_q = \bigvee_{q \in F} C_q = \bigvee_{q \in F} B_q = A$. Construct a section $v \in S(A, V^Q)$ by putting $v(p) = v_q(p)$ for each point $p \in A$, where q is a (unique) element of F such that $p \in C_q$. The section v is continuous, since $v = v_q$ on C_q $(q \in F)$. It is easily seen that $v \in u{\downarrow} = C(A, X)$.

Let q be an arbitrary element of A.

Consider an $x \in u(q){\downarrow}$, choose a section $w \in C(Q, V^Q)$ such that $w(q) = x$, and construct a section $\overline{w} \in S(A, V^Q)$ as follows:

$$\overline{w}(p) = \begin{cases} w(p) & \text{if } p \in [\![w \in u]\!], \\ v(p) & \text{if } p \in A \backslash [\![w \in u]\!]. \end{cases}$$

Obviously, the section $\overline{w}$ is continuous and $\overline{w} \in u{\downarrow} = C(A, X)$, whence $x = \overline{w}(q) \in X$ in view of the containment $q \in [\![w \in u]\!]$.

Now let $x \in X \cap V^q$. As before, choose a section $w \in C(Q, V^Q)$ such that $w(q) = x$. Consider the section $\overline{w} \in S(A, V^Q)$ defined as follows:

$$\overline{w}(p) = \begin{cases} w(p) & \text{if } p \in w^{-1}(X), \\ v(p) & \text{if } p \in A \backslash w^{-1}(X). \end{cases}$$

The obvious relations $\overline{w} \in C(A, X) = u{\downarrow}$ and $q \in w^{-1}(X)$ imply that $x = w(q) = \overline{w}(q) \in u(q){\downarrow}$.

(2)$\to$(4): Consider an arbitrary section $v \in C(Q, V^Q)$. If $q \in [\![v \in u]\!] = v^{-1}(\llcorner u \lrcorner)$ then $v(q) \in \llcorner u \lrcorner$; consequently, $v(q) \in u(q){\downarrow} = X \cap V^q$, i.e., $q \in v^{-1}(X)$.

If $q \in v^{-1}(X)$ then $v(q) \in X \cap V^q = u(q){\downarrow}$ and, hence, $V^q \models v(q) \in u(q)$ and $q \in [\![v \in u]\!]$.

(4)→(1): Observe that $v^{-1}(\llcorner u \lrcorner) = [\![v \in u]\!] = v^{-1}(X)$ for all $v \in C(Q, V^Q)$. Therefore, in view of Lemma 2.2.7, the equality $X = \llcorner u \lrcorner$ holds. ▷

Obviously, for every $X \sqsubset V^Q$, a section u satisfying conditions (1)–(4) is unique. We call this section the *pack* of the set X and denote it by $\ulcorner X \urcorner$.

It is easy to verify validity of the following

Proposition. *Let X be an open subset of V^Q. A section $\overline{u} \in C(Q, V^Q)$ coincides with $\ulcorner \mathrm{cl}\, X \urcorner$ if and only if $\overline{u}$ is pointwise the least section among $u \in C(Q, V^Q)$ satisfying the inclusion $X \cap V^q \subset u(q){\downarrow}$ for all $q \in Q$.*

2.3.7. Lemma. *If $u \in C(Q, V^Q)$ and $A \in \mathrm{Clop}(Q)$ then $\sqcup u(A) \in \mathrm{Clop}(V^Q)$.*

◁ For every section $v \in C(Q, V^Q)$, the set $v^{-1}(\sqcup u(A)) = A \cap [\![v \in u]\!]$ is clopen; whence, in view of 2.2.4, the set $\sqcup u(A)$ is clopen too. ▷

2.3.8. *Proposition.* *Every continuous section of V^Q defined on an open or closed subset of Q is extendible to a global continuous section.*

◁ Consider $A \subset Q$ and $u \in C(A, V^Q)$. For every point $q \in A$, there exist a section $u_q \in C(Q, V^Q)$ and a set $B_q \sqsubset Q$ such that $q \in B_q$ and $u_q = u$ on $B_q \cap A$.

Suppose that A is an open set. Without loss of generality, we may assume that $B_q \subset A$. Consider the open set $X = \bigcup_{q \in Q} u(q){\downarrow} = \bigcup_{q \in A} \sqcup u_q(B_q)$ and show that $(\mathrm{cl}\, X) \cap V^q = u(q){\downarrow}$ for all $q \in A$. We only establish the inclusion $(\mathrm{cl}\, X) \cap V^q \subset u(q){\downarrow}$ (the reverse inclusion follows from the obvious properties of closure). Take an $x \in \mathrm{cl}\, X \cap V^q$. There is a section $v \in C(Q, V^Q)$ such that $v(q) = x$. Evidently, for each neighborhood $B \sqsubset Q$ about q, the intersection $v(B) \cap X$ is nonempty and, thus, there exists a point $p \in B \cap B_q$ such that $v(p) \in u(p){\downarrow}$. On the other hand, $u(p) = u_q(p)$; consequently, $v(B) \cap \sqcup u_q(B_q) \neq \varnothing$. The set $\sqcup u_q(B_q)$ is closed and, therefore, $x \in \sqcup u_q(B_q)$, whence $x \in u_q(q){\downarrow} = u(q){\downarrow}$. Put $\overline{u} = \ulcorner \mathrm{cl}\, X \urcorner$. From what was established above it follows that $\overline{u}(q){\downarrow} = u(q){\downarrow}$ for all $q \in A$. Thus, $\overline{u}$ is a sought global extension of the section u.

Suppose now that the set A is closed. The family $(B_q)_{q \in A}$ forms an open cover of the compact set A and, therefore, we can refine a subcover $(B_q)_{q \in F}$, where F is a finite subset of A. Without loss of generality, we may assume that $\bigcup_{q \in F} B_q = Q$. By the exhaustion principle, there is an antichain $(C_q)_{q \in F}$ such that $C_q \subset B_q$ for all $q \in F$ and $\bigcup_{q \in F} C_q = Q$. Construct a section $\overline{u} \in S(Q, V^Q)$ by putting $\overline{u}(p) = u_q(p)$ for each point $p \in Q$, where q is a (unique) element of F such that $p \in C_q$. The section $\overline{u}$ is continuous, since $\overline{u} = u_q$ on C_q $(q \in F)$. Obviously, $\overline{u} = u$ on A. ▷

Corollary. *If A is an open or closed subset of Q then $C(A, V^Q) = \{u|_A : u \in C(Q, V^Q)\}$.*

Extension Principle. *For every section $u \in C(A, V^Q)$ defined on an open subset $A \subset Q$, there is a unique section $\overline{u} \in C(\operatorname{cl} A, V^Q)$ extending u.*

$\lhd$ Due to Proposition 2.3.8, there exists a section $u_1 \in C(Q, V^Q)$ such that $u_1 = u$ on A. Put $\overline{u} = u_1|_{\operatorname{cl} A}$.

Uniqueness of this extension is obvious. $\rhd$

The section $\overline{u}$ of the extension principle is called the *closure* of u and denoted by $\operatorname{ext}(u)$.

2.3.9. It is easy to verify validity of the following

Theorem. *Consider a family $(u_\xi)_{\xi \in \Xi}$ of global continuous sections of V^Q and an antichain $(B_\xi)_{\xi \in \Xi}$ in the algebra $\operatorname{Clop}(Q)$ and put $B = (\bigvee_{\xi \in \Xi} B_\xi)^\perp$. The continuous section*

$$u = \operatorname{ext}\left(\bigcup_{\xi \in \Xi} u_\xi|_{B_\xi} \cup \varnothing^\wedge|_B \right)$$

is the mixing of the family $(u_\xi)_{\xi \in \Xi}$ by $(B_\xi)_{\xi \in \Xi}$. In particular, the mixing principle is valid for the Boolean valued algebraic system $C(Q, V^Q)$.

Corollary. *The Boolean valued algebraic system $C(Q, V^Q)$ satisfies the maximum principle.*

2.3.10. *Pointwise Truth Value Theorem.* *The following equality holds*

$$[\![\varphi(u_1, \dots, u_n)]\!] = \{q \in Q : V^q \models \varphi(u_1(q), \dots, u_n(q))\} \qquad (*)$$

for an arbitrary formula $\varphi(t_1, \dots, t_n)$ and sections $u_1, \dots, u_n \in C(Q, V^Q)$.

$\lhd$ The proof is carried out by induction on the length of φ.

If φ is atomic, i.e., has the form $t_1 \in t_2$ or $t_1 = t_2$; then $(*)$ follows from the definitions of $[\![\cdot = \cdot]\!]$ and $[\![\cdot \in \cdot]\!]$.

Assume that the claim is proven for formulas of a "smaller" length. We restrict ourselves to the case in which the formula φ has the form $(\exists t_0)\ \varphi(t_0, \vec{t})$.

If $V^q \models (\exists t_0)\ \varphi(t_0, \vec{u}(q))$ then there exists an element $x \in V^q$ such that $V^q \models \varphi(x, \vec{u}(q))$. Choose a section $u_0 \in C(Q, V^Q)$ satisfying the equality $u_0(q) = x$. By the induction hypothesis, $q \in [\![\varphi(u_0, \vec{u})]\!] \subset [\![(\exists t_0)\ \varphi(t_0, \vec{u})]\!]$, which proves the inclusion $\supset$ in $(*)$.

Show the reverse inclusion. Suppose that $q \in [\![(\exists t_0)\ \varphi(t_0, \vec{u})]\!]$. By the maximum principle, there is a continuous section u_0 such that $[\![\varphi(u_0, \vec{u})]\!] = [\![(\exists t_0)\ \varphi(t_0, \vec{u})]\!]$. Therefore, by the induction hypothesis, $V^q \models \varphi(u_0(q), \vec{u}(q))$ and, hence, $V^q \models (\exists t_0)\ \varphi(t_0, \vec{u}(q))$. $\rhd$

2.3.11. *Lemma.* *For every subset $X \subset V^Q$, the following hold:*

(1) $\sqcup \operatorname{cl} X \subset \operatorname{cl} \sqcup X$;

(2) $\sqcup \operatorname{int} X \subset \operatorname{int} \sqcup X$;

(3) *if* $X \in \operatorname{Clop}(V^Q)$ *then* $\sqcup X \in \operatorname{Clop}(V^Q)$;

(4) *if* X *is open then* $\sqcup X$ *is an open subset of* V^Q;

(5) *if* X *is open then* $\sqcup \operatorname{cl} X = \operatorname{cl} \sqcup X$.

$\lhd$ (1): Suppose that $x \in \sqcup \operatorname{cl} X$. Then $x \in y{\downarrow}$ for some $y \in \operatorname{cl} X$. Consider sections $u, v \in C(Q, V^Q)$ such that $u(q) = x$ and $v(q) = y$, where $q = \operatorname{pr}(x)$. For every $A \in \operatorname{Clop}(q)$, we have $v(A) \cap X \neq \varnothing$. Put $B = A \cap [\![u \in v]\!] \sqsubset Q$. Since $q \in B$, there is a point $p \in B$ such that $v(p) \in X$. Obviously, $u(p) \in v(p){\downarrow} \subset \sqcup X$ and, hence, $u(A) \cap (\sqcup X) \neq \varnothing$. Consequently, $x \in \operatorname{cl} \sqcup X$.

(2): Suppose that $x \in \sqcup \operatorname{int} X$ and consider $y \in \operatorname{int} X$ and $u, v \in C(Q, V^Q)$ such that $x \in y{\downarrow}$, $u(q) = x$, and $v(q) = y$, where $q = \operatorname{pr}(x)$. It is clear that the set $B = v^{-1}(X) \cap [\![u \in v]\!]$ is a neighborhood about q and, hence, $u(B)$ is a neighborhood about x. Furthermore, $u(p) \in v(p){\downarrow} \subset \sqcup X$ for all $p \in B$, i.e., $u(B) \subset \sqcup X$. Thus, $x \in \operatorname{int} \sqcup X$.

(3): According to Lemma 2.2.4, it suffices to consider an arbitrary section $v \in C(Q, V^Q)$ and show that the set $v^{-1}(\sqcup X)$ is clopen. Put $u = \ulcorner X \urcorner$. Obviously, $v(q) \in \sqcup X$ if and only if

$$V^q \models (\exists t \in u(q))\, v(q) \in t.$$

By the Pointwise Truth Value Theorem,

$$v^{-1}(X) = \{q \in Q : V^q \models (\exists t \in u(q))\ v(q) \in t\} = [\![(\exists t \in u)\ v \in t]\!]$$

and, consequently, $v^{-1}(X) \sqsubset Q$.

(4): The claim follows readily from (2).

(5): Let X be an open set. Then its closure $\operatorname{cl} X$ is clopen and, according to (3), the set $\sqcup \operatorname{cl} X$ is clopen too. The obvious relation $\sqcup X \subset \sqcup \operatorname{cl} X$ implies $\operatorname{cl} \sqcup X \subset \sqcup \operatorname{cl} X$. The reverse inclusion holds by (1). $\rhd$

2.3.12. Theorem. *The Boolean valued algebraic system $C(Q, V^Q)$ satisfies the ascent principle.*

$\lhd$ Let $(u_\xi)_{\xi \in \Xi}$ be a family of global continuous sections of V^Q and let $(B_\xi)_{\xi \in \Xi}$ be a family of clopen subsets of Q. Consider the clopen set $X = \operatorname{cl} \bigcup_{\xi \in \Xi} u_\xi(B_\xi)$ and put $u = \ulcorner X \urcorner$. Show that the section $u \in C(Q, V^Q)$ thus constructed is an ascent of $(u_\xi)_{\xi\in\Xi}$ by $(B_\xi)_{\xi\in\Xi}$. Indeed, for every section $v \in C(Q, V^Q)$, the following hold:

$$\llbracket v \in u \rrbracket = v^{-1}(\llcorner u \lrcorner) = v^{-1}\Big(\operatorname{cl} \bigcup_{\xi\in\Xi} u_\xi(B_\xi)\Big) = \operatorname{cl} v^{-1}\Big(\bigcup_{\xi\in\Xi} u_\xi(B_\xi)\Big)$$

$$= \operatorname{cl} \bigcup_{\xi\in\Xi} v^{-1}(u_\xi(B_\xi)) = \operatorname{cl} \bigcup_{\xi\in\Xi} B_\xi \cap \llbracket v = u_\xi \rrbracket = \bigvee_{\xi\in\Xi} B_\xi \wedge \llbracket v = u_\xi \rrbracket.$$

Consider now an arbitrary section $u \in C(Q, V^Q)$ and show that u is an ascent of some family of elements in $C(Q, V^Q)$ by suitable weights. Put $X = \llcorner u \lrcorner$. For each $x \in X$, choose a section $u_x \in C(Q, V^Q)$ such that $x \in \operatorname{im} u_x$. Assign $B_x = \llbracket u_x \in u \rrbracket = u_x^{-1}(X)$. Obviously, $x \in u_x(B_x) \subset X$ for all $x \in X$, whence $X = \bigcup_{x\in X} u_x(B_x) = \operatorname{cl} \bigcup_{x\in X} u_x(B_x)$. As in the first part of the proof, we can establish the equality $\llbracket v \in u \rrbracket = \bigvee_{x\in X} B_x \wedge \llbracket v = u_x \rrbracket$ for all $v \in C(Q, V^Q)$. Thus, u is an ascent of $(u_x)_{x\in X}$ by $(B_x)_{x\in X}$. $\triangleright$

2.3.13. Consider a $D \sqsubset Q$ and suppose that $\mathscr{U}$ is a subset of $C(D, V^Q)$. Given a point $q \in D$, denote by $\mathscr{U}(q)$ the totality $\{u(q) : u \in \mathscr{U}\}$.

Proposition. *Consider a $D \sqsubset Q$ and suppose that $\mathscr{U}$ is a nonempty subset of $C(D, V^Q)$. The following properties of a section $\overline{u} \in C(Q, V^Q)$ are equivalent:*

(1) $\overline{u} = \ulcorner \operatorname{cl} \bigcup_{u\in\mathscr{U}} \operatorname{im} u \urcorner$;

(2) $\llbracket v \in \overline{u} \rrbracket = \operatorname{cl}\{q \in D : v(q) \in \mathscr{U}(q)\}$ *for all* $v \in C(Q, V^Q)$;

(3) $\llbracket v \in \overline{u} \rrbracket = \operatorname{cl} \bigcup_{u\in\mathscr{U}} \{v = u\}$ *for all* $v \in C(Q, V^Q)$;

(4) $\overline{u}{\downarrow} = \Big\{\operatorname{ext}\Big(\bigcup_{u\in\mathscr{U}} u|_{D_u}\Big) : (D_u)_{u\in\mathscr{U}}$ *is a partition of unity in the algebra* $\operatorname{Clop}(D)\Big\}$;

(5) $\overline{u}{\downarrow} = C(D, \operatorname{cl} \bigcup_{u\in\mathscr{U}} \operatorname{im} u)$;

(6) $\overline{u}$ *is pointwise the least section among* $u \in C(Q, V^Q)$ *satisfying the inclusion* $\mathscr{U}(q) \subset u(q){\downarrow}$ *for all* $q \in D$. *If* $\mathscr{U} \subset C(Q, V^Q)$ *then* $\llbracket v \in \overline{u} \rrbracket = \bigvee_{u\in\mathscr{U}} \llbracket v = u \rrbracket$ *for all* $v \in C(Q, V^Q)$.

$\triangleleft$ (1)→(2): Put $X = \bigcup_{u\in\mathscr{U}} \operatorname{im} u$. Then $\llcorner \overline{u} \lrcorner = \operatorname{cl} X$ and, therefore, $\llbracket v \in \overline{u} \rrbracket = v^{-1}(\llcorner u \lrcorner) = v^{-1}(\operatorname{cl} X) = \operatorname{cl} v^{-1}(X)$ for all $v \in C(Q, V^Q)$. It is easy to verify the relation $X = \bigcup_{q\in D} \mathscr{U}(q)$ and establish equivalence of the containments $v(q) \in \mathscr{U}(q)$ and $q \in v^{-1}\big(\bigcup_{q\in D} \mathscr{U}(q)\big)$.

(2)→(3): It suffices to show that $\{q \in D : v(q) \in \mathscr{U}(q)\} = \bigcup_{u\in\mathscr{U}} \{v = u\}$ for all $v \in C(Q, V^Q)$. Take an arbitrary point $q \in D$.

If $v(q) \in \mathscr{U}(q)$ then, for some element $u \in \mathscr{U}$, we have $v(q) = u(q)$ and, consequently, $q \in \{v = u\}$.

If $q \in \bigcup_{u\in\mathscr{U}}\{v = u\}$ then, for a suitable $u \in \mathscr{U}$, we have $q \in \{v = u\}$ and, hence, $v(q) = u(q) \in \mathscr{U}(q)$.

(3)→(4): Consider an arbitrary element $v \in C(D, V^Q)$ and define a section $\overline{v} \in C(Q, V^Q)$ as follows:

$$\overline{v}(q) = \begin{cases} v(q) & \text{if } q \in D, \\ \varnothing^{\wedge}(q) & \text{if } q \notin D. \end{cases}$$

Suppose that $v \in \overline{u}{\downarrow}$. Then

$$D = \{v \in \overline{u}\} \subset [\![\overline{v} \in \overline{u}]\!] = \operatorname{cl} \bigcup_{u\in\mathscr{U}} \{\overline{v} = u\} \subset D.$$

For all $u \in \mathscr{U}$, the set $\{\overline{v} = u\} = u^{-1}(\operatorname{im} \overline{v})$ is clopen. According to the exhaustion principle, there is an antichain $(D_u)_{u\in\mathscr{U}}$ in the algebra $\operatorname{Clop}(Q)$ such that $D_u \subset \{\overline{v} = u\}$ and $\bigvee_{u\in\mathscr{U}} D_u = \operatorname{cl}\bigcup_{u\in\mathscr{U}}\{\overline{v} = u\} = D$. Obviously, the section $w = \bigcup_{u\in\mathscr{U}} u|_{D_u}$ is continuous, the set $\operatorname{dom} w$ is open, $D = \operatorname{cl}\operatorname{dom} w$, and $\{w = v\} = \{w = \overline{v}\} = \operatorname{dom} w$. It is clear that $\operatorname{ext}(w) \in C(D, V^Q)$ and $\{\operatorname{ext}(w) = v\} = D$. Therefore, $\operatorname{ext}(w) = v$ and, thus, the inclusion $\subset$ holds.

We now establish the reverse inclusion. Let $(D_u)_{u\in\mathscr{U}}$ be a partition of unity in the algebra $\operatorname{Clop}(D)$ and let $v = \operatorname{ext}\big(\bigcup_{u\in\mathscr{U}} u|_{D_u}\big)$. Show that $v \in \overline{u}{\downarrow}$. Since $\operatorname{dom} v = D$, it suffices to establish the inclusion $\operatorname{im} v \subset \llcorner\overline{u}\lrcorner$. Obviously, $u(D_u) \subset \llcorner\overline{u}\lrcorner$ for all $u \in \mathscr{U}$ and, consequently, $\bigcup_{u\in\mathscr{U}} u(D_u) \subset \llcorner\overline{u}\lrcorner$. Observe that $\operatorname{im} v = \operatorname{cl}\bigcup_{u\in\mathscr{U}} u(D_u)$ and, hence, $\operatorname{im} v \subset \llcorner\overline{u}\lrcorner$.

(4)→(5): Put $X = \operatorname{cl}\bigcup_{u\in\mathscr{U}} \operatorname{im} u$. Let $(D_u)_{u\in\mathscr{U}}$ be a partition of unity in the algebra $\operatorname{Clop}(D)$ and let $v = \operatorname{ext}(\bigcup_{u\in\mathscr{U}} u|_{D_u})$. Obviously, $\operatorname{dom} v = D$. Show that $\operatorname{im} v \subset X$. The inclusion $u(D_u) \subset X$ implies $\bigcup_{u\in\mathscr{U}} u(D_u) \subset X$; whence, in view of the equality $\operatorname{im} v = \operatorname{cl}\bigcup_{u\in\mathscr{U}} u(D_u)$, the desired relation $\operatorname{im} v \subset X$ follows. Thus, $\overline{u}{\downarrow} \subset C(D, X)$.

For the reverse inclusion, consider an arbitrary section $v \in C(D, X)$ and prove that $v = \operatorname{ext}\big(\bigcup_{u\in\mathscr{U}} u|_{D_u}\big)$ for some partition of unity $(D_u)_{u\in\mathscr{U}}$ in the algebra $\operatorname{Clop}(D)$. Obviously, $v^{-1}(X) = D$. Since the section v is open, we have $D = \operatorname{cl} v^{-1}\big(\bigcup_{u\in\mathscr{U}} \operatorname{im} u\big)$. In addition, the set $A = v^{-1}\big(\bigcup_{u\in\mathscr{U}} \operatorname{im} u\big)$ is open and dense in D.

With each element $u \in \mathscr{U}$ we associate a clopen set $C_u = \{v = u\} = v^{-1}(\operatorname{im} u)$. The obvious equality $A = \bigcup_{u\in\mathscr{U}} C_u$ implies that $\bigvee_{u\in\mathscr{U}} C_u = D$. In view of the exhaustion principle, there is a partition of unity $(D_u)_{u\in\mathscr{U}}$ in the algebra $\operatorname{Clop}(D)$ such that $D_u \subset C_u$ for all $u \in \mathscr{U}$. Put $w = \bigcup_{u\in\mathscr{U}} u|_{D_u}$. It is clear that, for each $u \in \mathscr{U}$, the equalities $w|_{D_u} = u|_{D_u} = v|_{D_u}$ hold, since $D_u \subset \{v = u\}$. Consequently, by the extension principle, $\operatorname{ext}(w) = v$, which proves the desired inclusion.

(5)→(1): It is sufficient to observe that $D = \mathrm{pr}\big(\mathrm{cl}\bigcup_{u\in\mathscr{U}} \mathrm{im}\, u\big)$ and use Proposition 2.3.6 (3).

Equivalence of (1) and (6) is evident. ▷

Obviously, the section $\overline{u}$ of the proposition is unique. We call $\overline{u}$ the *ascent* of the set $\mathscr{U}$ and denote it by $\mathscr{U}\uparrow$. In case $\mathscr{U}$ is a nonempty subset of $C(Q, V^Q)$, the notion of the ascent of $\mathscr{U}$ coincides with the eponymized notion of 2.1.8.

2.4. Functional Representation

Throughout the section, we assume that Q is an extremally disconnected Hausdorff compact space and $\mathfrak{U}$ is a Boolean valued universe over $\mathrm{Clop}(Q)$.

2.4.1. For the further considerations we need the notion of the quotient class $X/{\sim}$ where X is a class (that need not be a set) and $\sim$ is an equivalence relation on X. The traditional definition of quotient class, for the case in which X is a set, cannot be always applied to the case of a class, since the elements of X equivalent to a given $x \in X$ form a class that need not be a set. We can overcome this difficulty with the help of the following fact:

Theorem (Frege–Russell–Scott). *To every equivalence $\sim$ on a class X, there is a function $F : X \to \mathbf{V}$ such that*

$$F(x) = F(y) \leftrightarrow x \sim y \quad \textit{for all } x, y \in X. \qquad (**)$$

As F we can take the function

$$F(x) = \big\{y \in X : y \sim x \ \&\ (\forall z \in X)\big(z \sim x \to \mathrm{rank}(y) \leqslant \mathrm{rank}(z)\big)\big\}.$$

This function F is conventionally called the *canonical projection* of $\sim$. In view of (**) we may regard $F(x)$ as an analog of the coset containing an element $x \in X$. In this connection, we denote $F(x)$ by ${\sim}(x)$.

2.4.2. For each point $q \in Q$, introduce the equivalence relation $\sim_q$ on the class $\mathfrak{U}$ as follows:

$$u \sim_q v \ \leftrightarrow\ q \in [\![u = v]\!].$$

Consider the bundle

$$V^Q = \big\{\big(q, {\sim_q}(u)\big) : q \in Q,\ u \in \mathfrak{U}\big\}$$

and make the convention to denote a pair $\big(q, {\sim_q}(u)\big)$ by $\widehat{u}(q)$. Obviously, for every element $u \in \mathfrak{U}$, the mapping

$$\widehat{u} : q \mapsto \widehat{u}(q)$$

is a section of the bundle V^Q. Note that, for each $x \in V^Q$, there exist $u \in \mathfrak{U}$ and $q \in Q$ such that $\widehat{u}(q) = x$. In addition, the equality $\widehat{u}(q) = \widehat{v}(q)$ holds if and only if $q \in [\![u = v]\!]$.

Make each stalk V^q of the bundle V^Q into an algebraic system of signature $\{\in\}$ by letting

$$V^q \models x \in y \ \leftrightarrow \ q \in [\![u \in v]\!],$$

where the elements $u, v \in \mathfrak{U}$ are such that $\widehat{u}(q) = x$ and $\widehat{v}(q) = y$. It is easy to verify that the above definition is sound. Indeed, if $\widehat{u}_1(q) = x$ and $\widehat{v}_1(q) = y$ for another pair u_1, v_1, then the containments $q \in [\![u \in v]\!]$ and $q \in [\![u_1 \in v_1]\!]$ are equivalent.

It is easily seen that the class $\{\widehat{u}(A) : u \in \mathfrak{U},\ A \sqsubset Q\}$ is a base for some open topology on V^Q, which allows us to regard V^Q as a continuous bundle.

2.4.3. Theorem. *The following hold:*

(1) *The bundle V^Q is a continuous polyverse.*

(2) *The mapping $u \mapsto \widehat{u}$ is an isomorphism between the Boolean valued universes $\mathfrak{U}$ and $C(Q, V^Q)$.*

We divide the proof of the last theorem into several steps.

2.4.4. Lemma. *If $u \in \mathfrak{U}$ and $A \sqsubset Q$ then $\widehat{u}(A) \sqsubset V^Q$.*

$\lhd$ For every element $x \in V^Q \backslash \widehat{u}(A)$, there exist $v \in \mathfrak{U}$ and $q \in Q$ such that $x = \widehat{v}(q)$.

If $q \in A$ then

$$\widehat{u}(q) \neq x = \widehat{v}(q), \quad q \in [\![u \neq v]\!],$$

and, thus, the set $\widehat{v}([\![u \neq v]\!])$ is a neighborhood about x disjoint from $\widehat{u}(A)$. If, otherwise, $q \notin A$, then the neighborhood $\widehat{v}(Q \backslash A)$ about x is disjoint from $\widehat{u}(A)$. $\rhd$

2.4.5. Lemma. *The classes $\{\widehat{u} : u \in \mathfrak{U}\}$ and $C(Q, V^Q)$ coincide.*

$\lhd$ Consider an arbitrary element $u \in \mathfrak{U}$ and show that the section $\widehat{u}$ is continuous. If $v \in \mathfrak{U}$ and $A \sqsubset Q$ then the set $\widehat{u}^{-1}\big(\widehat{v}(A)\big) = A \cap [\![u = v]\!]$ is open. Arbitrariness of v and A allows us to conclude that $\widehat{u} \in C(Q, V^Q)$.

We now establish the reverse inclusion. Take an $f \in C(Q, V^Q)$. For each point $q \in Q$, choose an element $u_q \in \mathfrak{U}$ such that $\widehat{u}_q(q) = f(q)$ and assign

$$A_q := \{p \in Q : \widehat{u}_q(p) = f(p)\} = f^{-1}\big(\widehat{u}(Q)\big) \sqsubset Q.$$

Thus, $(A_q)_{q \in Q}$ is an open cover of the compact space Q from which we can refine a subcover $(A_q)_{q \in F}$, where F is a finite subset of Q. By the exhaustion principle, there is an antichain $(B_q)_{q \in F}$ such that $B_q \subset A_q$ for all $q \in B$ and $\bigcup_{q \in F} B_q = Q$.

Since the Boolean valued algebraic system $\mathfrak{U}$ satisfies the mixing principle, we may consider

$$u = \operatorname*{mix}_{q \in F} B_q u_q \in \mathfrak{U}.$$

It is easy to become convinced that $\widehat{u} = f$. ▷

2.4.6. Lemma. *The topology of V^Q is extremally disconnected.*

◁ The claim follows from Lemmas 2.4.4 and 2.4.5 and Proposition 2.2.5. ▷

2.4.7. Lemma. *The mapping $(u \mapsto \widehat{u}) : \mathfrak{U} \to C(Q, V^Q)$ is bijective and, for all $u, v \in \mathfrak{U}$, the following hold:*

$$[\![u = v]\!]_{\mathfrak{U}} = [\![\widehat{u} = \widehat{v}]\!]_{C(Q,V^Q)},$$
$$[\![u \in v]\!]_{\mathfrak{U}} = [\![\widehat{u} \in \widehat{v}]\!]_{C(Q,V^Q)}.$$

◁ It is easily seen that, for all $u, v \in \mathfrak{U}$ and $q \in Q$, we have:

$$V^q \models \widehat{u}(q) \in \widehat{v}(q) \;\leftrightarrow\; q \in [\![u \in v]\!],$$
$$V^q \models \widehat{u}(q) = \widehat{v}(q) \;\leftrightarrow\; q \in [\![u = v]\!].$$

The desired equalities are thus established. In Lemma 2.4.6, it is shown that the mapping $u \mapsto \widehat{u}$ is surjective. We are left with proving its injectivity. Let elements $u, v \in \mathfrak{U}$ be such that $\widehat{u} = \widehat{v}$. Then $[\![u = v]\!] = [\![\widehat{u} = \widehat{v}]\!] = Q$, which implies the equality $u = v$ due to the fact that the system $\mathfrak{U}$ is separated. ▷

Thus, the triple

$$(C(Q, V^Q),\ [\![\cdot = \cdot]\!],\ [\![\cdot \in \cdot]\!])$$

is a Boolean valued algebraic system over $\operatorname{Clop}(Q)$ isomorphic to $\mathfrak{U}$ and, hence, $C(Q, V^Q)$ is a Boolean valued universe over $\operatorname{Clop}(Q)$.

2.4.8. Lemma. *If $u \in C(Q, V^Q)$ then $\llcorner u \lrcorner$ is a clopen subset of V^Q.*

◁ Take a $u \in C(Q, V^Q)$. Since $C(Q, V^Q)$ satisfies the ascent principle, $u = \operatorname{asc}_{\xi \in \Xi} B_\xi u_\xi$ for some family $(u_\xi)_{\xi \in \Xi}$ of continuous sections of V^Q and a family $(B_\xi)_{\xi \in \Xi}$ of clopen subsets of Q. For each $v \in C(Q, V^Q)$, the following relations hold:

$$v^{-1}\Big(\operatorname{cl} \bigcup_{\xi \in \Xi} u_\xi(B_\xi)\Big) = \operatorname{cl} \bigcup_{\xi \in \Xi} v^{-1}(u_\xi(B_\xi)) = \operatorname{cl} \bigcup_{\xi \in \Xi} B_\xi \cap [\![v = u_\xi]\!]$$
$$= \bigvee_{\xi \in \Xi} B_\xi \wedge [\![v = u_\xi]\!] = [\![v \in u]\!] = v^{-1}(\llcorner u \lrcorner).$$

Thus, in view of Lemma 2.2.7, the equality

$$\llcorner u \lrcorner = \operatorname{cl} \bigcup_{\xi \in \Xi} u_\xi(B_\xi)$$

is established. The set

$$\bigcup_{\xi \in \Xi} u_\xi(B_\xi)$$

is open; therefore, by Lemma 2.4.6, the class $\llcorner u \lrcorner$ is a clopen set. $\triangleright$

2.4.9. Lemma. *For every subset $X \sqsubset V^Q$, there exists a section $u \in C(Q, V^Q)$ such that $\llcorner u \lrcorner = X$.*

$\triangleleft$ With each element $x \in X$ we associate a section $u_x \in C(Q, V^Q)$ such that $x \in \operatorname{im} u_x$. Obviously, the set $B_x = u_x^{-1}(X)$ is clopen. Consider the ascent

$$u = \operatorname*{asc}_{x \in X} B_x u_x$$

and establish the equality $\llcorner u \lrcorner = X$. Since $x \in u_x(B_x) \subset X$ for all $x \in X$, we have

$$X = \bigcup_{x \in X} u_x(B_x) = \operatorname{cl} \bigcup_{x \in X} u_x(B_x).$$

For an arbitrary section $v \in C(Q, V^Q)$, the following relations hold:

$$v^{-1}(X) = \bigcup_{x \in X} v^{-1}\big(u_x(B_x)\big) = \operatorname{cl} \bigvee_{x \in X} B_x \wedge [\![v = u_x]\!] = [\![v \in u]\!] = v^{-1}(\llcorner u \lrcorner).$$

In view of Lemma 2.2.7, the desired equality is established. $\triangleright$

2.4.10. Lemma. *For every formula $\varphi(t_1, \dots, t_n)$ and arbitrary sections $u_1, \dots, u_n \in C(Q, V^Q)$, the following holds:*

$$[\![\varphi(u_1, \dots, u_n)]\!] = \big\{q \in Q : V^q \models \varphi\big(u_1(q), \dots, u_n(q)\big)\big\}.$$

$\triangleleft$ The proof of the lemma repeats that of the Pointwise Truth Value Theorem (see 2.3.10). $\triangleright$

The last lemma implies in particular that the axioms of extensionality and regularity are true in each stalk. Thus, Theorem 2.4.3 is completely proven.

In conclusion, we state a theorem that combines the basic results of Sections 2.3 and 2.4.

Theorem. *Let Q be the Stone space of a complete Boolean algebra B.*

(1) *The class $C(Q, V^Q)$ of continuous sections of a polyverse V^Q on Q is a Boolean valued universe.*

(2) *For an arbitrary Boolean valued universe $\mathfrak{U}$ over B, there exists a continuous polyverse V^Q on Q such that $C(Q, V^Q)$ is isomorphic to $\mathfrak{U}$.*

References

1. Arkhangel′skiĭ A. V. and Ponomarëv V. I., Fundamentals of General Topology: Problems and Exercises, Reidel, Dordrecht etc. (1984).
2. Bourbaki N., General Topology. Parts 1 and 2, Addison-Wesley, Reading (1966).
3. Gordon E. I. and Morozov S. F., Boolean Valued Models of Set Theory [in Russian], Gor′kiĭ State University, Gor′kiĭ (1982).
4. Kusraev A. G. and Kutateladze S. S., Nonstandard Methods in Analysis, Kluwer, Dordrecht (1994).
5. Sikorski R., Boolean Algebras, Springer-Verlag, Berlin (1964).
6. Solovay R. and Tennenbaum S., "Iterated Cohen extensions and Souslin's problem," Ann. Math., **94**, No. 2, 201–245 (1972).
7. Vladimirov D. A., Boolean Algebras [in Russian], Nauka, Moscow (1969).

CHAPTER 3

Dual Banach Bundles

BY

A. E. Gutman and A. V. Koptev

S.S. Kutateladze (ed.), *Nonstandard Analysis and Vector Lattices*, 105–159.

Bundles are traditionally employed for studying various algebraic systems in mathematical analysis. The technique of bundles is used in examining Banach spaces, Riesz spaces, C^*-algebras, Banach modules, etc. (see, for instance, [3, 6, 7, 13–15]). Representation of some objects of functional analysis as spaces of sections of corresponding bundles serves as a basis for some theories valuable in their own right. One of these theories in [8–12] is devoted to the notion of a continuous Banach bundle (CBB) and its applications to lattice normed spaces (LNSs). Within this theory, in particular, a representation is obtained for an arbitrary LNS as a space of sections of a suitable CBB.

In some sense, a CBB over a topological space Q formally reflects the intuitive notion of a family of Banach spaces $(X_q)_{q \in Q}$ varying continuously from point to point in the space Q. To be more precise, a Banach bundle $\mathscr{X}$ over Q is a mapping associating with each point $q \in Q$ a Banach space $\mathscr{X}(q)$ the so-called stalk of $\mathscr{X}$ at q. Furthermore, the bundle $\mathscr{X}$ is endowed with some structure that allows us to speak about continuity of sections of the bundle (a section is a function u defined on a subset of Q and taking values $u(q) \in \mathscr{X}(q)$ for all $q \in \operatorname{dom} u$). The notion of a section can be regarded as a generalization of the notion of a vector valued function: if X is a Banach space then X-valued functions are sections of the Banach bundle whose stalks are all equal to X.

In many questions of analysis, an essential role is played by duality theory, one of whose basic tools is the concept of a dual space (see, for instance, [17]). Existence of a functional representation for the initial space by means of sections of some bundle allows us to construct an analogous representation for the dual space. In particular, the problem of representing a dual LNS leads to the notion of a dual Banach bundle.

Which CBB $\mathscr{X}'$ should be considered dual to a given bundle $\mathscr{X}$ (discussed, for instance, in [7–9, 12, 19]) is a question closely connected with the notion of a homomorphism. A homomorphism v of a continuous Banach bundle $\mathscr{X}$ over Q is a functional valued mapping $v : q \mapsto v(q) \in \mathscr{X}(q)'$ taking every continuous section u of the bundle $\mathscr{X}$ into the continuous real-valued function $\langle u|v \rangle : q \mapsto \langle u(q)|v(q) \rangle$. When we try to define a dual CBB $\mathscr{X}'$, the following two requirements are worth to be imposed: first, homomorphisms should be continuous sections of the bundle $\mathscr{X}'$ and, second, all continuous sections of $\mathscr{X}'$ should be homomorphisms.

In the case of ample bundles over extremally disconnected compact spaces, the problem of defining a dual CBB is solved in [8] (see also [12]). However, the approach to the definition of a dual bundle presented in that article rests essentially on the specific properties of ample bundles and extremally disconnected compact spaces and, thus, cannot be extended to a wider class of bundles.

The natural intention to extend the domain of application for duality theory leads to the problem of constructing a dual CBB for an arbitrary Banach bundle over an arbitrary topological space. The study of this problem is the main topic of the present chapter, where, in particular, a definition of a dual bundle is presented, with the above-formulated requirements fulfilled, and a number of necessary and sufficient conditions is suggested for existence of a dual bundle.

In Section 3.1, auxiliary results are collected on topological spaces, Banach spaces, and functions acting in them.

Section 3.2 is devoted to studying the notion of a homomorphism of a Banach bundle. In particular, description of homomorphisms is suggested therein for a wide class of bundles and the question is examined of continuity of the pointwise norm of a homomorphism.

The question about the possibility of representing the space of all homomorphisms from a CBB $\mathscr{X}$ into a CBB $\mathscr{Y}$ as the space of continuous sections of some Banach bundle leads to the notion of an operator bundle $B(\mathscr{X}, \mathscr{Y})$. In Section 3.3, some necessary and sufficient conditions are given for existence of such a bundle.

In Section 3.4, the notion of a dual Banach bundle is introduced and studied. This bundle is a particular case of an operator bundle (considered in the previous section). The definition of a dual bundle therein generalizes that of [8, 12] where the case is considered of an ample bundle over an extremally disconnected compact space. In the same articles it is established in particular that every ample CBB has the dual bundle. In the general case, dual bundles may fail to exist. Nevertheless, the above generalization is justified by the fact that new classes arise of CBBs that have dual bundles. In Section 3.4, various necessary and sufficient conditions are presented for existence of a dual bundle, the norming duality relations are established between the bundles $\mathscr{X}$ and $\mathscr{X}'$, and the questions are studied of existence of the second dual bundle and embedding of a bundle into its second dual.

In examining the notion of a dual bundle, one of the natural steps is consideration of weakly continuous sections (these are sections continuous with respect to the duality between a bundle and its dual). The notion of a weakly continuous section is introduced and studied in Section 3.5. In particular, the question is discussed about continuity of weakly continuous sections for various classes of Banach bundles and conditions are suggested for coincidence of the space of weakly continuous sections of a trivial CBB and the space of weakly continuous vector valued functions acting into the corresponding stalk.

When speaking about Banach bundles, we use the terminology and notation of [8] (see also [12]). In particular, we distinguish the notion of a Banach bundle and that of a continuous Banach bundle and employ the approach to the definition of continuity for sections by means of the notion of a continuity structure. All necessary information on the theory of Banach bundles can be found in [3, 7–12].

If $\mathscr{X}$ and $\mathscr{Y}$ are some CBBs over a topological space Q then we denote by $\mathrm{Hom}(\mathscr{X}, \mathscr{Y})$ the set of all Q-homomorphisms from $\mathscr{X}$ into $\mathscr{Y}$ (which is denoted by

$\mathrm{Hom}_Q(\mathscr{X},\mathscr{Y})$ in [8, 12]). As usual, the symbol $\mathrm{Hom}_D(\mathscr{X},\mathscr{Y})$ is used for denoting the set of D-homomorphisms from $\mathscr{X}|_D$ into $\mathscr{Y}|_D$, where $D \subset Q$. Instead of "Q-homomorphism" we just say "homomorphism." Analogous convention is effective concerning the terms "Q-isometric embedding" and "Q-isometry."

In contrast to [8, 12], we use the symbol X_Q for denoting the trivial Banach bundle with stalk X over a topological space Q. The symbol $\mathscr{R}$ denotes the trivial CBB with stalk $\mathbb{R}$ over the topological space under consideration.

Let $\mathscr{X}$ be a continuous Banach bundle over a topological space Q, let u be a section of $\mathscr{X}$ defined on an $A \subset Q$, and let v be a section of $\mathscr{X}$ defined on a $B \subset Q$ such that $v(q) \in \mathscr{X}(q)'$ for all $q \in B$. The symbol $\langle u|v\rangle$ denotes the function acting from $A \cap B$ into $\mathbb{R}$ by the rule $\langle u|v\rangle(q) = \langle u(q)|v(q)\rangle$.

All vector spaces under consideration are assumed over $\mathbb{R}$, the field of reals.

3.1. Auxiliary Results

This section contains facts to be used in the sequel about topological and Banach spaces as well as functions acting in this spaces. The collected results are auxiliary and do not involve the notion of a Banach bundle.

3.1.1. *Lemma.* *Let X be a normed space and let x and y be norm-one vectors in X. Then either of the intervals $[x,y]$ or $[x,-y]$ does not intersect the open ball with radius $1/2$ centered at the origin, i.e..*

$$\inf_{\lambda\in[0,1]} \|\lambda x + (1-\lambda)y\| \geqslant 1/2 \quad \text{or} \quad \inf_{\lambda\in[0,1]} \|\lambda x + (1-\lambda)(-y)\| \geqslant 1/2.$$

$\triangleleft$ Assume that there are vectors $u = \lambda x + (1-\lambda)(-y)$ and $v = \mu x + (1-\mu)y$ such that $\|u\| < 1/2$ and $\|v\| < 1/2$. Obviously, $0 < \lambda, \mu < 1$ and $x \neq \pm y$. Moreover, the vectors u and v are linearly independent. Hence, $x = \alpha u + \beta v$ and $y = \gamma u + \delta v$ for some $\alpha, \beta, \gamma, \delta \in \mathbb{R}$. Linear independence of (u, v) and (x, y), together with the equalities

$$\begin{pmatrix} \alpha & \beta \\ \gamma & \delta \end{pmatrix} \begin{pmatrix} u \\ v \end{pmatrix} = \begin{pmatrix} x \\ y \end{pmatrix}, \qquad \begin{pmatrix} \lambda & \lambda - 1 \\ \mu & 1-\mu \end{pmatrix} \begin{pmatrix} x \\ y \end{pmatrix} = \begin{pmatrix} u \\ v \end{pmatrix},$$

implies that

$$\begin{pmatrix} \alpha & \beta \\ \gamma & \delta \end{pmatrix} = \begin{pmatrix} \lambda & \lambda - 1 \\ \mu & 1-\mu \end{pmatrix}^{-1},$$

i.e.,

$$\begin{pmatrix} \alpha & \beta \\ \gamma & \delta \end{pmatrix} = \frac{1}{\lambda + \mu - 2\lambda\mu} \begin{pmatrix} 1-\mu & 1-\lambda \\ -\mu & \lambda \end{pmatrix}.$$

The relations

$$1 = \|x\| \leqslant |\alpha|\|u\| + |\beta|\|v\| < \frac{|\alpha| + |\beta|}{2}$$

and

$$1 = \|y\| \leqslant |\gamma|\|u\| + |\delta|\|v\| < \frac{|\gamma| + |\delta|}{2}$$

allow us to conclude that

$$\frac{1}{|\alpha| + |\beta|} + \frac{1}{|\gamma| + |\delta|} < 1,$$

i.e., $|\alpha| + |\beta| + |\gamma| + |\delta| < (|\alpha| + |\beta|)(|\gamma| + |\delta|)$. It is easy to see that $\lambda + \mu - 2\lambda\mu > \lambda^2 + \mu^2 - 2\lambda\mu \geqslant 0$. Furthermore, $|\alpha| + |\beta| = (2 - \lambda - \mu)/(\lambda + \mu - 2\lambda\mu)$ and $|\gamma| + |\delta| = (\lambda + \mu)/(\lambda + \mu - 2\lambda\mu)$, whence

$$\frac{2}{\lambda + \mu - 2\lambda\mu} < \frac{2 - \lambda - \mu}{\lambda + \mu - 2\lambda\mu} \, \frac{\lambda + \mu}{\lambda + \mu - 2\lambda\mu}.$$

Consequently, $2(\lambda + \mu - 2\lambda\mu) < 2(\lambda + \mu) - (\lambda + \mu)^2$ and, finally, $(\lambda - \mu)^2 < 0$. This contradiction completes the proof. ▷

3.1.2. The following statement may be found, for instance, in [21, Proposition 1 (SP1)].

Lemma. *If a Banach space X possesses the Schur property then every weakly Cauchy sequence in X is norm convergent.*

◁ Consider a norm divergent sequence $(x_n) \subset X$ and show that it is not a weakly Cauchy sequence. There exist a number $\varepsilon > 0$ and a strictly increasing sequence $(n_k) \subset \mathbb{N}$ such that $\|x_{n_k} - x_{n_{k+1}}\| > \varepsilon$ for all odd $k \in \mathbb{N}$. Since the sequence $(x_{n_k} - x_{n_{k+1}})$ does not vanish in norm and X possesses the Schur property, there is a functional $x' \in X'$ such that the numerical sequence $\langle x_{n_k} - x_{n_{k+1}} \mid x' \rangle$ does not vanish. Consequently, the subsequence (x_{n_k}), together with the initial sequence (x_n), is not a weakly Cauchy sequence. ▷

3.1.3. ***Lemma.*** *Let X be an infinite-dimensional separable Banach space. Then every infinite-dimensional Banach subspace of X' includes a weakly* null sequence of norm-one functionals.*

◁ Let Y be an infinite-dimensional Banach subspace of X'. Consider a sequence (y_n) of norm-one vectors in Y such that $\|y_i - y_j\| \geqslant 1/2$ whenever $i \neq j$ (see, for instance, [18, 8.4.2]). By [4, XIII], from (y_n) we can extract a subsequence (y_{n_m}) convergent weakly* to an element $y \in X'$. It is clear that $y \in Y$. For every $m \in \mathbb{N}$, put $z_m := y_{n_m} - y$. Let $\varepsilon > 0$ and let (z_{m_k}) be a subsequence of (z_m) such that $\|z_{m_k}\| > \varepsilon$ for all $k \in \mathbb{N}$. Then $(z_{m_k} \,/\, \|z_{m_k}\|)$ is a sought sequence. ▷

3.1.4. *Lemma.* *Let X be an infinite-dimensional Banach space. Then there exist a weakly vanishing net $(x_\alpha)_{\alpha\in\aleph} \subset X$ and a norm vanishing net $(x'_\alpha)_{\alpha\in\aleph} \subset X'$ such that $\langle x_\alpha | x'_\alpha\rangle = 1$ for all $\alpha \in \aleph$.*

$\triangleleft$ As $\aleph$ we consider the set of all finite subsets of X' ordered by inclusion.

Fix an $\alpha = \{x'_1, \dots, x'_n\} \in \aleph$ and, employing the fact that X is infinite-dimensional, take an element $x_\alpha \in \bigcap_{i=1}^n \ker x'_i$ with norm $\|x_\alpha\| = n$. Next, choose a functional $x'_\alpha \in X'$ satisfying the equalities $\langle x_\alpha | x'_\alpha\rangle = 1$ and $\|x'_\alpha\| = 1/n$.

Obviously, the net $(x'_\alpha)_{\alpha\in\aleph}$ vanishes in norm. Show that the net $(x_\alpha)_{\alpha\in\aleph}$ is weakly vanishing. Let U be an arbitrary weak neighborhood about zero in X. Choose functionals $x'_1, \dots, x'_n \in X'$ so that $\bigcap_{i=1}^n \ker x'_i \subset U$. Then $x_\alpha \in \bigcap_{i=1}^n \ker x'_i \subset U$ for all $\alpha \in \aleph$, $\alpha \geqslant \{x'_1, \dots, x'_n\}$. $\triangleright$

3.1.5. Let (x_n) be a sequence in a Banach space X.

Lemma. *The following are equivalent:*

(a) *for every sequence $(x'_n) \subset X'$ and every element $x' \in X'$, weak* convergence $x'_n \to x'$ implies $\langle x_n | x'_n\rangle \to 0$;*

(b) *for every sequence $(x'_m) \subset X'$ and every element $x' \in X'$, weak* convergence $x'_m \to x'$ implies $\langle x_n | x'_m\rangle \to 0$ as $n, m \to \infty$;*

(c) *(x_n) is weakly null and $\langle x_n | x'_n\rangle \to 0$ for every weakly* null sequence $(x'_n) \subset X'$;*

(d) *(x_n) is weakly null and $\langle x_n | x'_m\rangle \to 0$ as $n, m \to \infty$ for every weakly* null sequence $(x'_m) \subset X'$;*

(e) *$\sup_{m\in\mathbb{N}} |\langle x_n | x'_m\rangle| \to 0$ as $n \to \infty$ for every weakly* null sequence $(x'_m) \subset X'$;*

(f) *for every operator $T \in B(X, c_0)$, the sequence (Tx_n) vanishes in norm.*

The proof of equivalence of the above assertions is a routine and quite simple exercise.

DEFINITION. Say that a sequence is *w-w^*-vanishing* if (x_n) satisfies one of the conditions (a)–(f) of the above lemma. If $x \in X$ and the sequence $(x_n - x)$ is w-w^*-vanishing then we say that (x_n) *w-w^*-converges* to x.

A Banach space X is said to possesses the WS *property* (or the *weak Schur property*) if every w-w^*-convergent sequence in X converges in norm (or, which is the same, every w-w^*-vanishing sequence vanishes in norm).

We list some evident facts concerning the above notions.

Proposition. *The following are true:*

(1) *Each norm convergent sequence is w-w^*-convergent.*

(2) *Every subsequence of a w-w^*-convergent sequence is also w-w^*-convergent.*

(3) *If X and Y are Banach spaces, $T \in B(X,Y)$, and a sequence $(x_n) \subset X$ is w-w^*-convergent to an $x \in X$, then the sequence (Tx_n) is w-w^*-convergent to Tx.*

(4) *If a Banach space possesses the* WS *property then this property is also enjoyed by every Banach subspace.*

(5) *If a Banach space X possesses the* WS *property then this property is also enjoyed by every Banach space isomorphic to X.*

(6) *If a Banach space contains a copy of a space which does not possess the* WS *property, then the space does not possess the* WS *property either.*

3.1.6. Lemma. *If a Banach space X has weakly* sequentially compact dual ball then X possesses the* WS *property. The converse fails to be true.*

$\triangleleft$ Suppose that X does not possess the WS property. Then there exists a w-w^*-vanishing sequence $(x_n) \subset X$ which does not vanish in norm. Without loss of generality, we may assume that $\|x_n\| > \varepsilon$ for all $n \in \mathbb{N}$ and a suitable $\varepsilon > 0$. Since X has weakly* sequentially compact dual ball, from a sequence of functionals $(x'_n) \subset X'$ satisfying the conditions $\|x'_n\| = 1$ and $\langle x_n | x'_n \rangle > \varepsilon$ for all $n \in \mathbb{N}$ we can extract a weakly* convergent subsequence x'_{n_k}. However, $\langle x_{n_k} | x'_{n_k} \rangle > \varepsilon$, which contradicts the fact that (x_{n_k}) is w-w^*-vanishing.

The space $\ell^1(\mathbb{R})$ can be considered as a counterexample to the converse assertion. Indeed, this space possesses the Schur property and, therefore, the WS property. On the other hand, as is shown in [4, XIII], the dual ball of the space $\ell^1(\mathbb{R})$ is not weakly* sequentially compact. $\triangleright$

Each of the following properties of a Banach space X implies the WS property:

(1) X possesses the Schur property;
(2) X is separable;
(3) X' does not contain a copy of ℓ^1;
(4) X is reflexive;
(5) X is a subspace of a weakly compactly generated Banach space;
(6) for every separable subspace Y of X, the space Y' is separable.

Property (1) obviously implies the WS property, and the other properties guarantee that X has weakly* sequentially compact closed dual ball (see [4, XIII]), which allows us to apply the last lemma. Recall that a Banach space Y is said to be *weakly compactly generated* if Y contains a weakly compact absolutely convex set whose linear span is dense in Y.

3.1.7. A Banach space X is said to possess the *Dunford–Pettis property* if

$$\langle x_n | x'_n \rangle \to 0 \quad \text{for all weakly null sequences } (x_n) \subset X \text{ and } (x'_n) \subset X'.$$

In Section 3.5, within the study of weakly continuous sections of Banach bundles, the important role is clarified of the question whether a Banach space under consideration possesses the following property close to the Dunford–Pettis property.

DEFINITION. Say that a Banach space X possesses the DP* *property* if

$$\langle x_n | x'_n \rangle \to 0 \quad \begin{array}{l} \text{for every weakly null sequence } (x_n) \subset X \\ \text{and every weakly* null sequence } (x'_n) \subset X'. \end{array}$$

(Note that there is no reason to consider the analog of the DP* property for nets, since, in view of Lemma 3.1.4, only finite-dimensional spaces possess such a property.)

It is clear that X possesses the DP* property if and only if the sets of weakly convergent and w-w^*-convergent sequences in X coincide.

A Banach space X with the property that weakly* null sequences in X' are weakly null is called a *Grothendieck space* (see [4, VII, p. 121]). Obviously, every reflexive Banach space is a Grothendieck space.

The following assertions are easy to verify.

Lemma. *Let X be a Banach space.*

(1) *If X possesses the Schur property then X possesses the* DP* *property.*

(2) *If X possesses the* DP* *property then X possesses the Dunford–Pettis property.*

(3) *The space X possesses the* WS *and* DP* *properties if and only if X possesses the Schur property.*

(4) *For a Grothendieck space, the* DP* *property is equivalent to the Dunford–Pettis property.*

It is worth noting that assertion (2) does not admit conversion. Indeed, the space c_0 does not possess the Schur property and possesses the WS property, since c_0 is separable; therefore, by (3), c_0 does not possess the DP* property. At the same time, c_0 enjoys the Dunford–Pettis property, since $c'_0 \simeq \ell^1$ possesses the Schur property.

Recall that the intersection (union) of countably many open (closed) subsets of a topological space is called a σ-*open* (σ-*closed*) set.

Let K be a quasiextremally disconnected compact Hausdorff space (i.e. a compact Hausdorff space in which the closure of every open σ-closed subset is open). The spaces ℓ^∞ and $C(K)$ are Grothendieck spaces enjoying the Dunford–Pettis property and not the Schur property (see, for instance, [4, VII, Theorem 15, Exercise 1 (ii), XI, Exercise 4 (ii)], [1, Theorem 13.13], and [20, Theorem V.2.1]).

Corollary. *Let K be a quasiextremally disconnected compact Hausdorff space.*

(1) *The Banach spaces ℓ^∞ and $C(K)$ possess the* DP* *property.*

(2) *Every Banach space containing a copy of ℓ^∞ does not possess the* WS *property.*

◁ The claim follows immediately from the above-indicated properties of ℓ^∞ and $C(K)$, assertions (4) and (3) of the last lemma, and Proposition 3.1.5 (6). ▷

3.1.8. Lemma. *Given an arbitrary topological space Q, the following are equivalent:*

(a) *all functions in $C(Q)$ are locally constant;*

(b) *for every sequence of functions $(f_n) \subset C(Q)$ and every point $q \in Q$, there exists a neighborhood about q such that all functions f_n, $n \in \mathbb{N}$, are constant on the neighborhood;*

(c) *for every sequence of functions $(f_n) \subset C(Q)$, there is a partition of Q into clopen sets such that all functions f_n, $n \in \mathbb{N}$, are constant on every element of the partition.*

◁ (a)→(b): It is sufficient to find a neighborhood about q on which all functions $g_n = |f_n - f_n(q)| \wedge 1$, $n \in \mathbb{N}$, vanish. Since, the sum $g = \sum_{n=1}^{\infty} g_n/2^n$ is a continuous function and $g(q) = 0$, by (a) there is a neighborhood about q on which $g \equiv 0$. It is clear that all functions g_n, $n \in \mathbb{N}$, vanish too.

(b)→(c): According to (b), for every point $q \in Q$, the intersection $\bigcap_{n \in \mathbb{N}} \{f_n = f_n(q)\}$ of closed sets is a neighborhood about its every point; therefore, this intersection is clopen. All intersections of this kind form a sought partition of Q.

The implication (c)→(a) is evident. ▷

Definition. A topological space Q satisfying one of the equivalent conditions (a)–(c) of Lemma 3.1.8 is called *functionally discrete*.

3.1.9. A point of a topological space is *σ-isolated* or a *P-point* if the intersection of every sequence of neighborhoods about this point is again a neighborhood.

Remark. A Hausdorff topological space containing a single nonisolated point is a normal and Baire space.

Proposition. *Let Q be a completely regular topological space.*

(1) *The following are equivalent:*

   (a) *Q is functionally discrete;*

   (b) *all points in Q are σ-isolated;*

   (c) *every σ-open subset of Q is open;*

   (d) *every σ-closed subset of Q is closed.*

(2) *If Q is functionally discrete then all countable subsets of Q are closed.*

(3) *The converse of (2) is false.*

◁ (1): (a)→(b): Consider an arbitrary point $q \in Q$, a sequence (U_n) of neighborhoods about q, and put $V = \bigcap_{n\in\mathbb{N}} U_n$. Since the space Q is completely regular; for every $n \in \mathbb{N}$, there is a continuous function $f_n : Q \to [0,1]$ such that $f_n(q) = 0$ and $f_n \equiv 1$ on $Q \setminus U_n$. The sum

$$f = \sum_{n=1}^{\infty} f_n/2^n : Q \to [0,1]$$

is a continuous function and, by (a), vanishes on some neighborhood U_0 about q. Since $f > 0$ outside V, the neighborhood U_0 is a subset of V; therefore, V is a neighborhood about q too.

(b)→(c): By (b), the intersection of a sequence of open subsets of Q is a neighborhood about its every point and, hence, is open.

(c)→(a): By (c), for every function $f \in C(Q)$ and a point $q \in Q$, the intersection

$$\bigcap_{n\in\mathbb{N}} \{p \in Q : |f(p) - f(q)| < 1/n\}$$

is a neighborhood about q on which the function f is constant.

Equivalence of the mutually dual assertions (c) and (d) is evident.

(2): It is sufficient to observe that countable subsets of Q are σ-closed and to apply (1).

(3): Construct a completely regular topological space Q whose all countable subsets are closed and choose a function in $C(Q)$ which is not locally constant.

Make the interval $[0,1]$ into a topological space Q by taking as a base for open sets all subsets of $(0,1]$ and all subsets of the form $[0,t] \setminus S$, where $t \in (0,1]$ and S is a countable subset of $(0,1]$. The topological space Q is constructed. It is clear that all countable subsets of Q are closed. Since Q is a Hausdorff space and contains a single nonisolated point, it is normal (see the remark above the proposition); therefore, Q is completely regular. It is easy to see that the identity mapping of $[0,1]$ is continuous and is not constant on every neighborhood about 0. ▷

3.1.10. Recall that a topological space is *countably compact* if from every countable open cover of this space we can refine a finite subcover. A topological space is *perfectly normal* if it is normal and its every closed subset is σ-open.

Proposition. *Let Q be a completely regular topological space. Under each of the following conditions, the space Q includes a nonclosed countable subset (hence, Q is not functionally discrete):*

(1) *Q includes a nondiscrete countable compact subspace;*
(2) *Q includes an infinite compact subspace;*
(3) *Q includes a nondiscrete subspace that is a Fréchet–Urysohn space;*
(4) *Q includes a convergent sequence of pairwise distinct elements;*
(5) *Q contains a nonisolated point at which there is a countable base.*

Furthermore, a perfectly normal topological space is functionally discrete only if it is discrete.

$\lhd$ It is known (see, for example, [2, III, assertion 189]) that a topological space is countably compact if and only if its every infinite subset has a limit point. Using this criterion, we easily prove that condition (1) is sufficient for existence of a nonclosed countable subset of Q. Sufficiency of conditions (2), (4), and (5) is easily validated. Condition (3) is equivalent to (4).

For a nondiscrete perfectly normal topological space, existence of a not locally constant function follows from the Vedenisov Theorem (see [5, 1.5.19]). $\rhd$

3.1.11. If a topological space Q is functionally discrete and completely regular then Q satisfies none of the conditions 3.1.10 (1)–(5). In particular, if Q is nondiscrete then Q cannot be compact, first-countable, or metrizable. These observations essentially restrict the class of topological spaces in which Q may fall. Therefore, it is worth verifying that a completely regular functionally discrete topological space need not be discrete.

First, for an arbitrary upward-directed set $\aleph$ without greatest element, define a nondiscrete normal topological space $\aleph^\bullet$. As the underlying set we take $\overline{\aleph} = \aleph \cup \{\infty\}$, where $\infty \notin \aleph$. Endow $\overline{\aleph}$ with an order, regarding $\aleph$ as an ordered subset of $\overline{\aleph}$ and assuming $\infty > \alpha$ for all $\alpha \in \aleph$. Consider open the subsets of $\aleph$ and all intervals of the form $(\alpha, \infty] := \{\beta \in \overline{\aleph} : \alpha < \beta \leqslant \infty\}$, where $\alpha \in \aleph$ to be open. Therefore, $\aleph^\bullet$ becomes a topological space. Since $\aleph$ has no greatest element, the point $\infty \in \aleph^\bullet$ is nonisolated; hence, the topology of $\aleph^\bullet$ is nondiscrete. The space $\aleph^\bullet$ is normal, since it is Hausdorff and contains a single nonisolated point (see Remark 3.1.9).

REMARK. **(1)** If all countable subsets of $\aleph$ have upper bounds, every continuous function $f : \aleph^\bullet \to \mathbb{R}$ takes a constant value $f(\infty)$ on some neighborhood about ∞. (For instance, the intersection

$$\bigcap_{n\in\mathbb{N}} \{\alpha \in \aleph^\bullet : \ |f(\alpha) - f(\infty)| < 1/n\}$$

is such a neighborhood.)

(2) For an arbitrary topological space P, continuity of a function $f : \aleph^{\bullet} \to P$ is equivalent to the fact that the net $\big(f(\alpha)\big)_{\alpha \in \aleph}$ converges to $f(\infty)$.

EXAMPLE. There exists a functionally discrete normal topological space that is not discrete.

$\lhd$ Let $\aleph$ be an upward-directed set without greatest element and let all countable subsets of $\aleph$ have upper bounds. For instance, an arbitrary uncountable cardinal or the set of all countable subsets of an uncountable set (ordered by inclusion) is such an upward-directed set. Then, by the above remark, $\aleph^{\bullet}$ is a sought space. $\rhd$

3.1.12. Lemma. *Let Y be a locally convex space and let a sequence $(y_n) \subset Y$ converge to some $y \in Y$. Suppose that a vector valued function $u : [0,1] \to Y$ satisfies the equality $u(0) = y$ and, for every $n \in \mathbb{N}$, maps the interval $[\frac{1}{n+1}, \frac{1}{n}]$ onto the interval $[y_{n+1}, y_n]$ by the formula*

$$u\left(\frac{\lambda}{n+1} + \frac{1-\lambda}{n}\right) = \lambda y_{n+1} + (1-\lambda) y_n, \quad 0 \leqslant \lambda \leqslant 1.$$

Then u is continuous.

$\lhd$ It is clear that u is continuous on the half-open interval $(0,1]$. Take an arbitrary neighborhood V about $y = u(0)$, take an arbitrary convex neighborhood $W \subset V$ about the same element, and consider a number n_0 such that $y_n \in W$ for $n \geqslant n_0$. Then, in view of convexity of W, the inclusion $u\big([0, \frac{1}{n_0}]\big) \subset W$ holds. $\rhd$

3.1.13. Lemma. *Let X be an infinite-dimensional Banach space, whereas Q is not a functionally discrete topological space. Then there exists a weakly* continuous function from Q into X' whose pointwise norm is bounded and discontinuous.*

$\lhd$ By the Josefson–Nissenzweig Theorem [4, XII], there exists a weakly* null sequence (x'_n) of norm-one vectors in X'. Put $y_1 = x'_1$ and

$$y_{n+1} = \begin{cases} x'_{n+1}, & \|\lambda y'_n + (1-\lambda) x'_{n+1}\| \geqslant 1/2 \text{ for all } \lambda \in [0,1], \\ -x'_{n+1}, & \text{otherwise} \end{cases}$$

for every $n \in \mathbb{N}$. Obviously, the sequence (y_n) is weakly* null and, by Lemma 3.1.1, every interval $[y'_{n+1}, y'_n]$, $n \in \mathbb{N}$, does not intersect the open ball with radius $1/2$ centered at the origin. Then the vector valued function $u : [0,1] \to X'$ defined in Lemma 3.1.12 (where Y is equal to the space X' endowed with the weak* topology and y equals to 0) is weakly* continuous. At the same time, $|\!|\!|u|\!|\!|(0) = 0$ and $|\!|\!|u|\!|\!|\big((0,1]\big) \subset [1/2, 1]$.

Now consider a function $f \in C(Q)$ such that f is not constant on each neighborhood about a point $q \in Q$ and put $g = |f - f(q)| \wedge 1$. It is clear that $g : Q \to [0,1]$, $g(q) = 0$, and $q \in \mathrm{cl}\{g > 0\}$. Consequently, the composition $u \circ g : Q \to X'$ is a sought vector valued function. ▷

3.1.14. Let X be a Banach space. A subset $F \subset X'$ is called *total* (or *separating*) if, for every nonzero element $x \in X$, there is a functional $x' \in F$ such that $\langle x | x' \rangle \neq 0$.

REMARK. In each of the following cases, the dual X' of a Banach space includes a countable total subset:

(1) X is separable;

(2) X is isomorphic to the dual of a separable Banach space.

◁ (1): Consider a set $\{x_n : n \in \mathbb{N}\}$ everywhere dense in X. With each number $n \in \mathbb{N}$, associate a norm-one functional $x'_n \in X'$ such that $\langle x_n | x'_n \rangle = \|x_n\|$. Then, for an arbitrary nonzero element $x \in X$, there is an $n \in \mathbb{N}$ for which $\|x - x_n\| \leqslant \|x\|/3$ and, consequently,

$$
\begin{aligned}
&|\langle x | x'_n \rangle| \geqslant |\langle x_n | x'_n \rangle| - |\langle x_n - x \mid x'_n \rangle| \\
&\geqslant \|x_n\| - \|x\|/3 \geqslant \|x\| - \|x\|/3 - \|x\|/3 > 0.
\end{aligned}
$$

(2): Without loss of generality, we may assume that $X = Y'$, where Y is a separable Banach space. It remains to observe that the image of a countable everywhere dense subset of Y under the canonical embedding of Y into Y'' is total. ▷

Given a topological space Q and a Banach space X, the symbol $C_w(Q, X)$ denotes the totality of all weakly continuous functions from Q into X.

Lemma. *Let X be a Banach space and let Q be a functionally discrete topological space. Suppose that X' includes a countable total subset. Then $C(Q, X) = C_w(Q, X)$.*

◁ Consider an arbitrary vector valued function $u \in C_w(Q, X)$. It is sufficient to show that, for some partition of Q into clopen subsets, the function u is constant on each element of the partition.

Let $\{x'_n : n \in \mathbb{N}\}$ be a total subset of X'. Since u is weakly continuous, $\langle u | x'_n \rangle \in C(Q)$ for all $n \in \mathbb{N}$. According to 3.1.8 (c), there is a partition of Q into clopen subsets such that all functions $\langle u | x'_n \rangle$, $n \in \mathbb{N}$, are constant on each element of the partition. Since the set $\{x'_n : n \in \mathbb{N}\}$ is total, the function u is constant on each element of the partition. ▷

3.2. Homomorphisms of Banach Bundles

The current section, as follows from its title, is devoted to studying homomorphisms of Banach bundles. Some of the facts below are of interest in their own right, but usefulness of the majority of the results in the section reveals itself later, in studying operator bundles (see Sections 3.3 and 3.4).

The first group of results, 3.2.1–3.2.4, suggests a number of conditions guaranteeing that continuous sections of a Banach bundle with operator stalks are homomorphisms.

Subsections 3.2.5–3.2.7 provide a repeatedly employed useful way of constructing sections, homomorphisms, and Banach bundles.

In 3.2.8 and 3.2.9, the notion of the dimension of a Banach bundle is studied. The results obtained, concerning domains of constancy for the dimension, are, to our opinion, of interest in their own right.

In 3.2.10, a description is given for homomorphisms of Banach bundles over a first-countable topological space. This result is supplied with examples (3.2.11) which justify essence of the restrictions imposed on the topological space.

Closing this section, we study the question of continuity for the pointwise norm of a homomorphism acting from a CBB with constant finite dimension into an arbitrary CBB (3.2.12). A number of examples (see 3.2.13) demonstrates that the constancy of dimension is an essential requirement.

3.2.1. *Proposition.* *Let $\mathscr{X}$, $\mathscr{Y}$, and $\mathscr{Z}$ be CBBs over a topological space Q, with $\mathscr{Z}(q) \subset B(\mathscr{X}(q), \mathscr{Y}(q))$ for all $q \in Q$, and let sets of sections $\mathscr{U} \subset C(Q, \mathscr{X})$ and $\mathscr{W} \subset C(Q, \mathscr{Z})$ be stalkwise dense in $\mathscr{X}$ and $\mathscr{Z}$. Suppose that the global section $w \otimes u$ of $\mathscr{Y}$ is continuous for every $u \in \mathscr{U}$ and $w \in \mathscr{W}$. Then, for every $D \subset Q$, the inclusion $C(D, \mathscr{Z}) \subset \mathrm{Hom}_D(\mathscr{X}, \mathscr{Y})$ holds.*

$\lhd$ Fix an arbitrary subset $D \subset Q$, elements $\overline{u} \in C(D, \mathscr{X})$ and $\overline{w} \in C(D, \mathscr{Z})$, and a point $q \in D$. We prove that the section $\overline{w} \otimes \overline{u}$ of $\mathscr{Y}$ is continuous at q. By [8, Proposition 1.3.2], it is sufficient to show upper semicontinuity of the function $\|\overline{w} \otimes \overline{u} - v\| : D \to \mathbb{R}$ at the point q for every $v \in C(D, \mathscr{Y})$. Let $\varepsilon > 0$ and $v \in C(D, \mathscr{Y})$. We find a neighborhood about q on which

$$\|\overline{w} \otimes \overline{u} - v\| < \|\overline{w} \otimes \overline{u} - v\|(q) + \varepsilon.$$

Take an element $u \in \mathscr{U}$ such that $\|\overline{w}\|(q)\|\overline{u} - u\|(q) < \varepsilon/8$. By continuity of the real-valued functions $\|\overline{u} - u\|$ and $\|\overline{w}\|$, we may find a neighborhood U_1 about q on which $\|\overline{w}\|\|\overline{u} - u\| \leqslant \varepsilon/4$. Similarly, we take an element $w \in \mathscr{W}$ and a neighborhood U_2 about q such that $\|\overline{w} - w\|(q)\|u\|(q) < \varepsilon/8$ and $\|\overline{w} - w\|\|u\| \leqslant \varepsilon/4$ on U_2. Then, on the intersection $U_1 \cap U_2$, the following hold:

$$\begin{aligned} \|\overline{w} \otimes \overline{u} - w \otimes u\| &\leqslant \|\overline{w} \otimes \overline{u} - \overline{w} \otimes u\| + \|\overline{w} \otimes u - w \otimes u\| \\ &\leqslant \|\overline{w}\|\|\overline{u} - u\| + \|\overline{w} - w\|\|u\| \leqslant \varepsilon/4 + \varepsilon/4 = \varepsilon/2. \end{aligned}$$

The same calculations yield the inequality $\|\overline{w}\otimes\overline{u}-w\otimes u\|(q)<\varepsilon/4$. Now we take a neighborhood U_3 about q, on which $\|w\otimes u-v\|\leqslant\|w\otimes u-v\|(q)+\varepsilon/4$. On the neighborhood $U_1\cap U_2\cap U_3$ about q, the following hold:

$$\begin{aligned}\|\overline{w}\otimes\overline{u}-v\| &\leqslant \|\overline{w}\otimes\overline{u}-w\otimes u\|+\|w\otimes u-v\|\\ &\leqslant \varepsilon/2+\|w\otimes u-v\|(q)+\varepsilon/4\\ &\leqslant \varepsilon/2+\|w\otimes u-\overline{w}\otimes\overline{u}\|(q)+\|\overline{w}\otimes\overline{u}-v\|(q)+\varepsilon/4\\ &< \varepsilon/2+\varepsilon/4+\|\overline{w}\otimes\overline{u}-v\|(q)+\varepsilon/4\\ &= \|\overline{w}\otimes\overline{u}-v\|(q)+\varepsilon,\end{aligned}$$

which completes the proof. ▷

3.2.2. Corollary. *Let $\mathscr{X}$, $\mathscr{Y}$, and $\mathscr{Z}$ be CBBs over a topological space Q, with $\mathscr{Z}(q)\subset B(\mathscr{X}(q),\mathscr{Y}(q))$ at every point $q\in Q$. Suppose that $C(Q,\mathscr{Z})\subset \operatorname{Hom}(\mathscr{X},\mathscr{Y})$. Then, for every $D\subset Q$, the inclusion $C(D,\mathscr{Z})\subset\operatorname{Hom}_D(\mathscr{X},\mathscr{Y})$ holds.*

◁ The claim follows from 3.2.1 with $\mathscr{U}=C(Q,\mathscr{X})$ and $\mathscr{W}=C(Q,\mathscr{Z})$. ▷

3.2.3. Corollary. *The inclusion $C\big(Q,B(X,Y)\big)\subset\operatorname{Hom}(X_Q,Y_Q)$ holds for arbitrary Banach spaces X and Y.*

◁ Put $\mathscr{U}$ and $\mathscr{W}$ equal to the sets of all constant X-valued and $B(X,Y)$-valued functions and apply Proposition 3.2.1. ▷

One of the natural questions which may arise when considering the above corollary is as follows: When does the equality

$$C\big(Q,B(X,Y)\big)=\operatorname{Hom}(X_Q,Y_Q)$$

hold? This question is addressed in Section 3.3.

3.2.4. Corollary. *Let $\mathscr{X}$, $\mathscr{Y}$, and $\mathscr{Z}$ be CBBs over a topological space Q and let $\mathscr{Z}(q)\subset B\big(\mathscr{X}(q),\mathscr{Y}(q)\big)$ at every point $q\in Q$. Suppose that the space $\operatorname{Hom}(\mathscr{X},\mathscr{Y})$ includes a continuity structure for $\mathscr{Z}$. Then $C(Q,\mathscr{Z})\subset\operatorname{Hom}(\mathscr{X},\mathscr{Y})$.*

◁ Taking $C(Q,\mathscr{X})$ as $\mathscr{U}$, the above-mentioned continuity structure for $\mathscr{Z}$ as $\mathscr{W}$, and applying Proposition 3.2.1, we obtain the claim. ▷

3.2.5. In the sequel, we use the following auxiliary result.

Lemma. *Let Q be a completely regular topological space. Suppose that $q\in Q$ is a limit point for a countable discrete set $\{q_n : n\in\mathbb{N}\}$, with $q_i\neq q_j$ whenever $i\neq j$.*

(1) *There is a sequence (W_n) of open subsets of Q such that $q_n\in W_n$, $\operatorname{cl}W_n\cap\operatorname{cl}\bigcup_{k\neq n}W_k=\varnothing$, and $q\notin\operatorname{cl}W_n$ for all $n\in\mathbb{N}$.*

Consider continuous functions $f_n : Q \to [0,1]$, $f_n \equiv 0$ on $Q \setminus W_n$ for all $n \in \mathbb{N}$. Furthermore, let (ε_n) be a vanishing numerical sequence.

If at q there is a countable base then we may additionally stipulate that $\big(\operatorname{cl} \bigcup_{n\in\mathbb{N}} W_n\big) \setminus \bigcup_{n\in\mathbb{N}} \operatorname{cl} W_n = \{q\}$.

(2) *The function $f : Q \to [0,1]$ defined by the formula*

$$f(p) = \begin{cases} \varepsilon_n f_n(p), & p \in W_n, \\ 0, & p \notin \bigcup_{n\in\mathbb{N}} W_n \end{cases}$$

is continuous.

(3) *Let $\mathscr{X}$ be a CBB over Q. Given a sequence $(u_n)_{n\in\mathbb{N}} \subset C(Q, \mathscr{X})$ such that $\|u_n\| \leqslant M$ on W_n, from some index on, the section u over Q defined by the formula*

$$u(p) = \begin{cases} \varepsilon_n f_n(p) u_n(p), & p \in W_n, \\ 0, & p \notin \bigcup_{n\in\mathbb{N}} W_n \end{cases}$$

is continuous.

(4) *Let $\mathscr{X}$ and $\mathscr{Y}$ be CBBs over Q. If $(H_n)_{n\in\mathbb{N}} \subset \operatorname{Hom}(\mathscr{X}, \mathscr{Y})$ and $\|H_n\| \leqslant K$ on W_n for all n from some index on, then the mapping $H : p \in Q \mapsto H(p) \in B\big(\mathscr{X}(p), \mathscr{Y}(p)\big)$ defined by the formula*

$$H(p) = \begin{cases} \varepsilon_n f_n(p) H_n(p), & p \in W_n, \\ 0, & p \notin \bigcup_{n\in\mathbb{N}} W_n \end{cases}$$

is a homomorphism from $\mathscr{X}$ into $\mathscr{Y}$.

(5) *If X is a topological vector space and a sequence $(x_n) \subset X$ converges to an $x \in X$, then the vector valued function $u : Q \to X$ defined by the formula*

$$u(p) = \begin{cases} f_n(p) x_n + \big(1 - f_n(p)\big) x, & p \in W_n, \\ x, & p \notin \bigcup_{n\in\mathbb{N}} W_n \end{cases}$$

is continuous.

(6) *If X is a Banach space and a sequence of functionals $(x'_n) \subset X'$ converges weakly* to an $x' \in X'$, then the vector valued function $H : Q \to X'$ defined by the formula*

$$H(p) = \begin{cases} f_n(p) x'_n + \big(1 - f_n(p)\big) x', & p \in W_n, \\ x', & p \notin \bigcup_{n\in\mathbb{N}} W_n \end{cases}$$

is a homomorphism from X_Q into $\mathscr{R}$.

◁ (1): By induction, for every $n \in \mathbb{N}$, we construct open sets $W_n, V_n \subset Q$. Since the space Q is regular, the point q_1 and the closed set $\operatorname{cl}\{q_k : k \geqslant 2\}$ have disjoint open neighborhoods W_1 and V_1. We may assume that $\operatorname{cl} W_1 \cap \operatorname{cl} V_1 = \varnothing$. If W_k and V_k are chosen for all $k \leqslant n$ then we take W_{n+1} and V_{n+1} so that V_n contain W_{n+1} and V_{n+1}, and the sets $\operatorname{cl} W_{n+1}$ and $\operatorname{cl} V_{n+1}$ separate the point q_{n+1} and the closed set $\operatorname{cl}\{q_k : k \geqslant n+2\}$. It is easy to see that (W_n) is a sought sequence.

Finally, let (U_n) be a countable base for open neighborhoods about q, with $U_1 = Q$ and $U_n \supset U_{n+1}$ for all $n \in \mathbb{N}$. Then, when constructing the sequence of sets W_n, we may take $W_n \subset U_{k(n)}$, where $k(n) = \max\{k \in \mathbb{N} : q_n \in U_k\}$. This provides the desired relation, $(\operatorname{cl} \bigcup_{n\in\mathbb{N}} W_n) \setminus \bigcup_{n\in\mathbb{N}} \operatorname{cl} W_n = \{q\}$.

(2): It is obvious that the function f is the pointwise sum of the uniformly convergent series $\sum_{n=1}^{\infty} \varepsilon_n f_n$; therefore, f is continuous.

Assertions (3)–(5) may be proven in much the same way by using Proposition [8, 1.3.6] for (3) and [8, 1.4.11] for (4).

(6): By (5) the function H is weakly* continuous; therefore, $H \otimes u \in C(Q)$ for all constant functions $u : Q \to X$. It remains to observe that the pointwise norm of H is bounded by construction and to apply [8, Theorem 1.4.9]. ▷

3.2.6. Corollary. *Let $\mathscr{X}$ and $\mathscr{Y}$ be CBBs over a completely regular topological space Q. Suppose that a sequence $(q_n)_{n\in\mathbb{N}}$, $q_i \neq q_j$ $(i \neq j)$ converges to a point q and $q \neq q_k$ for all $k \in \mathbb{N}$.*

(1) *Let $x_n \in \mathscr{X}(q_n)$ $(n \in \mathbb{N})$, let $x \in \mathscr{X}(q)$, and let the convergence $(q_n, x_n) \to (q, x)$ as $n \to \infty$ hold in the topological space $Q \otimes \mathscr{X}$ (see [8, 1.1.4]). (For $x = 0$, this is equivalent to the equality $\lim_{n\to\infty} \|x_n\| = 0$.) Then there exists a bounded section $u \in C(Q, \mathscr{X})$ such that $u(q_n) = x_n$ for all $n \in \mathbb{N}$ and $u(q) = x$.*

(2) *Let $H_n \in \operatorname{Hom}(\mathscr{X}, \mathscr{Y})$ $(n \in \mathbb{N})$ and let the sequence $(\|H_n\|)_{n\in\mathbb{N}}$ be uniformly vanishing. Then there exists a bounded homomorphism $H \in \operatorname{Hom}(\mathscr{X}, \mathscr{Y})$ such that $H(q_n) = H_n(q_n)$ for all $n \in \mathbb{N}$ and $H(q) = 0$.*

(3) *Let X be a topological vector space. Suppose that the sequence $(x_n) \subset X$ converges to an $x \in X$. Then there is a continuous vector valued function $u : Q \to X$ such that $u(q_n) = x_n$ for all $n \in \mathbb{N}$ and $u(q) = x$.*

(4) *Let X be a Banach space. Suppose that the sequence $(x'_n) \subset X'$ is convergent weakly* to an $x' \in X'$. Then there exists a homomorphism $H \in \operatorname{Hom}(X_Q, \mathscr{R})$ such that $H(q_n) = x'_n$ for all $n \in \mathbb{N}$ and $H(q) = x'$.*

◁ We only need to explain assertion (1). If $x = 0$, this assertion follows directly from Lemma 3.2.5 (3) and Dupré's Theorem (see [8, 1.3.5]). Dealing with the general case, use Dupré's Theorem again and consider a bounded section $v \in C(Q, \mathscr{X})$ taking the value x at q. From [8, Proposition 1.3.8] it follows that $\|x_n - v(q_n)\| \to 0$ as $n \to \infty$. Since the assertion under proof is true for the case $x = 0$, there is a bounded section $w \in C(Q, \mathscr{X})$ satisfying the equalities $w(q_n) = x_n - v(q_n)$ $(n \in \mathbb{N})$ and $w(q) = 0$. It remains to put $u = v + w$. $\triangleright$

3.2.7. Lemma. *Let $X_1 \subset X_2 \subset \cdots$ be Banach spaces, let Q be a completely regular topological space, and let $(U_n)_{n\in\mathbb{N}}$ be a partition of Q such that the sets $U_1 \cup \cdots \cup U_n$ are closed for all $n \in \mathbb{N}$. Then there is a* CBB *$\mathscr{X}$ over Q satisfying the following conditions:*

(a) *$\mathscr{X}|_{U_n} \equiv X_n$ for all $n \in \mathbb{N}$;*

(b) *if the sequence of functionals $x'_n \in X'_n$ $(n \in \mathbb{N})$ is such that x'_{n+1} extends x'_n and $\|x'_n\| \leqslant 1$ for all $n \in \mathbb{N}$, then the mapping H satisfying the relations $H|_{U_n} \equiv x'_n$ $(n \in \mathbb{N})$ belongs to* $\mathrm{Hom}(\mathscr{X}, \mathscr{R})$.

$\triangleleft$ Consider a (discrete) Banach bundle $\mathscr{X}$ satisfying condition (a) and define a continuity structure in $\mathscr{X}$ as follows: Put

$$C_0 = C(Q);$$
$$C_n = \{f \in C(Q) : f \equiv 0 \text{ on } U_1 \cup \cdots \cup U_n\}, \quad n \in \mathbb{N}.$$

It is clear that the set of sections

$$\mathscr{C} = \{f_1 x_1 + \cdots + f_n x_n : f_i \in C_i,\ x_i \in X_i,\ i = 1, \dots, n,\ n \in \mathbb{N}\}$$

of the bundle $\mathscr{X}$ is a subspace of the space of all global sections of $\mathscr{X}$. Moreover, the set $\mathscr{C}$ is stalkwise dense in $\mathscr{X}$. Indeed, let $q \in Q$, $x \in \mathscr{X}(q)$, and let a number $n \in \mathbb{N}$ be such that $q \in U_n$. Since the space Q is completely regular, there is a function $f \in C_{n-1}$ such that $f(q) = 1$. Therefore, fx belongs to $\mathscr{C}$ and passes through x at q. Consequently, $\mathscr{C}$ is a continuity structure in $\mathscr{X}$ which makes $\mathscr{X}$ a CBB.

Let H satisfy condition (b). Verify that $H \in \mathrm{Hom}(\mathscr{X}, \mathscr{R})$. By Theorem [8, 1.4.9], it is sufficient to show that $H \otimes u \in C(Q)$ for all $u \in \mathscr{C}$. If $u = f_1 x_1 + \cdots + f_n x_n \in \mathscr{C}$, where $f_i \in C_i$, $x_i \in X_i$, $i = 1, \dots, n$, then, for all $q \in Q$, the equality $(H \otimes u)(q) = \langle u(q) | x'_n \rangle$ holds. Next,

$$\langle u(q) | x'_n \rangle = f_1(q) \langle x_1 | x'_n \rangle + \cdots + f_n(q) \langle x_n | x'_n \rangle.$$

Therefore, the function $H \otimes u$ is continuous. $\triangleright$

3.2.8. DEFINITION. Let $\mathscr{X}$ be an arbitrary Banach bundle over a set Q. The function $\dim \mathscr{X}$ which, with every point $q \in Q$, associates the dimension $\dim \mathscr{X}(q)$ of the stalk $\mathscr{X}(q)$ is the *dimension of* $\mathscr{X}$.

We say that $\mathscr{X}$ *has constant dimension* n if $\dim \mathscr{X}(q) = n$ for all $q \in Q$.

Lemma. *Let $\mathscr{X}$ be a CBB with finite-dimensional stalks over an arbitrary topological space. For every $n = 0, 1, 2, \dots$, consider the following conditions:*

(a) *the set $\{\dim \mathscr{X} = n\}$ is open;*

(b) *the set $\{\dim \mathscr{X} < n\}$ is open;*

(c) *the set $\{\dim \mathscr{X} \leqslant n\}$ is open;*

(d) *the set $\{\dim \mathscr{X} > n\}$ is closed;*

(e) *the set $\{\dim \mathscr{X} \geqslant n\}$ is closed.*

If one of the conditions (a)–(e) *holds for every $n = 0, 1, 2, \dots$, then each of the conditions holds for every $n = 0, 1, 2, \dots$. In this case, all sets mentioned in* (a)–(e) *are clopen.*

$\lhd$ It suffices to observe that, due to [7, 18.1], the sets of the form $\{\dim \mathscr{X} > n\}$ and $\{\dim \mathscr{X} \geqslant n\}$ are open and, therefore, the sets of the form $\{\dim \mathscr{X} < n\}$ and $\{\dim \mathscr{X} \leqslant n\}$ are closed. $\rhd$

3.2.9. Proposition. *The following hold:*

(1) *Let Q be a Baire topological space. Then, for every* CBB $\mathscr{X}$ *over Q with finite-dimensional stalks, the union $\bigcup_{n \geqslant 0} \operatorname{int} \{\dim \mathscr{X} = n\}$ is everywhere dense in Q.*

(2) *If the space Q is completely regular and, for every* CBB $\mathscr{X}$ *over Q with finite-dimensional stalks, the set $\bigcup_{n \geqslant 0} \operatorname{int} \operatorname{cl} \{\dim \mathscr{X} = n\}$ is everywhere dense, then Q is a Baire space.*

$\lhd$ (1): For proving that the union under consideration is everywhere dense, it is sufficient, given a nonempty open set $U \subset Q$, to find an open nonempty subset $W \subset U$ such that the dimension of $\mathscr{X}$ is constant on W.

Since Q is a Baire space, there is a number $n \geqslant 0$ such that

$$V := \operatorname{int} \operatorname{cl} \{\dim \mathscr{X} = n\} \neq \varnothing.$$

Consequently, from [7, 18.1] we easily infer that the set $\{\dim \mathscr{X} \leqslant n\}$ is closed; therefore, $V \subset \operatorname{cl} \{\dim \mathscr{X} = n\} \subset \{\dim \mathscr{X} \leqslant n\}$, i.e., $\dim \mathscr{X} \leqslant n$ on V. The relation $V \subset \operatorname{cl} \{\dim \mathscr{X} = n\}$ and the fact that the set V is open imply that there exists a point $q \in V \cap \{\dim \mathscr{X} = n\}$. Since the set $\{\dim \mathscr{X} \geqslant n\}$ is open, $\dim \mathscr{X} \geqslant n$ on some open neighborhood $W \subset V$ about q. Thus, the dimension of $\mathscr{X}$ is constant on the open nonempty set $W \subset V \subset U$.

(2): Let Q be a completely regular space that is not a Baire space. We will construct a CBB $\mathscr{X}$ over Q such that $\mathscr{X}$ has finite-dimensional stalks while the set $\bigcup_{n\geqslant 0} \operatorname{int}\operatorname{cl}\{\dim \mathscr{X} = n\}$ is not everywhere dense.

Since Q is not a Baire space, there exist an open nonempty set $U \subset Q$ and a cover $(V_n)_{n\in\mathbb{N}}$ consisting of nowhere dense subsets $V_n \subset U$. Put $U_1 = Q\backslash U$ and $U_{n+1} = \operatorname{cl} V_n \setminus (U_1 \cup \dots \cup U_n)$ for all $n \in \mathbb{N}$. It is clear that, for all $n \in \mathbb{N}$, the set U_n is nowhere dense, the union $U_1 \cup \dots \cup U_n$ is closed, and $\bigcup_{n\in\mathbb{N}} U_n = Q$.

Consider a sequence $X_1 \subset X_2 \subset \cdots$ of finite-dimensional Banach spaces with strictly increasing dimensions: $\dim X_n < \dim X_{n+1}$ for all $n \in \mathbb{N}$. By Lemma 3.2.7, there exists a CBB $\mathscr{X}$ over Q such that $\mathscr{X}|_{U_n} \equiv X_n$ for all $n \in \mathbb{N}$. It is easy to see that

$$\bigcup_{n\geqslant 0} \operatorname{int}\operatorname{cl}\{\dim \mathscr{X} = n\} = \bigcup_{m\geqslant 0} \operatorname{int}\operatorname{cl} U_m = \operatorname{int} U_1,$$

where the latter set is not everywhere dense. ▷

Corollary. *If $\mathscr{X}$ is a CBB with finite-dimensional stalks over a Baire space Q then, for every $m = 0, 1, 2, \dots$, the equality holds*

$$\operatorname{cl}\{\dim \mathscr{X} \geqslant m\} = \operatorname{cl} \bigcup_{n\geqslant m} \operatorname{int}\{\dim \mathscr{X} = n\}.$$

◁ Fix a number $0 \leqslant m \in \mathbb{Z}$. The inclusion $\supset$ is obvious. Prove the reverse inclusion. Let $q \in Q$ and $\dim \mathscr{X}(q) \geqslant m$. The union $\bigcup_{n\geqslant 0} \operatorname{int}\{\dim \mathscr{X} = n\}$ is everywhere dense by Proposition 3.2.9 (1),

$$\bigcup_{n<m} \operatorname{int}\{\dim \mathscr{X} = n\} \subset \{\dim \mathscr{X} < m\},$$

and the latter set is closed.

Hence, the point q belongs to the closure of $\bigcup_{n\geqslant m} \operatorname{int}\{\dim \mathscr{X} = n\}$. Therefore,

$$\{\dim \mathscr{X} \geqslant m\} \subset \operatorname{cl} \bigcup_{n\geqslant m} \operatorname{int}\{\dim \mathscr{X} = n\},$$

which implies the required inclusion. ▷

3.2.10. The following assertion differs from [8, Theorem 1.4.7] only in the conditions on Q.

Theorem. *Let $\mathscr{X}$ and $\mathscr{Y}$ be CBBs over a first-countable completely regular topological space Q. A mapping $H : q \in Q \mapsto H(q) \in B(\mathscr{X}(q), \mathscr{Y}(q))$ is a homomorphism from $\mathscr{X}$ into $\mathscr{Y}$ if and only if $H \otimes u \in C(Q, \mathscr{Y})$ for all $u \in C(Q, \mathscr{X})$.*

◁ Necessity follows from [8, Theorem 1.4.4]. Prove sufficiency. In view of [8, Theorem 1.4.4], it is enough to prove that H is locally bounded. Suppose that the function $\|H\|$ is not bounded in any neighborhood about a point $q \in Q$. In this case, since Q is first-countable, there is a sequence $(q_n) \subset Q \setminus \{q\}$, $q_i \neq q_j$ $(i \neq j)$, convergent to q such that $\|H\|(q_n) > (\|H\|(q) + n)^2$ for all $n \in \mathbb{N}$. For every $n \in \mathbb{N}$, we take an element $x_n \in \mathscr{X}(q_n)$ so that $\|H(q_n)x_n\| = \|H(q_n)\|$ and $\|x_n\| \leqslant 2$. By Corollary 3.2.6 (1) there exists a bounded section $u \in C(Q, \mathscr{X})$ taking values $u(q_n) = \frac{1}{n} x_n$ for all $n \in \mathbb{N}$ and $u(q) = 0$. Then

$$\|H \otimes u\|(q_n) = \tfrac{1}{n}\,\|H(q_n)\| \geqslant \tfrac{1}{n}(\|H\|(q) + n)^2 > n.$$

This contradicts continuity of $H \otimes u$, since $q_n \to q$ and $(H \otimes u)(q) = 0$. ▷

REMARK. From the above proof and the proof of 3.2.5 (3), it is clear that, in the last theorem, the condition $H \otimes u \in C(Q, \mathscr{Y})$ for all $u \in C(Q, \mathscr{X})$ can be replaced by a "weaker" condition: $H \otimes u \in C(Q, \mathscr{Y})$ for all u in a stalkwise dense $C^b(Q)$-submodule of $C^b(Q, \mathscr{X})$ closed with respect to the uniform norm. For instance, we may take as such a submodule $C^b(Q, \mathscr{X})$.

3.2.11. Thus, Theorem 3.2.10 is stated for the case of a first-countable topological space Q. In the literature, the class of Fréchet–Urysohn spaces is usually the smallest class of topological spaces under consideration which includes the class of first-countable spaces (cf. [5, 1.6.14]). (Recall that a topological space Q is said to be a *Fréchet–Urysohn space* if, for every point $q \in Q$ and every $P \subset Q$, the condition $p \in \operatorname{cl} P$ implies existence of a sequence in P convergent to q.) Show that Theorem 3.2.10 cannot be generalized to the class of Fréchet–Urysohn spaces Q.

EXAMPLE. We construct a topological space Q with the following properties:

(a) Q is a Fréchet–Urysohn space;
(b) Q is a normal space;
(c) Q is not first-countable;
(d) Q is not locally pseudocompact;
(e) Q is a Baire space;
(f) there exist a CBB $\mathscr{X}$ over Q with finite-dimensional stalks and a mapping $H : q \in Q \mapsto H(q) \in \mathscr{X}(q)'$ such that $H \otimes u \in C(Q)$ for all $u \in C(Q, \mathscr{X})$, but $H \notin \operatorname{Hom}(\mathscr{X}, \mathscr{R})$;
(g) for every infinite-dimensional Banach space X, there is a mapping $H : Q \to X'$ such that $H \otimes u \in C(Q)$ for all $u \in C(Q, \mathscr{X})$, but $H \notin \operatorname{Hom}(X_Q, \mathscr{R})$.

Consider the set $Q = (\mathbb{N} \times \mathbb{N}) \cup \{\infty\}$, where $\infty \notin \mathbb{N} \times \mathbb{N}$, and endow Q with a topology in the following way. We regard all elements of $\mathbb{N} \times \mathbb{N}$ as isolated points and all subsets $U \subset Q$, for which $\infty \in U$ and

$$(\forall m \in \mathbb{N})\ (\exists n_m \in \mathbb{N})\ (\forall n \geqslant n_m)\quad (m, n) \in U,$$

as neighborhoods of ∞. It is clear that

$$C(Q) = \Big\{f : Q \to \mathbb{R} \ :\ \lim_{n\to\infty} f\big((m,n)\big) = f(\infty) \text{ for all } m \in \mathbb{N}\Big\}. \tag{1}$$

Verify that the topological space Q possesses properties (a)–(g).

(a): It is sufficient to consider a subset $P \subset Q$ that does not contain a sequence convergent to ∞ and show that $\infty \notin \operatorname{cl} P$. Obviously, for every $m \in \mathbb{N}$, there is a number n_m such that $\big\{(m,n) \in P : n \in \mathbb{N}\big\} \subset \big\{(m,1), \dots, (m, n_m)\big\}$. Hence, the set P and the neighborhood $\big\{(m,n) : m \in \mathbb{N},\ n > n_m\big\} \cup \{\infty\}$ about ∞ are disjoint; therefore, $\infty \notin \operatorname{cl} P$.

(b), (e): See Remark 3.1.9.

Conditions (c) and (d) immediately follow from assertion (f) proven below and Theorems 3.2.10 and [8, 1.4.7] respectively.

(f): Consider a CBB $\mathscr{X}$ over Q such that $\mathscr{X}(q) = \mathbb{R}$ for all $q \in \mathbb{N}\times\mathbb{N}$, $\mathscr{X}(\infty) = \{0\}$, and $C(Q, \mathscr{X}) = \{u \in C(Q) : u(\infty) = 0\}$. Define an H by the equalities $H(\infty) = 0$ and $H\big((m,n)\big) = m$ for all $(m,n) \in \mathbb{N} \times \mathbb{N}$. It is easy to verify that $H \otimes u \in C(Q)$ for all $u \in C(Q, \mathscr{X})$ (see (1)). Nevertheless, the pointwise norm of H is not locally bounded; therefore, by [8, Theorem 1.4.4], $H \notin \operatorname{Hom}(\mathscr{X}, \mathscr{R})$.

(g): By the Josefson–Niessenzweig Theorem [4, XII], there is a weakly* null sequence (x'_n) of norm-one vectors in X'. Define $H(\infty) = 0 \in X'$ and $H\big((m,n)\big) = mx'_n$ for all $(m,n) \in \mathbb{N} \times \mathbb{N}$. Then $H \otimes u \in C(Q)$ for an arbitrary section $u \in C(Q, X_Q)$. Indeed, for every $m \in \mathbb{N}$, the relation $\lim_{n\to\infty} (H \otimes u)\big((m,n)\big) = 0$ holds, since $\big(H\big((m,n)\big)\big)_{n\in\mathbb{N}}$ is a weakly* null sequence and $\|u\big((m,n)\big) - u(\infty)\| \to 0$ as $n \to \infty$. It remains to observe that the pointwise norm of H is not locally bounded and to apply [8, Theorem 1.4.4].

3.2.12. *Theorem.* *Let a* CBB $\mathscr{X}$ *over a topological space* Q *have constant finite dimension, let* $\mathscr{Y}$ *be an arbitrary* CBB *over* Q*, and let* $\mathscr{U}$ *be a subset of* $C(Q, \mathscr{X})$ *stalkwise dense in* $\mathscr{X}$*. If a mapping* $H : p \in Q \mapsto H(p) \in B\big(\mathscr{X}(p), \mathscr{Y}(p)\big)$ *is such that* $H \otimes u \in C(Q, \mathscr{Y})$ *for every* $u \in \mathscr{U}$*, then* $H \in \operatorname{Hom}(\mathscr{X}, \mathscr{Y})$ *and the pointwise norm* $|\!|\!|H|\!|\!|$ *is continuous.*

$\lhd$ Fix an arbitrary point $q \in Q$ and prove continuity of $|\!|\!|H|\!|\!|$ at this point. Due to the relation

$$\begin{aligned}\|H(p)\| &= \sup\Big\{\Big\|H(p)\Big(\frac{1}{\max\{\|u(p)\|, 1\}}\, u(p)\Big)\Big\| : u \in \mathscr{U}\Big\} \\ &= \sup\Big\{\Big(\frac{1}{|\!|\!|u|\!|\!| \vee 1} |\!|\!|H \otimes u|\!|\!|\Big)(p) : u \in \mathscr{U}\Big\}\end{aligned}$$

valid for all $p \in Q$, the function $\|H\|$ is lower-semicontinuous. It remains to prove that the function $\|H\|$ is upper-semicontinuous at q. Take an arbitrary $\varepsilon > 0$ and prove that, in some neighborhood U about q, the inequality $\|H\| \leqslant \|H\|(q) + \varepsilon$ holds.

Since the stalk $\mathscr{X}(q)$ is finite-dimensional, there is a collection of sections $\mathbf{u} = (u_1, \dots, u_n) \subset \operatorname{lin} \mathscr{U}$ such that the values $u_1(q), \dots, u_n(q)$ lie on the unit sphere and constitute a basis for $\mathscr{X}(q)$. Since the set

$$\Lambda = \{\boldsymbol{\lambda} \in \mathbb{R}^n : \|\boldsymbol{\lambda}\mathbf{u}\|(q) = 1\}$$

is bounded in $\mathbb{R}^n$, the number

$$\|\Lambda\|_1 := \sup\{|\lambda_1| + \dots + |\lambda_n| : (\lambda_1, \dots, \lambda_n) \in \Lambda\}$$

is finite. (Here and in the sequel, we denote by $\boldsymbol{\lambda}\mathbf{u}$ the sum $\lambda_1 u_1 + \dots + \lambda_n u_n$.) Choose some number $\delta \in (0,1)$ such that $\frac{1}{1-\delta}(\delta + \|H\|(q)) < \|H\|(q) + \varepsilon$.

By [16, Lemma 7], there exists a neighborhood U_δ about q, where $1 - \delta \leqslant \|\boldsymbol{\lambda}\mathbf{u}\| \leqslant 1 + \delta$ for all $\boldsymbol{\lambda} \in \Lambda$. Without loss of generality, we may assume that the collection $\mathbf{u}(p) = (u_1(p), \dots, u_n(p))$ is linearly independent for every element $p \in U_\delta$ (see [7, 18.1]). In particular, an arbitrary vector $x \in \mathscr{X}(p)$ can be represented as

$$x = \frac{\|x\|}{\|\boldsymbol{\lambda}_x \mathbf{u}\|(p)} (\boldsymbol{\lambda}_x \mathbf{u})(p)$$

with a suitable $\boldsymbol{\lambda}_x \in \Lambda$. Since the sections $H \otimes u_i$, $i = 1, \dots, n$, are continuous, there exists a neighborhood $U \subset U_\delta$ about q such that

$$\|\Lambda\|_1 \max\big\{\big|\|H \otimes u_i\|(p) - \|H \otimes u_i\|(q)\big| \ : \ i = 1, \dots, n\big\} < \delta$$

for all $p \in U$. At a point $p \in U$, the value of the norm $\|H(p)\|$ is attained at some vector $x(p) \in \mathscr{X}(p)$, $\|x(p)\| = 1$. Hence,

$$\begin{aligned}
\|H\|(p) = \|H(p)x(p)\| &= \frac{1}{\|\boldsymbol{\lambda}_{x(p)}\mathbf{u}\|(p)} \|H \otimes (\boldsymbol{\lambda}_{x(p)}\mathbf{u})\|(p) \\
&\leqslant \frac{1}{1-\delta}\Big(\Big|\|H \otimes (\boldsymbol{\lambda}_{x(p)}\mathbf{u})\|(p) - \|H \otimes (\boldsymbol{\lambda}_{x(p)}\mathbf{u})\|(q)\Big| \\
&\qquad\qquad + \|H \otimes (\boldsymbol{\lambda}_{x(p)}\mathbf{u})\|(q)\Big) \\
&\leqslant \frac{1}{1-\delta}\Big(\|\Lambda\|_1 \max\big\{\big|\|H \otimes u_i\|(p) - \|H \otimes u_i\|(q)\big| : \\
&\qquad\qquad i = 1, \dots, n\big\} + \|H\|(q)\Big) \\
&\leqslant \frac{1}{1-\delta}(\delta + \|H\|(q)) < \|H\|(q) + \varepsilon.
\end{aligned}$$

The fact that $H \in \operatorname{Hom}(\mathscr{X}, \mathscr{Y})$ now follows from continuity of $\|H\|$ and [8, Theorem 1.4.4]. $\triangleright$

Corollary. *Let $\mathscr{X}$ and $\mathscr{Y}$ be CBBs over the same topological space. If $\mathscr{X}$ has constant finite dimension then the pointwise norm of every homomorphism from $\mathscr{X}$ into $\mathscr{Y}$ is continuous.*

3.2.13. As we see from the examples below, the constant dimension requirement for a bundle $\mathscr{X}$ in Corollary 3.2.12 is essential.

Intending to emphasize diversity of situations in which a homomorphism $H \in \mathrm{Hom}(\mathscr{X}, \mathscr{R})$ with a discontinuous norm arises for a CBB $\mathscr{X}$ with finite-dimensional stalks, we give three different examples. In the first case, the dimension of $\mathscr{X}$ is equal to 0 at a unique discontinuity point of the function $|\!|\!| H |\!|\!|$ and the dimension of $\mathscr{X}$ is equal to 1 at other points. In the second case, the dimension of $\mathscr{X}$ takes two distinct (possibly, nonzero) values and, in the third case, the dimension of $\mathscr{X}$ takes infinitely many distinct values and the function $|\!|\!| H |\!|\!|$ is discontinuous at every point.

EXAMPLES. **(1)** Let $Q = [0,1]$. Define $\mathscr{X}(q) = \mathbb{R}$ whenever $0 < q \leqslant 1$ and $\mathscr{X}(0) = \{0\}$. Consider the set $\{u \in C[0,1] : u(0) = 0\}$ as a continuity structure in $\mathscr{X}$. Then the pointwise norm of the homomorphism H identically equal to values $\mathrm{id}_{\mathbb{R}}$ on the half-open interval $(0,1]$ is not continuous at the point $0 \in Q$. It is easy to verify that, in this case, $\mathrm{Hom}(\mathscr{X}, \mathscr{R})$ can be identified in a natural way with the space of real-valued continuous functions defined on the interval $[0,1]$ bounded on the half-open interval $(0,1]$ and vanishing at the point $0 \in Q$. However, such functions are far from being always continuous.

(2) This time, consider a completely regular topological space Q and let q be a nonisolated point of Q. Define $U_1 = \{q\}$, $U_2 = Q \backslash U_1$, and $U_3 = U_4 = \dots = \varnothing$. Let X be a finite-dimensional Banach space, let X_1 be a proper subspace of X, and let $X_2 = X_3 = \dots = X$. Fix a norm-one functional $x' \in X'$ vanishing on X_1 and define $x'_1 = 0$, $x'_2 = x'_3 = \dots = x'$. Consider the CBB $\mathscr{X}$ of Lemma 3.2.7 and a homomorphism H satisfying condition (b) of the lemma. It is clear that $|\!|\!| H |\!|\!|(q) = 0$ and $|\!|\!| H |\!|\!| \equiv 1$ outside $\{q\}$. Therefore, since the point q is nonisolated, the function $|\!|\!| H |\!|\!|$ is discontinuous.

(3) Let $Q = \mathbb{Q}$ be the set of rationals with the natural topology and let $n \mapsto q_n$ be an arbitrary bijection from $\mathbb{N}$ onto Q. Define $U_n = \{q_n\}$ for all $n \in \mathbb{N}$ and consider an arbitrary sequence of Banach spaces $X_1 \subset X_2 \subset \cdots$ and an arbitrary sequence of functionals x'_n satisfying condition 3.2.7 (b). We additionally require that the dimensions of X_n and the norms of x'_n be strictly monotone increasing. Let $\mathscr{X}$ be the CBB of Lemma 3.2.7 and let H be a homomorphism satisfying condition 3.2.7 (b). It is obvious that the stalks of $\mathscr{X}$ have pairwise distinct dimensions and the pointwise norm of H is discontinuous at every point of Q.

The authors are unaware of an answer to the following question: Given a bundle, is the requirement that the dimension be constant on some neighborhood about

q sufficient for continuity of the pointwise norms of all homomorphisms at q? Theorem 3.3.5 (2) in the next section gives a positive answer to this question in some particular case.

3.3. An Operator Bundle

In this section, we suggest a number of necessary and sufficient conditions for existence of a Banach bundle $B(\mathscr{X},\mathscr{Y})$ whose continuous sections are homomorphisms from a CBB $\mathscr{X}$ into a CBB $\mathscr{Y}$. Separately treated are the cases of arbitrary bundles $\mathscr{X}$ and $\mathscr{Y}$, bundles with finite-dimensional stalks, and the case of trivial CBBs and CBBs with constant finite dimension.

3.3.1. Let $\mathscr{X}$, $\mathscr{Y}$, and $\mathscr{Z}$ be CBBs over a topological space Q, with $\mathscr{Z}(q) \subset B(\mathscr{X}(q),\mathscr{Y}(q))$ for all $q \in Q$.

Lemma. *The following assertions are equivalent:*

(a) $C(Q,\mathscr{Z}) = \mathrm{Hom}(\mathscr{X},\mathscr{Y})$;

(b) $\mathrm{Hom}(\mathscr{X},\mathscr{Y})$ *is a stalkwise dense subset of* $C(Q,\mathscr{Z})$ *in* $\mathscr{Z}$ (*in other words,* $\mathrm{Hom}(\mathscr{X},\mathscr{Y})$ *is a continuity structure in* $\mathscr{Z}$).

$\lhd$ Equivalence of (a) and (b) follows immediately from Corollary 3.2.4. $\rhd$

Obviously, a bundle $\mathscr{Z}$ satisfying condition (a) or (b) of the lemma is unique. This allows us to introduce the following notion.

DEFINITION. The Banach bundle $\mathscr{Z}$ satisfying condition (a) or (b) of the above lemma (if such a bundle exists) is called the *operator bundle* for the CBBs $\mathscr{X}$ and $\mathscr{Y}$ and denoted by the symbol $B(\mathscr{X},\mathscr{Y})$.

The above definition of operator bundle generalizes the analogous notion introduced in [8, 1.2.3] for the case of bundles over extremally disconnected compact Hausdorff spaces.

3.3.2. The following result, repeatedly used throughout the article, provides the basic criterion for existence of an operator bundle.

Theorem. *Let* $\mathscr{X}$ *and* $\mathscr{Y}$ *be* CBBs *over a topological space* Q. *For existence of the bundle* $B(\mathscr{X},\mathscr{Y})$, *it is necessary and sufficient that the pointwise norm of every homomorphism from* $\mathscr{X}$ *into* $\mathscr{Y}$ *be continuous.*

$\lhd$ Necessity for continuity of pointwise norms is evident. Sufficiency of this condition may be explained by using the equivalent definition 3.3.1 (b) of an operator bundle. The stalk $B(\mathscr{X},\mathscr{Y})(q)$ for each point $q \in Q$ is the closure of the subspace $\{H(q) : H \in \mathrm{Hom}(\mathscr{X},\mathscr{Y})\}$ in the Banach space $B(\mathscr{X}(q),\mathscr{Y}(q))$. $\rhd$

By [8, Corollary 2.2.2], in the case of an ample CBB $\mathscr{X}$ over an extremally disconnected compact Hausdorff space Q, the pointwise norm of every homomorphism from $\mathscr{X}$ into an arbitrary CBB $\mathscr{Y}$ over Q is continuous. Proven by using this lemma, by [8, 2.2.3] we see that, in the case indicated, the operator bundle $B(\mathscr{X},\mathscr{Y})$ exists. This allows us to regard criterion 3.3.2 as a generalization of [8, Theorem 2.2.3] to the case of an arbitrary CBB over an arbitrary topological space.

3.3.3. Proposition. *If a CBB $\mathscr{X}$ over a topological space Q has constant finite dimension then, for every CBB $\mathscr{Y}$ over Q, the operator bundle $B(\mathscr{X},\mathscr{Y})$ exists.*

$\lhd$ The claim follows from Corollary 3.2.12 and Theorem 3.3.2. $\rhd$

Examples 3.2.13, with 3.3.2 taken into account, demonstrate that the constant dimension requirement for a bundle $\mathscr{X}$ in the last proposition is essential.

3.3.4. Proposition. *Suppose that a CBB $\mathscr{X}$ over a completely regular topological space Q has constant finite dimension. Then, for every CBB $\mathscr{Y}$ over Q, the equality $B(\mathscr{X},\mathscr{Y})(q) = B(\mathscr{X}(q),\mathscr{Y}(q))$ holds at every point $q \in Q$. In particular, if $\mathscr{X}$ and $\mathscr{Y}$ have constant finite dimension, then $B(\mathscr{X},\mathscr{Y})$ has the same property.*

$\lhd$ Fix a point $q \in Q$ and a linear operator $S \in B(\mathscr{X}(q),\mathscr{Y}(q))$. If we construct a homomorphism $H \in \mathrm{Hom}(\mathscr{X},\mathscr{Y})$ such that $H(q) = S$ then the claim will be proven.

First, observe that if W is a closed neighborhood about q, a section w over W is continuous (locally bounded), and a function $f \in C(Q)$ vanishes outside W, then the global section $f * w$, defined by the formula

$$(f * w)(p) = \begin{cases} f(p)w(p), & p \in W, \\ 0, & p \notin W, \end{cases}$$

is continuous (locally bounded). Hence, in view of [8, Theorem 1.4.4], given a homomorphism $G \in \mathrm{Hom}_W(\mathscr{X},\mathscr{Y})$, the mapping

$$H = f * G : p \in Q \mapsto H(p) \in B(\mathscr{X}(p),\mathscr{Y}(p))$$

is a homomorphism from $\mathscr{X}$ into $\mathscr{Y}$ because the pointwise norm of H is locally bounded and $H \otimes u = f * (G \otimes u) \in C(Q,\mathscr{Y})$ for all $u \in C(Q,\mathscr{X})$.

Recalling the fact that the space Q is completely regular, we can require $f(q) = 1$. Then $H(q) = G(q)$. Therefore, for proving the claim, it suffices to define a homomorphism $G \in \mathrm{Hom}_W(\mathscr{X},\mathscr{Y})$ on any closed neighborhood W about q taking value S at the point q. By [16, Lemma 7], there exists a linear operator $T : \mathscr{X}(q) \to C(Q,\mathscr{X})$ such that, for every $x \in \mathscr{X}(q)$, the inequality $\|x\| \leqslant \|Tx\|$ holds on some neighborhood U about q. Since, for every point $p \in U$, the operator

$$T_p : x \in \mathscr{X}(q) \mapsto (Tx)(p) \in \mathscr{X}(p)$$

is invertible and the dimension of $\mathscr{X}$ is constant, we conclude that the range of T is stalkwise dense in $\mathscr{X}$ on U. By Dupré's Theorem (see [8, 1.3.5]), there exists a collection of sections $\mathscr{V} \subset C(Q, \mathscr{Y})$ such that $\{v(q) : v \in \mathscr{V}\}$ is a basis for the subspace $\operatorname{Im} S \subset \mathscr{Y}(q)$ on the unit sphere. Therefore, by [16, Lemma 7], there is a linear operator $R : \operatorname{Im} S \to C(Q, \mathscr{Y})$ such that the range of R coincides with the linear span of $\mathscr{V}$ and $|\!|\!| Ry |\!|\!| \leqslant 2 \|y\|$ for every $y \in \operatorname{Im} S$ on some neighborhood V about q. By analogy to the definition of the operators T_p, we consider a linear operator $R_r : \operatorname{Im} S \to \mathscr{Y}(p)$ for every point $r \in V$. It is obvious that the operator R_q is invertible and $\|R_r\| \leqslant 2$ for all $r \in V$. At the same time, for all $p \in U$, the estimate $\|T_p^{-1}\| \leqslant 1$ holds.

Finally, take a closed neighborhood $W \subset U \cap V$ about q and, with each element $p \in W$, associate the linear operator

$$G(p) = R_p \circ R_q^{-1} \circ S \circ T_q \circ T_p^{-1} : \mathscr{X}(p) \to \mathscr{Y}(p).$$

By [8, Theorem 1.4.9], the mapping $G : p \in W \mapsto G(p) \in B(\mathscr{X}(p), \mathscr{Y}(p))$ thus obtained is a sought homomorphism, because $G(q) = S$, $|\!|\!| G |\!|\!| \leqslant 2 \|R_q^{-1}\| \|S\| \|T_q\|$, and $G \otimes u \in C(W, \mathscr{Y})$ for all $u \in \operatorname{Im} T$. $\rhd$

3.3.5. Assertion (1) of the following theorem under the assumption $\mathscr{Y} = \mathscr{R}$ presents a particular answer to G. Gierz's question [7, 19, Problem 1, p. 231].

Theorem. *Let $\mathscr{X}$ be a CBB with finite-dimensional stalks over a completely regular Baire space Q and let $\mathscr{Y}$ be a CBB over Q.*

(1) *Given a point q of the everywhere dense set $\bigcup_{n \geqslant 0} \operatorname{int}\{\dim \mathscr{X} = n\}$ (see Proposition 3.2.9) and an operator $T \in B(\mathscr{X}(q), \mathscr{Y}(q))$, there exists a homomorphism $H \in \operatorname{Hom}(\mathscr{X}, \mathscr{Y})$ such that $H(q) = T$ and $|\!|\!| H |\!|\!| \leqslant \|T\|$.*

(2) *Suppose that there is a countable base at a point $q \in Q$ and the bundle $\mathscr{Y}$ has nonzero stalks on an everywhere dense set. The pointwise norms of all elements in $\operatorname{Hom}(\mathscr{X}, \mathscr{Y})$ are continuous at q if and only if the dimension of $\mathscr{X}$ is constant on some neighborhood about q.*

$\lhd$ (1): Let $0 \leqslant n \in \mathbb{Z}$, $q \in \operatorname{int}\{\dim \mathscr{X} = n\}$, let $U \subset \operatorname{int}\{\dim \mathscr{X} = n\}$ be a closed neighborhood about q, and let $T \in B(\mathscr{X}(q), \mathscr{Y}(q))$. From Proposition 3.3.4 and [8, Lemma 1.3.9] we easily infer that there is a homomorphism $G \in \operatorname{Hom}_U(\mathscr{X}, \mathscr{Y})$ such that $G(q) = T$ and $|\!|\!| G |\!|\!| \leqslant \|T\|$. Since the space Q is completely regular, there exists a continuous function $f : Q \to [0,1]$ satisfying the equalities $f(q) = 1$ and $f \equiv 0$ on $Q \backslash U$. It remains to put $H = f * G$ (see the proof of 3.3.4).

(2): Theorem 3.2.12 implies the sufficiency part of the assertion. For proving necessity, suppose that, in every neighborhood about q, there are points at which the dimension of $\mathscr{X}$ is greater than $\dim \mathscr{X}(q) =: m$, and construct a homomorphism $H \in \mathrm{Hom}(\mathscr{X}, \mathscr{Y})$ with discontinuous pointwise norm at q.

By Corollary 3.2.9, the point q belongs to the closure of the open set

$$\bigcup_{n>m} \mathrm{int}\{\dim \mathscr{X} = n\};$$

moreover, the hypotheses imply that the set $\{\dim \mathscr{Y} > 0\}$ is open. Since at the point q there is a countable base, we may take a sequence $(q_n) \subset \bigcup_{n>m} \mathrm{int}\{\dim \mathscr{X} = n\} \cap \{\dim \mathscr{Y} > 0\}$, $q_i \neq q_j$ $(i \neq j)$, convergent to q.

According to Dupré's Theorem [8, 1.3.5], there exist bounded sections $u_1, \dots, u_m \in C(Q, \mathscr{X})$ with linearly independent values $u_1(q), \dots, u_m(q)$. From [7, Proposition 18.1] it follows that the sections are pointwise linearly independent on an open neighborhood U about q. Without loss of generality, we may assume that $q_n \in U$ for all $n \in \mathbb{N}$.

For every $n \in \mathbb{N}$, the inequality $\dim \mathscr{X}(q_n) > m$ and nondegeneracy of the stalk $\mathscr{Y}(q_n)$ allow us to find an operator $T_n \in B(\mathscr{X}(q_n), \mathscr{Y}(q_n))$ such that $T_n \equiv 0$ on $\mathrm{lin}\{u_1(q_n), \dots, u_m(q_m)\}$ and $\|T_n\| = 1$. By (1), for every number $n \in \mathbb{N}$, there is a homomorphism $H_n \in \mathrm{Hom}(\mathscr{X}, \mathscr{Y})$ satisfying the relations $H_n(q_n) = T_n$ and $\|\!|H_n|\!\| \leqslant 1$.

Let $\mathscr{X}_0$ be the CBB over U with continuity structure $\mathrm{lin}\{u_1|_U, \dots, u_m|_U\}$, let $\mathscr{Y}_0 = \mathscr{Y}|_U$, and let $n \in \mathbb{N}$. By Theorem 3.2.12, the mapping $p \in U \mapsto H_n(p)\big|_{\mathrm{lin}\{u_1(p), \dots, u_m(p)\}} \in B(\mathscr{X}_0(p), \mathscr{Y}_0(p))$ has continuous pointwise norm. Therefore, we can take an open neighborhood $V_n \subset U$ about q_n such that

$$\|H_n(p)\big|_{\mathrm{lin}\{u_1(p), \dots, u_m(p)\}}\| < 1/n$$

for all $p \in V_n$.

By Lemma 3.2.5 (1), there exists a sequence (W_n) of open subsets of Q satisfying the conditions $\mathrm{cl}\, W_n \cap \mathrm{cl} \bigcup_{k \neq n} W_k = \varnothing$, $q_n \in W_n$, and

$$\Big(\mathrm{cl} \bigcup_{n \in \mathbb{N}} W_n\Big) \setminus \bigcup_{n \in \mathbb{N}} \mathrm{cl}\, W_n = \{q\}.$$

We additionally require that $W_n \subset V_n$ for all $n \in \mathbb{N}$. Moreover, consider a sequence of continuous functions $f_n : Q \to [0, 1]$ such that $f_n(q_n) = 1$ and $f_n \equiv 0$ in $Q \setminus W_n$. Define

$$H(p) = \begin{cases} f_n(p) H_n(p), & p \in W_n, \\ 0, & p \notin \bigcup_{n \in \mathbb{N}} W_n \end{cases}$$

for all $p \in Q$. It is obvious that $\|\!|H|\!\| \leqslant 1$. Since the space Q is completely regular, the set $N_q = \{u \in C(Q, \mathscr{X}) : u(q) = 0\}$ enlarges the linear span of $\mathrm{lin}\{u_1, \dots, u_m\}$

to a subset of $C(Q, \mathscr{X})$ stalkwise dense in $\mathscr{X}$. By applying [8, Theorem 1.4.9] to this subset, we show that H is a homomorphism from $\mathscr{X}$ into $\mathscr{Y}$.

If $u \in \operatorname{lin}\{u_1, \dots, u_m\}$ then the series $\sum_{n=1}^{\infty} f_n H_n \otimes u$ uniformly converges. Indeed, its terms have disjoint supports, the pointwise norm of u is bounded, and $\|f_n H_n \otimes u\| \leqslant \frac{1}{n}\|u\|$ for all $n \in \mathbb{N}$. Then, by [8, Theorem 1.3.6], the section $H \otimes u$ is continuous as the sum of the series.

Now let $u \in N_q$. The section $H \otimes u$ is continuous on every set $\operatorname{cl} W_n$, $n \in \mathbb{N}$, since $\operatorname{cl} W_n$ is a subset of an open set $Q \setminus \operatorname{cl} \bigcup_{k \neq n} W_k$, and $H \otimes u = f_n H_n \otimes u$ on this subset. If

$$p \in \left(\operatorname{cl} \bigcup_{n \in \mathbb{N}} W_n\right) \setminus \bigcup_{n \in \mathbb{N}} \operatorname{cl} W_n$$

then $p = q$ and the section $H \otimes u$ is continuous at p, since $\|H\| \leqslant 1$ and the function $\|u\|$ is continuous and vanishes at q. Finally, the set $Q \setminus \operatorname{cl} \bigcup_{n \in \mathbb{N}} W_n$ is open and the equality $\|H \otimes u\| \equiv 0$ holds on this set.

Thus, $H \in \operatorname{Hom}(\mathscr{X}, \mathscr{Y})$. Furthermore, $\|H\|(q) = 0$, $\|H\|(q_n) = 1$ for all $n \in \mathbb{N}$, and $q_n \to q$; therefore, the function $\|H\|$ is discontinuous at q. $\rhd$

3.3.6. *Theorem.* *Let $\mathscr{X}$ and $\mathscr{Y}$ be CBBs over a first-countable completely regular Baire space Q. Suppose that all stalks of $\mathscr{X}$ are finite-dimensional and the bundle $\mathscr{Y}$ has nonzero stalks on an everywhere dense subset of Q. Then the operator bundle $B(\mathscr{X}, \mathscr{Y})$ exists if and only if the sets $\{\dim \mathscr{X} = n\}$ are clopen for all $n = 0, 1, 2, \dots$.*

$\lhd$ Sufficiency of the indicated condition for existence of the bundle $B(\mathscr{X}, \mathscr{Y})$ follows from Proposition 3.3.3.

For proving necessity, observe that, by Theorem 3.3.2 and assertion (2) of Theorem 3.3.5, existence of the bundle $B(\mathscr{X}, \mathscr{Y})$ implies that the sets $\{\dim \mathscr{X} = n\}$ are open for all $n = 0, 1, 2, \dots$. It remains to use Lemma 3.2.8. $\rhd$

3.3.7. The following assertion follows immediately from Theorem 3.3.6.

Corollary. *Let $\mathscr{X}$ be a CBB with finite-dimensional stalks over a first-countable connected completely regular Baire topological space Q and let $\mathscr{Y}$ be a CBB over Q with nonzero stalks on an everywhere dense subset of Q. Then existence of the bundle $B(\mathscr{X}, \mathscr{Y})$ is equivalent to the fact that the dimension of $\mathscr{X}$ is constant.*

Observe that the space Q satisfying the hypotheses of the above corollary may fail to be metrizable. It is easy to verify that the Nemytskiĭ plane is such a nonmetrizable space (see [5, 1.2.4, 1.4.5, 2.1.10]).

3.3.8. In the rest of this section, we mainly deal with trivial CBBs. For these CBBs, the existence of the bundle $B(X_Q, Y_Q)$ is closely connected with the question whether the inclusion

$$C(Q, B(X,Y)) \subset \mathrm{Hom}(X_Q, Y_Q)$$

is strict (considered in 3.2.3).

Proposition. *Given Banach spaces X and Y, the bundle $B(\mathscr{X}, \mathscr{Y})$ exists if and only if $C(Q, B(X,Y)) = \mathrm{Hom}(X_Q, Y_Q)$. Moreover, if the bundle $B(X_Q, Y_Q)$ exists then it is equal to the trivial* CBB *with stalk $B(X,Y)$.*

$\triangleleft$ We first prove the second assertion. Let $B(X_Q, Y_Q)$ exist. Since the relations $B(X_Q, Y_Q)(q) \subset B(X_Q(q), Y_Q(q)) = B(X,Y)$ are true at each point $q \in Q$ and the relation $C(Q, B(X,Y)) \subset \mathrm{Hom}(X_Q, Y_Q) = C(Q, B(X_Q, Y_Q))$ holds, every stalk of $B(X_Q, Y_Q)$ coincides with the space $B(X,Y)$. In this case, $C(Q, B(X,Y))$ is a continuity structure for both $B(X,Y)_Q$ and $B(X_Q, Y_Q)$; therefore, these two CBBs coincide (see [8, 2.1.8, 2.1.9]). Whence it is immediate that the equality $C(Q, B(X,Y)) = \mathrm{Hom}(X_Q, Y_Q)$ is necessary for existence of $B(X_Q, Y_Q)$. Sufficiency is evident by Theorem 3.3.2. $\triangleright$

3.3.9. Corollary. *Let X and Y be Banach spaces and let X be finite-dimensional. Then the bundle $B(X_Q, Y_Q)$ exists and, moreover, $B(X_Q, Y_Q) = B(X,Y)_Q$ and $\mathrm{Hom}(X_Q, Y_Q) = C(Q, B(X,Y))$.*

$\triangleleft$ The claim follows from 3.3.2 and 3.3.8. $\triangleright$

3.3.10. Theorem. *Let X be an infinite-dimensional Banach space and let Q be a topological space. Suppose that, for some* CBB *$\mathscr{Y}$ with nonzero stalks, the bundle $B(X_Q, \mathscr{Y})$ exists. Then the space Q is functionally discrete.*

$\triangleleft$ Assume that there exists a not locally constant function in $C(Q)$ and construct a homomorphism H from X_Q into $\mathscr{Y}$ with discontinuous pointwise norm. By the theorem of 3.3.2, the theorem will be thus proven.

Due to Lemma 3.1.13, there exists a weakly* continuous vector valued function $w : Q \to X'$ with bounded and discontinuous pointwise norm. Let q be a discontinuity point of $\|w\|$. Consider a section $v \in C(Q, \mathscr{Y})$ with nonzero value $v(q)$ and define a mapping $H : q \in Q \mapsto H(q) \in B(X, \mathscr{Y}(q))$ by the rule $H(q) : x \in X \mapsto \langle x | w(q) \rangle v(q)$ for all $q \in Q$. Then, for every constant section $u \in C(Q, X_Q)$, the equality $H \otimes u = \langle u | w \rangle v \in C(Q, \mathscr{Y})$ holds. Moreover, $\|H\| = \|w\| \|v\|$. Boundedness of $\|w\|$ implies local boundedness of $\|H\|$. Therefore, $H \in \mathrm{Hom}(X_Q, \mathscr{Y})$ by [8, Theorem 1.4.9]. Finally, since the function $\|w\|$ is discontinuous at q, and the function $\|v\|$ is continuous and nonzero at this point, $\|H\| = \|w\| \|v\| \notin C(Q)$. $\triangleright$

Below (see 3.3.13) we show that, in the last theorem, the necessary condition for existence of the operator bundle $B(X_Q, \mathscr{Y})$ (namely, functional discreteness of Q) is also sufficient in case the Banach space X is separable. In general, this condition is not sufficient (cf. Proposition 3.3.14 as applied to the Banach space X' and the bundle $\mathscr{Y} = \mathscr{R}$).

3.3.11. Proposition. *Let X and Y be Banach spaces, $Y \neq \{0\}$, and let Q be a topological space that is not functionally discrete. The following are equivalent:*

(a) *the Banach bundle $B(X_Q, Y_Q)$ exists;*

(b) $B(X,Y)_Q = B(X_Q, Y_Q)$;

(c) $\mathrm{Hom}(X_Q, Y_Q) = C(Q, B(X,Y))$;

(d) *X is finite-dimensional.*

$\lhd$ Equivalence of (a), (b), and (c) is proven in 3.3.8, (d) follows from (a) by 3.3.10, and (a) follows from (d) by 3.3.9. $\rhd$

3.3.12. Proposition. *Let $\mathscr{X}$ be a CBB over a functionally discrete topological space Q. Suppose that $C(Q, \mathscr{X})$ includes a countable subset stalkwise dense in $\mathscr{X}$. Then, for every CBB $\mathscr{Y}$ over Q, the bundle $B(\mathscr{X}, \mathscr{Y})$ exists.*

$\lhd$ Let $\mathscr{U} \subset C(Q, \mathscr{X})$ be a countable subset stalkwise dense in $\mathscr{X}$. Consider an arbitrary CBB $\mathscr{Y}$ over Q, a homomorphism $H \in \mathrm{Hom}(\mathscr{X}, \mathscr{Y})$, and a point $q \in Q$ and prove continuity for the pointwise norm of H at q. Since the space Q is functionally discrete, there is a neighborhood U about q on which all functions $|\!|\!|u|\!|\!|$, $|\!|\!|H \otimes u|\!|\!|$, $u \in \mathscr{U}$, are constant. In view of stalkwise denseness of $\mathscr{U}$ in $\mathscr{X}$, the equality $|\!|\!|H|\!|\!|(p) = \sup\{|\!|\!|H \otimes u|\!|\!|(p) : u \in \mathscr{U},\ |\!|\!|u|\!|\!|(p) \leqslant 1\}$ holds for every point $p \in Q$; therefore, the function $|\!|\!|H|\!|\!|$ is constant on U and, in particular, $|\!|\!|H|\!|\!|$ is continuous at q. It remains to use Theorem 3.3.2. $\rhd$

3.3.13. Corollary. *Let Q be an arbitrary topological space and let X be a separable infinite-dimensional Banach space. The following are equivalent:*

(a) *for every CBB $\mathscr{Y}$ over Q, the bundle $B(X_Q, \mathscr{Y})$ exists;*

(b) *the bundle $B(X_Q, \mathscr{R})$ exists;*

(c) *the space Q is functionally discrete.*

$\lhd$ The implication (a)$\to$(b) is evident, (c) follows from (b) by 3.3.10, and (a) follows from (c) by 3.3.12. $\rhd$

3.3.14. Proposition. *Let X be a nonseparable Banach space. There exists a functionally discrete normal topological space Q such that, for every CBB $\mathscr{Y}$ over Q with nonzero stalks, the bundle $B(X_Q, \mathscr{Y})$ does not exist.*

$\lhd$ Given a subset $F \subset X$, denote by the symbol $F^{\perp}$ the annihilator of F, i.e., $F^{\perp} = \{x' \in X' : \langle x | x' \rangle = 0 \text{ for all } x \in F\}$. Consider the set

$$\aleph = \{F^{\perp} : F \text{ is a countable subset of } X\}$$

ordered by the rule

$$F_1^\perp \leqslant F_2^\perp \Leftrightarrow F_1^\perp \supset F_2^\perp.$$

It is easy to see that all countable subsets of $\aleph$ have upper bounds. Moreover, $\aleph$ has no greatest element. Indeed, since the space X is nonseparable, for every annihilator $F^\perp \in \aleph$, there exists a nonzero element $x \in X$ outside the closure of the linear span of F. On the other hand, there is a functional in $F^\perp$ with nonzero value at x. Whence, $F^\perp < (F \cup \{x\})^\perp$.

As is shown in 3.1.11, the space $Q := \aleph^\bullet$ is normal and functionally discrete.

Let $\mathscr{Y}$ be an arbitrary CBB over Q with nonzero stalks. Construct a homomorphism $H \in \mathrm{Hom}(X_Q, \mathscr{Y})$ with discontinuous pointwise norm. To this end, consider a section $v \in C(Q, \mathscr{Y})$ taking nonzero value at the point $\infty \in Q$. Since $\{0\} \notin \aleph$, for every element $\alpha \in \aleph$, we may take a norm-one functional $x'_\alpha \in \alpha$. Let $H(\alpha) = v(\alpha) \otimes x'_\alpha$ for all $\alpha \in \aleph$ and let $H(\infty) = 0$. Then, by [8, Theorem 1.4.9], the mapping H is a homomorphism, since, for every constant section $u_x \equiv x$, $x \in X$, the section $H \otimes u_x$ vanishes on the interval $(\{x\}^\perp, \infty]$; therefore, $H \otimes u_x$ is continuous. At the same time, the pointwise norm of H is discontinuous at ∞. Consequently, by Theorem 3.3.2 the bundle $B(X_Q, \mathscr{Y})$ does not exist. $\rhd$

3.3.15. Lemma. *Let $\aleph$ be an upward-directed set without greatest element and let $\mathscr{X}$ be a CBB over $\aleph^\bullet$ (see 3.1.11). Suppose that in $C(\aleph^\bullet, \mathscr{X})$ there is a stalkwise dense subset such that every subset of $\aleph$ of the same cardinality has an upper bound. Then, for every CBB $\mathscr{Y}$ over $\aleph^\bullet$, the bundle $B(\mathscr{X}, \mathscr{Y})$ exists.*

$\lhd$ Let $\mathscr{U}$ be a subset of $C(\aleph^\bullet, \mathscr{X})$ satisfying the hypotheses of the lemma.

Consider an arbitrary CBB $\mathscr{Y}$ over $\aleph^\bullet$ and verify continuity for the pointwise norm of an arbitrary homomorphism $H \in \mathrm{Hom}(\mathscr{X}, \mathscr{Y})$. Hence, by Theorem 3.3.2, the assertion will be proven.

For every element $u \in \mathscr{U}$, take an $\alpha_u \in \aleph$ such that $|\!|\!|u|\!|\!|(\alpha) = |\!|\!|u|\!|\!|(\infty)$ and $|\!|\!|H \otimes u|\!|\!|(\alpha) = |\!|\!|H \otimes u|\!|\!|(\infty)$ for all $\alpha \geqslant \alpha_u$ (see Remark 3.1.11 (1)). Then, for every $u \in \mathscr{U}$, the two latter equalities hold for $\alpha \geqslant \beta$, where β is an upper bound for the set $\{\alpha_u : u \in \mathscr{U}\}$. Since $\mathscr{U}$ is stalkwise dense in $\mathscr{X}$, the value of the norm $|\!|\!|H|\!|\!|$ can be calculated at every point $\alpha \in \aleph^\bullet$ by the formula $|\!|\!|H|\!|\!|(\alpha) = \sup\{|\!|\!|H \otimes u|\!|\!|(\alpha) : u \in \mathscr{U}, |\!|\!|u|\!|\!|(\alpha) \leqslant 1\}$. From this formula we readily see that, for $\alpha \geqslant \beta$, the pointwise norm of H takes the value $|\!|\!|H|\!|\!|(\alpha) = |\!|\!|H|\!|\!|(\infty)$ and, therefore, is continuous. $\rhd$

Corollary. *Given a Banach space X, there is a nondiscrete normal topological space Q such that, for every CBB $\mathscr{Y}$ over Q, the bundle $B(X_Q, \mathscr{Y})$ exists.*

$\lhd$ It is sufficient to take $Q = \aleph^\bullet$, where $\aleph$ is a cardinal greater than the cardinality of X, and use Lemma 3.3.15. $\rhd$

3.4. The Dual of a Banach Bundle

In this section, we consider the problem of existence and the properties of the bundle $\mathscr{X}'$ dual to a Banach bundle $\mathscr{X}$.

In 3.4.2, we state various necessary and sufficient conditions for existence of a dual bundle. All assertions in the subsection are direct consequences of results of the preceding section. Proposition 3.4.3 asserts existence for a dual bundle of a CBB with Hilbert stalks.

One of the natural steps in studying the notion of a dual bundle is establishing norming duality relations between the bundles $\mathscr{X}$ and $\mathscr{X}'$. Item 3.4.5 is devoted to this subject. As a preliminary, in 3.4.4, we discuss the condition that the stalks of a CBB are stalkwise normed by the values of the corresponding homomorphisms. Unfortunately, we have to leave open the question whether this condition always holds, restricting ourselves to listing certain situations in which the condition is satisfied.

In 3.4.6–3.4.9, the interrelation is considered between separability of a distinguished stalk of a CBB and finiteness of the dimension of the stalks of the bundle or of the stalks of its dual.

The rest of the section (3.4.10–3.4.15) is devoted to studying the second dual bundle, $\mathscr{X}''$. Among the topics considered here, are existence of $\mathscr{X}''$, isometry between the bundles under study, and embedding of a Banach bundle into its second dual.

3.4.1. DEFINITION. Let $\mathscr{X}$ be a continuous Banach bundle. The bundle $B(\mathscr{X},\mathscr{R})$ (whenever the latter exists) is called the *dual* of $\mathscr{X}$ and denoted by the symbol $\mathscr{X}'$. If the bundle $\mathscr{X}'$ exists then we say that $\mathscr{X}$ *has the dual bundle.*

By Theorem 3.3.2, the dual $\mathscr{X}'$ exists if and only if the pointwise norms of all homomorphisms from $\mathscr{X}$ into $\mathscr{R}$ are continuous.

3.4.2. *Proposition.* *The following are true:*

(1) *Every* CBB $\mathscr{X}$ *with constant finite dimension over a topological space Q has the dual bundle. Moreover, if Q is completely regular then $\mathscr{X}'(q) = \mathscr{X}(q)'$ for all $q \in Q$.*

(2) *A* CBB $\mathscr{X}$ *with finite-dimensional stalks over a first-countable completely regular Baire topological space has the dual bundle if and only if $\{\dim \mathscr{X} = n\}$ is a clopen set for every $n = 0, 1, 2, \dots$.*

(3) *Suppose that a trivial* CBB *with stalk X has the dual bundle. Then the latter is the trivial* CBB *with stalk X'.*

(4) *If a trivial* CBB *with infinite-dimensional stalk over a topological space Q has the dual bundle then Q is functionally discrete (if, in addition, Q is completely regular then all of its countable subsets are closed).*

(5) *For every nonseparable Banach space X, there exists a functionally discrete topological space Q such that the* CBB *X_Q has no dual bundle.*

(6) *A trivial* CBB *with infinite-dimensional separable stalk over a topological space Q has the dual bundle if and only if Q is functionally discrete.*

(7) *For every Banach space X, there exists a nondiscrete normal topological space Q such that the* CBB *X_Q has the dual bundle.*

(8) *If a topological space Q is not functionally discrete then, for every Banach space X, the following are equivalent:*
   (a) *the dual $(X_Q)'$ exists;*
   (b) $(X')_Q = (X_Q)'$;
   (c) $C(Q, X') = \mathrm{Hom}(X_Q, \mathscr{R})$;
   (d) *X is finite-dimensional.*

◁ Assertions (1)–(8) follow directly from 3.3.3 and 3.3.4, 3.3.6, 3.3.8, 3.3.10, 3.3.14, 3.3.13, Corollary 3.3.15, and 3.3.11. ▷

REMARK. Examples 3.2.13 (1)–(3), with 3.3.2 taken into account, imply that the constant dimension requirement in assertion (1) of the above proposition is essential for existence of a dual bundle.

3.4.3. Lemma. *Let Q be a topological space and let $\mathscr{X}$ be a* CBB *over Q with Hilbert stalks (i.e., all stalks of $\mathscr{X}$ are Hilbert spaces). For every global section u of $\mathscr{X}$ and every point $q \in Q$ put*

$$h(u)(q) = \langle \cdot, u(q) \rangle \in \mathscr{X}(q)'.$$

Then $h[C(Q, \mathscr{X})] \subset \mathrm{Hom}(\mathscr{X}, \mathscr{R})$. Moreover, $h[C(Q, \mathscr{X})]$ is a continuity structure in the (discrete) Banach bundle with stalks $\mathscr{X}(q)'$ $(q \in Q)$.

◁ By [8, 1.4.4], the inclusion $h[C(Q, \mathscr{X})] \subset \mathrm{Hom}(\mathscr{X}, \mathscr{R})$ follows from the relations

$$\langle u_1 | h(u_2) \rangle = \langle u_1(\cdot), u_2(\cdot) \rangle = \frac{1}{2}\left(|\!|\!| u_1 |\!|\!|^2 + |\!|\!| u_2 |\!|\!|^2 - |\!|\!| u_1 - u_2 |\!|\!|^2 \right) \in C(Q),$$

$$|\!|\!| h(u_2) |\!|\!| = |\!|\!| u_2 |\!|\!|$$

valid for all $u_1, u_2 \in C(Q, \mathscr{X})$. The second assertion follows from the Riesz Theorem. ▷

Proposition. *Let $\mathscr{X}$ be a CBB with Hilbert stalks. If the dual $\mathscr{X}'$ exists then $\mathscr{X}'$ is isometric to $\mathscr{X}$* (see [8, 1.4.12]).

$\triangleleft$ Let Q be a topological space and let $\mathscr{X}$ be a CBB over Q with Hilbert stalks. Consider a CBB $\mathscr{Y}$ with stalks $\mathscr{Y}(q) = \mathscr{X}(q)'$ $(q \in Q)$ and continuity structure $\mathscr{C} = h[C(Q, \mathscr{X})]$ (see the previous lemma). By [8, Theorem 1.4.12 (3)], the bundles $\mathscr{X}$ and $\mathscr{Y}$ are isometric. Stalkwise denseness of $\mathscr{C}$ in $\mathscr{Y}$ and the relations $\mathscr{C} \subset \mathrm{Hom}(\mathscr{X}, \mathscr{R}) = C(Q, \mathscr{X}')$ imply that, at every point $q \in Q$, the stalks $\mathscr{X}'(q)$ and $\mathscr{Y}(q)$ coincide and $\mathscr{C}$ is a continuity structure in $\mathscr{X}'$, i.e., $\mathscr{X}' = \mathscr{Y}$. $\triangleright$

3.4.4. DEFINITION. Let $\mathscr{X}$ be a CBB over a topological space Q. Say that $\mathrm{Hom}(\mathscr{X}, \mathscr{R})$ *norms* $\mathscr{X}$ *on a subset* $D \subset Q$ if, for every point $q \in D$ and every $x \in \mathscr{X}(q)$, the equality $\|x\| = \sup\{|\langle x|H(q)\rangle| : H \in \mathrm{Hom}(\mathscr{X}, \mathscr{R}),\ \|\!|H|\!\| \leqslant 1\}$ holds. Say that $\mathrm{Hom}(\mathscr{X}, \mathscr{R})$ *norms* $\mathscr{X}$ if $\mathrm{Hom}(\mathscr{X}, \mathscr{R})$ norms $\mathscr{X}$ on Q.

We are not aware of an example of a CBB $\mathscr{X}$ for which $\mathrm{Hom}(\mathscr{X}, \mathscr{R})$ does not norm $\mathscr{X}$. (Moreover, we do not know if there exists a nonzero Banach bundle whose dual is zero.) At present, we can only indicate some classes of Banach bundles $\mathscr{X}$ for which $\mathrm{Hom}(\mathscr{X}, \mathscr{R})$ does norm $\mathscr{X}$. The following bundles fall in such a class:

(1) a CBB $\mathscr{X}$ over a topological space Q such that, for every $q \in Q$, the set $\{H(q) : H \in \mathrm{Hom}(\mathscr{X}, \mathscr{R})\} \subset \mathscr{X}(q)'$ norms $\mathscr{X}(q)$ and, for every homomorphism $H \in \mathrm{Hom}(\mathscr{X}, \mathscr{R})$, there is a homomorphism $G \in \mathrm{Hom}(\mathscr{X}, \mathscr{R})$ such that $G(q) = H(q)$ and $\|\!|G|\!\| \in C(Q)$;

(2) a CBB $\mathscr{X}$ over a completely regular topological space Q satisfying the following conditions: for every $q \in Q$, the set $\{H(q) : H \in \mathrm{Hom}(\mathscr{X}, \mathscr{R})\} \subset \mathscr{X}(q)'$ norms $\mathscr{X}(q)$ and, for every homomorphism $H \in \mathrm{Hom}(\mathscr{X}, \mathscr{R})$, there is a homomorphism $G \in \mathrm{Hom}(\mathscr{X}, \mathscr{R})$ such that $G(q) = H(q)$ and the pointwise norm of G is continuous at q;

(3) a trivial CBB;

(4) a CBB with constant finite dimension over a completely regular topological space;

(5) a CBB over a compact topological space or a locally compact Hausdorff topological space which admits a countable stalkwise dense set of continuous sections;

(6) a CBB with finite-dimensional stalks over a metrizable locally compact space;

(7) a CBB with Hilbert stalks;

(8) a CBB over a regular extremally disconnected topological space;

(9) a CBB over $\overline{\mathbb{N}}$ with separable stalk at ∞;

(10) a CBB $\mathscr{X}$ over a Hausdorff topological space with finitely many nonisolated points such that the stalks of $\mathscr{X}$ at these points are separable;

(11) the dual of a CBB.

◁ A proof of the fact that $\mathrm{Hom}(\mathscr{X},\mathscr{R})$ norms $\mathscr{X}$ in cases (1) and (2) can be easily obtained by multiplying the homomorphism G by a suitable element of $C(Q)$.

Cases (3), (4), and (7) are easily reduces to case (1) with the help of Corollary 3.2.3, Proposition 3.4.2 (1), and Lemma 3.4.3 respectively.

Case (5) for a compact topological space is considered in [7, 19.16], and the case of a locally compact Hausdorff (and, hence, completely regular) space is reduced to the case of a compact space by employing a compact neighborhood about an arbitrary point q and multiplying the homomorphism by a continuous real-valued function equal to unity at q and vanishing outside the neighborhood under consideration. By analogous reasoning, case (6) can be reduced to (5) with the help of assertion [7, 19.5 (iii)].

(8): Let $\mathscr{X}$ be a CBB over a regular extremally disconnected topological space D. Consider an extremally disconnected compact space Q that includes D as an everywhere dense subset, and let $\beta\mathscr{X}$ be the Stone–Čech extension of $\mathscr{X}$ onto Q (see [8, 1.1.4, 2.5.10]). Denote by $\overline{\beta\mathscr{X}}$ the ample hull of $\beta\mathscr{X}$ (see [8, 2.1.5]). With every homomorphism $\overline{H} \in \mathrm{Hom}(\overline{\beta\mathscr{X}},\mathscr{R})$ associate the mapping $H : q \in Q \mapsto \overline{H}(q)|_{\beta\mathscr{X}(q)}$, $q \in Q$. From [8, 1.4.4] it follows that $H \in \mathrm{Hom}(\beta\mathscr{X},\mathscr{R})$. Applying [8, Theorem 2.3.3 (1)] to the bundle $\overline{\beta\mathscr{X}}$, we conclude that $\mathrm{Hom}(\beta\mathscr{X},\mathscr{R})$ norms $\beta\mathscr{X}$. It remains to observe that $\{H|_D : H \in \mathrm{Hom}(\beta\mathscr{X},\mathscr{R})\} \subset \mathrm{Hom}_D(\mathscr{X},\mathscr{R})$.

(10): If a Hausdorff topological space Q has finitely many nonisolated points then, as is easily seen, each of these points is separated from the other nonisolated points by a clopen neighborhood. Consequently, without loss of generality, we may assume that Q has a single nonisolated point q.

Let $\mathscr{X}$ be a CBB over Q with the stalk $\mathscr{X}(q)$ separable. It is sufficient, given an $x' \in \mathscr{X}(q)'$, $\|x'\| < 1$, to construct a homomorphism $H \in \mathrm{Hom}(\mathscr{X},\mathscr{R})$ taking the value $H(q) = x'$ and satisfying the inequality $|\!|\!|H|\!|\!| \leqslant 1$.

Consider a countable system $\{x_n : n \in \mathbb{N}\}$ of linearly independent elements in $\mathscr{X}(q)$ whose linear span is everywhere dense in $\mathscr{X}(q)$ and, employing Dupré's Theorem (see [8, 1.3.5]), with each number $n \in \mathbb{N}$ associate a section $u_n \in C(Q,\mathscr{X})$ passing through x_n at q. By [7, Proposition 18.1], for every $n \in \mathbb{N}$, there exists a neighborhood U_n about q such that the sections $u_1,\dots,u_n$ are pointwise linearly independent over U_n. For all $n \in \mathbb{N}$ and $p \in U_n$, define a functional $y_n(p) :$ $\mathrm{lin}\{u_1(p),\dots,u_n(p)\} \to \mathbb{R}$ by the formula $\langle u_i(p)|y_n(p)\rangle = \langle u_i(q)|x'\rangle$, $i = 1,\dots,n$.

Since $\|x'\| < 1$, in view of [16, Lemma 7], each neighborhood U_n about q can be replaced by a smaller neighborhood V_n so that the inequalities $\|y_n(p)\| \leqslant 1$ be valid for all $p \in V_n$. Without loss of generality, we may assume that $V_n \supset V_{n+1}$ for all $n \in \mathbb{N}$.

The fact that the set $\{u_n : n \in \mathbb{N}\}$ is pointwise linearly independent over

$$V_\infty = \bigcap_{n\in\mathbb{N}} V_n$$

allows us, for every point $p \in V_\infty$, to define a functional $y_\infty(p) : \mathrm{lin}\{u_n(p) : n \in \mathbb{N}\} \to \mathbb{R}$ as a common extension of the functionals $y_n(p)$, $n \in \mathbb{N}$, i.e., to put $\langle u_n(p)|y_\infty(p)\rangle = \langle u_n(q)|x'\rangle$ for all $n \in \mathbb{N}$. Observe that $\|y_\infty(p)\| \leqslant 1$ for $p \in V_\infty$.

Define

$$H(p) := \begin{cases} 0, & p \notin V_1; \\ \overline{y}_n(p), & p \in V_n \backslash V_{n+1}; \\ \overline{y}_\infty(p), & p \in V_\infty, \end{cases}$$

where $\overline{y}_n(p)$, $1 \leqslant n \leqslant \infty$, is an arbitrary extension of $y_n(p)$ onto the entire stalk $\mathscr{X}(p)$ with norm preserved. It is clear that $H(q) = x'$ and $|\!|\!|H|\!|\!| \leqslant 1$.

Denote by $\mathscr{U}$ the set $\mathrm{lin}\{u_n : n \in \mathbb{N}\}$ complemented by all sections with singleton supports. Obviously, the set $\mathscr{U}$ is stalkwise dense in $\mathscr{X}$ and, for each $u \in \mathscr{U}$, the function $\langle u|H\rangle$ is constant on some neighborhood about q and, hence, continuous. Consequently, by Theorem [8, 1.4.4], the mapping H is a homomorphism.

(9): This is a particular case of (10).

(11): Let Q be a topological space and let $\mathscr{X}$ be a CBB over Q which has the dual bundle. From [8, 1.3.9] it follows that, for every point $q \in Q$ and every functional $x' \in \mathscr{X}'(q)$, the relation

$$\|x'\| = \sup\{\langle u(q)|x'\rangle : u \in C(Q, \mathscr{X})\}$$

holds. On the other hand, by [8, Theorem 1.4.4], for each section $u \in C(Q, \mathscr{X})$, the mapping

$$u'' : q \in Q \mapsto u(q)|_{\mathscr{X}'(q)}$$

belongs to $\mathrm{Hom}(\mathscr{X}', \mathscr{R})$ and, moreover, $\|u''\| \leqslant \|u\|$. Consequently, $\mathrm{Hom}(\mathscr{X}', \mathscr{R})$ norms $\mathscr{X}'$. ▷

3.4.5. Assertion (3) of the following proposition gives a positive answer to G. Gierz's question [7, 19, Problem 2, p. 231] for the bundles 3.4.4 (1)–(11) as well as for bundles with finite-dimensional stalks over completely regular Baire spaces (see Theorem 3.3.5 (1)).

Proposition. *Let $\mathscr{X}$ be a CBB over a topological space Q.*

(1) *Suppose that $\mathscr{X}$ has the dual bundle. Then, for every point $q \in Q$ and every element $x' \in \mathscr{X}'(q)$, the equality*

$$\|x'\| = \sup\left\{\left|\langle u(q)|x'\rangle\right| : u \in C(Q,\mathscr{X}),\ |\!|\!|u|\!|\!| \leqslant 1\right\}$$

holds. In particular, for every section $u' \in C(Q,\mathscr{X}')$, the relation

$$|\!|\!|u'|\!|\!| = \sup\left\{\left|\langle u|u'\rangle\right| : u \in C(Q,\mathscr{X}),\ |\!|\!|u|\!|\!| \leqslant 1\right\}$$

holds in the vector lattice $C(Q)$.

Suppose that $\mathrm{Hom}(\mathscr{X},\mathscr{R})$ *norms* $\mathscr{X}$ *on an everywhere dense subset of* Q.

(2) *For every section $u \in C(Q,\mathscr{X})$, the relation*

$$|\!|\!|u|\!|\!| = \sup\left\{\left|\langle u|H\rangle\right| : H \in \mathrm{Hom}(\mathscr{X},\mathscr{R}),\ |\!|\!|H|\!|\!| \leqslant 1\right\}$$

holds in the vector lattice $C(Q)$.

(3) *The uniform norm of every section $u \in C^b(Q,\mathscr{X})$ is calculated by the formula*

$$\|u\|_\infty = \sup\left\{\left\|\langle u|H\rangle\right\|_\infty : H \in \mathrm{Hom}(\mathscr{X},\mathscr{R}),\ |\!|\!|H|\!|\!| \leqslant 1\right\}.$$

$\lhd$ (1): Since $x' \in \mathscr{X}(q)'$ and the set $C(Q,\mathscr{X})$ is stalkwise dense in $\mathscr{X}$, there is a sequence of sections $(u_n) \subset C(Q,\mathscr{X})$ such that $|\!|\!|u_n|\!|\!|(q) \leqslant 1$ and $\|x'\| - 1/n \leqslant \langle u_n(q)|x'\rangle \leqslant \|x'\|$ for all $n \in \mathbb{N}$. It remains to observe that, by [8, Lemma 1.3.9], for every n, there is a section $v_n \in C(Q,\mathscr{X})$ satisfying the relations $v_n(q) = u_n(q)$ and $|\!|\!|v_n|\!|\!| \leqslant 1$.

(2): Let D be an everywhere dense subset of Q on which $\mathrm{Hom}(\mathscr{X},\mathscr{R})$ norms $\mathscr{X}$ and consider an arbitrary section $u \in C(Q,\mathscr{X})$ and put

$$\mathscr{F} = \left\{\langle u|H\rangle : H \in \mathrm{Hom}(\mathscr{X},\mathscr{R}),\ |\!|\!|H|\!|\!| \leqslant 1\right\}.$$

It is clear that $|\!|\!|u|\!|\!|$ is an upper bound for $\mathscr{F}$. If $g \in C(Q)$ is an arbitrary upper bound of $\mathscr{F}$ then it is easy to see that, for every point $q \in D$,

$$g(q) \geqslant \sup_{f\in\mathscr{F}} f(q) = |\!|\!|u|\!|\!|(q);$$

hence, $g \geqslant |\!|\!|u|\!|\!|$.

(3): Let $u \in C^b(Q,\mathscr{X})$. It is clear that

$$\|u\|_\infty \geqslant \sup\left\{\left\|\langle u|H\rangle\right\|_\infty : H \in \mathrm{Hom}(\mathscr{X},\mathscr{R}),\ |\!|\!|H|\!|\!| \leqslant 1\right\}.$$

To prove the assertion, for an arbitrary $\varepsilon > 0$, find a homomorphism H belonging to $\mathrm{Hom}(\mathscr{X},\mathscr{R})$ with $|||H||| \leqslant 1$ and such that $\|u\|_\infty - \varepsilon < \|\langle u|H\rangle\|_\infty$.

Consider a point $q \in Q$ satisfying the inequality $|||u|||(q) > \|u\|_\infty - \varepsilon$ and a neighborhood U about this point on which $|||u||| > \|u\|_\infty - \varepsilon$. Since $\mathrm{Hom}(\mathscr{X},\mathscr{R})$ norms $\mathscr{X}$ on an everywhere dense subset of Q, there is a $p \in U$ such that

$$|||u|||(p) = \|u(p)\| = \sup\left\{\left|\langle u(p)|H(p)\rangle\right| : H \in \mathrm{Hom}(\mathscr{X},\mathscr{R}),\ |||H||| \leqslant 1\right\};$$

therefore, $\|u\|_\infty - \varepsilon < \left|\langle u(p)|H(p)\rangle\right|$ for some homomorphism $H \in \mathrm{Hom}(\mathscr{X},\mathscr{R})$, $|||H||| \leqslant 1$. Consequently, $\|u\|_\infty - \varepsilon < \|\langle u|H\rangle\|_\infty$. ▷

3.4.6. Theorem. *Let Q be a completely regular topological space and let $q \in Q$ be a nonisolated point at which there is a countable base. Suppose that a CBB $\mathscr{X}$ over Q has the dual bundle. Then separability of the stalk $\mathscr{X}(q)$ implies that the stalk $\mathscr{X}'(q)$ is finite-dimensional.*

◁ Suppose that the stalk $\mathscr{X}(q)$ is separable and the stalk $\mathscr{X}'(q)$ is infinite-dimensional. We will construct a homomorphism H from $\mathscr{X}$ into $\mathscr{R}$ with discontinuous norm and thus, according to Theorem 3.3.2, obtain a contradiction with the hypotheses.

Let a set $\{x_n : n \in \mathbb{N}\}$ be everywhere dense in $\mathscr{X}(q)$ and let (x'_n) be a weakly* null sequence of elements in $\mathscr{X}'(q)$ such that $\|x'_n\| = 1$ for every $n \in \mathbb{N}$ (see 3.1.3). We assume that $|\langle x_i|x'_n\rangle| < 1/n$ for $i = 1,\dots,n$, since this can be fulfilled by passing to a subsequence. Making use of Dupré's Theorem (see [8, 1.3.5]), for every $n \in \mathbb{N}$, consider sections $u_n \in C(Q,\mathscr{X})$ and $v_n \in C(Q,\mathscr{X}')$ such that $u_n(q) = x_n$ and $v_n(q) = x'_n$.

Let $(U_n)_{n\in\mathbb{N}}$ be a neighborhood base at q. Since Q is a Hausdorff space, by induction we can construct a new neighborhood base $(V_n)_{n\in\mathbb{N}}$ at q such that, for every $n \in \mathbb{N}$, the following conditions hold: $V_{n+1} \subset V_n \cap U_1 \cap \dots \cap U_n$, the difference $V_n \setminus V_{n+1}$ contains a point q_n together with an open neighborhood W_n about q_n, and the estimates $1/2 < |||v_n||| < 2$ and $|\langle u_i|v_n\rangle| < 1/n$, $i = 1,\dots,n$, hold on V_n. Show that, for every continuous section $u \in C(Q,\mathscr{X})$ and an arbitrary $\varepsilon > 0$, for n large enough, the inequality $|\langle u|v_n\rangle| < \varepsilon$ holds on V_n. Indeed, let $\|u(q) - x_k\| < \varepsilon/4$ and $1/l < \varepsilon/2$ for some $k, l \in \mathbb{N}$. Take an element V_m of the constructed neighborhood base about q on which $|||u - u_k||| < \varepsilon/4$. Then, for every $n \geqslant \max\{k, l, m\}$, the following relations hold on V_n:

$$\begin{aligned}|\langle u|v_n\rangle| &\leqslant |\langle u - u_k \mid v_n\rangle| + |\langle u_k|v_n\rangle| \\ &< |||u - u_k|||\,|||v_n||| + 1/n < \frac{\varepsilon}{4}\cdot 2 + \frac{\varepsilon}{2} = \varepsilon.\end{aligned}$$

Now define a mapping $H : p \in Q \mapsto H(p) \in \mathscr{X}'(p)$. Put $H(p) = 0 \in \mathscr{X}(p)'$ whenever $p \notin \bigcup_{n\in\mathbb{N}} W_n$ and, for every $n \in \mathbb{N}$, put $H|_{W_n} = (f_n v_n)|_{W_n}$, where $f_n : Q \to [0,1]$ is a continuous function equal to 1 at q_n and vanishing outside W_n.

The function $\langle u|v\rangle : Q \to \mathbb{R}$ is continuous as the pointwise sum of the series $\sum_{n=1}^{\infty} f_n\langle u|v_n\rangle$ that uniformly converges due to pairwise disjointness of the sets W_n $(n \in \mathbb{N})$ and the relations $\operatorname{supp} f_n \subset W_n$ and $\sup_{W_n} |\langle u|v_n\rangle| \leqslant \sup_{V_n} |\langle u|v_n\rangle| \to 0$ as $n \to \infty$.

Thus, H is a homomorphism, since $\|\!|H|\!\| \leqslant 2$ (see [8, 1.4.4]). At the same time, $\|\!|H|\!\|(q_n) = |f_n(q_n)|\|\!|v_n|\!\|(q_n) = \|\!|v_n|\!\|(q_n) > 1/2$ for every $n \in \mathbb{N}$. Moreover, $q_n \to q$ and $\|\!|H|\!\|(q) = 0$. Consequently, the homomorphism H has discontinuous pointwise norm. $\triangleright$

3.4.7. Corollary. *Let Q be a completely regular topological space and let $q \in Q$ be a nonisolated point at which there is a countable base. Suppose that a CBB $\mathscr{X}$ over Q with Hilbert stalks has the dual bundle. Then the stalk $\mathscr{X}(q)$ is separable if and only if it is finite-dimensional.*

Thus, if a CBB $\mathscr{X}$ with Hilbert stalks over a completely regular topological space has the dual bundle, then the stalk of $\mathscr{X}$ at a nonisolated point with a countable base cannot be isometric to ℓ^2.

3.4.8. Proposition. *Let $Q = \overline{\mathbb{N}}$ be the one-point compactification of the set of naturals. A CBB $\mathscr{X}$ over Q with the stalk $\mathscr{X}(\infty)$ separable has the dual bundle if and only if the dimension of $\mathscr{X}$ is finite and constant on some neighborhood about ∞.*

$\triangleleft$ Sufficiency follows from Proposition 3.4.2 (2). Establish necessity. Suppose that the bundle $\mathscr{X}$ under consideration has dual bundle. Then, due to 3.4.6, the space $\mathscr{X}'(\infty)$ is finite-dimensional, whence, in view of 3.4.4 (10), it follows that the stalk $\mathscr{X}(\infty)$ is finite-dimensional too. Put $m = \dim \mathscr{X}(\infty)$ and consider sections $u_1, \dots, u_m \in C(Q, \mathscr{X})$ with linearly independent values $u_1(\infty), \dots, u_m(\infty)$ which exist by the Dupré Theorem (see [8, 1.3.5]). According to [7, 18.1], the sections $u_1, \dots, u_m$ are pointwise linearly independent over some neighborhood U about ∞ and, hence, $\dim \mathscr{X} \geqslant m$ on U.

Assume that there is no neighborhood about ∞ on which the dimension of $\mathscr{X}$ is constant. Then there exists a strictly increasing sequence of naturals n_k such that $\dim \mathscr{X}(n_k) > m$ for all $k \in \mathbb{N}$. Given a $k \in \mathbb{N}$, choose a functional $x'_k \in \mathscr{X}(n_k)'$ satisfying the equalities $\|x'_k\| = 1$ and $\langle u_1(n_k)|x'_k\rangle = \dots = \langle u_m(n_k)|x'_k\rangle = 0$. Introduce a mapping $H : q \in Q \mapsto H(q) \in \mathscr{X}(q)'$ as follows:

$$H(q) = \begin{cases} x'_k, & q = n_k; \\ 0, & q \notin \{n_k : k \in \mathbb{N}\}. \end{cases}$$

It is clear that $\|\!|H|\!\| \leqslant 1$. Denote by $\mathscr{U}$ the set $\operatorname{lin}\{u_1, \dots, u_m\}$ supplemented by all sections with singleton supports. Obviously, $\mathscr{U}$ is stalkwise dense in $\mathscr{X}$ and, for every $u \in \mathscr{U}$, the function $\langle u|H\rangle$ vanishes on a neighborhood about ∞

and, hence, continuous. Consequently, by [8, Theorem 1.4.4], the mapping H is a homomorphism, which, with 3.3.2 taken into account, contradicts existence of $\mathscr{X}'$ in view of the fact that the pointwise norm of H is discontinuous. ▷

3.4.9. The one-point compactification Q of the set of naturals can be regarded as the simplest topological space which is, on the one hand, classical (completely regular, metrizable, compact, etc.) and, on the other hand, nontrivial (nondiscrete, not antidiscrete, etc.). As Proposition 3.4.8 asserts, a CBB $\mathscr{X}$ over Q with the stalk $\mathscr{X}(\infty)$ separable has the dual bundle if and only if the dimension of $\mathscr{X}$ is finite and constant on some neighborhood about ∞. Moreover, due to Proposition 3.4.2 (4), every trivial bundle over Q with infinite-dimensional stalk has no dual bundle. Show that, nevertheless, there exists a CBB over Q with infinite-dimensional stalk at ∞ which has the dual bundle.

EXAMPLE. We construct a CBB $\mathscr{X}$ over $Q = \overline{\mathbb{N}}$ possessing the following properties:

(a) all stalks of $\mathscr{X}$ on $\mathbb{N}$ are finite-dimensional and $\mathscr{X}(\infty)$ is nonseparable;
(b) $\mathscr{X}'$ exists;
(c) the inclusion $\mathscr{X}'(\infty) \subset \mathscr{X}(\infty)'$ is strict;
(d) $\mathrm{Hom}(\mathscr{X}, \mathscr{R}) = C(Q, \mathscr{X}')$ norms $\mathscr{X}$.

For every natural n, consider the element $e_n = \chi_{\{n\}} \in \ell^\infty$ and the coordinate functional $\delta_n \in (\ell^\infty)'$, $\langle x | \delta_n \rangle = x(n)$ for all $x \in \ell^\infty$.

Denote by $\bar{\ell^1}$ the image of ℓ^1 under the natural isometric embedding of this space into $(\ell^\infty)'$. It is clear that $\delta_n \in \bar{\ell^1}$ for all $n \in \mathbb{N}$. Put $\mathscr{X}(\infty) = \ell^\infty$ and $\mathscr{X}(n) = \mathrm{lin}\{e_1, \ldots, e_n\}$, $n \in \mathbb{N}$.

Given an element $x \in \ell^\infty$, define a section u_x of $\mathscr{X}$ as follows:

$$u_x(q) = \begin{cases} (x(1), \ldots, x(q), 0, 0 \ldots), & q \in \mathbb{N}, \\ x, & q = \infty. \end{cases}$$

It is easy to see that the totality $\mathscr{C} = \{u_x : x \in \ell^\infty\}$ is a continuity structure in $\mathscr{X}$ which makes $\mathscr{X}$ a CBB.

By construction it is immediate that $\mathscr{X}$ possesses property (a).

(b), (c): For all $n \in \mathbb{N}$ and $f \in \mathscr{X}(n)'$, put

$$\langle x \mid \bar{f} \rangle = \langle (x(1), \ldots, x(n), 0, 0, \ldots) \mid f \rangle, \quad x \in \ell^\infty.$$

It is clear that, for each $n \in \mathbb{N}$, the correspondence $f \mapsto \bar{f}$ performs an isometric embedding of $\mathscr{X}(n)'$ into $\bar{\ell^1}$.

Let H be an arbitrary homomorphism from $\mathscr{X}$ into $\mathscr{R}$. For every $x \in \ell^\infty$, the following relations hold:

$$\begin{aligned}\langle x \mid \overline{H(n)}\,\rangle &= \big\langle (x(1), \dots, x(n), 0, 0, \dots) \mid H(n)\,\big\rangle \\ &= (H \otimes u_x)(n) \to (H \otimes u_x)(\infty) = \langle x \mid H(\infty)\rangle\end{aligned}$$

as $n \to \infty$. Therefore, $\big(\,\overline{H(n)}\,\big) \subset \ddot{\ell}^1$ is a weakly Cauchy sequence and, hence, converges in norm, since the space $\ddot{\ell}^1$ possesses the Schur property (see Lemma 3.1.2). Whence it follows that $H(\infty)$ is the norm limit of the sequence $\big(\,\overline{H(n)}\,\big)$; in particular, $H(\infty) \in \ddot{\ell}^1$ and $\|H\| \in C(Q)$. Thus, the CBB $\mathscr{X}$ has the dual bundle and $\mathscr{X}'(\infty) \neq \mathscr{X}(\infty)'$ due to the inclusion $\mathscr{X}'(\infty) \subset \ddot{\ell}^1$.

(d): According to 3.4.4 (1), it is sufficient, given an arbitrary functional $y \in \ddot{\ell}^1$, to present a homomorphism $H_y \in \mathrm{Hom}(\mathscr{X}, \mathscr{R})$ such that $H_y(\infty) = y$. The sought homomorphism can be defined as follows:

$$H_y(q) = \begin{cases} y|_{\mathscr{X}(q)}, & q \in \mathbb{N}, \\ y, & q = \infty. \end{cases}$$

The containment $H_y \in \mathrm{Hom}(\mathscr{X}, \mathscr{R})$ is justified by [8, Theorem 1.4.9] (with $\mathscr{V} = \mathscr{C}$).

3.4.10. The CBB $\mathscr{X}'' = (\mathscr{X}')'$ (if the latter exists) is called the *second dual* of a continuous Banach bundle $\mathscr{X}$.

It is clear that, for every CBB over a discrete topological space, the second dual exists. Ample CBBs over extremally disconnected compact Hausdorff spaces (see [8, 1.3]) form an important available class of continuous Banach bundles for which the second dual bundles exist.

First of all, we note that existence of $\mathscr{X}'$ does not imply existence of $\mathscr{X}''$.

Proposition. *Let X be a separable Banach space with nonseparable dual (for instance, $X = \ell^1$). Then there exists a topological space Q such that the trivial CBB X_Q has the dual bundle and has no second dual bundle.*

$\lhd$ By Proposition 3.4.2 (5), there exists a functionally discrete topological space Q such that the CBB $(X')_Q$ has no dual bundle. By 3.4.2 (6), the CBB X_Q has the dual bundle. By assertion 3.4.2 (3), the bundle $(X_Q)'$ coincides with $(X')_Q$ and, thereby, $(X_Q)'$ has no dual bundle, i.e., the bundle $(X_Q)''$ does not exist. $\rhd$

REMARK. The CBB $\mathscr{X}$ constructed in 3.4.9 is also an example of a Banach bundle which has the dual but not the second dual bundle. Indeed, with each element $n \in \mathbb{N}$ associate the functional $e_n'' \in \mathscr{X}'(n)'$ related to the element $e_n \in \mathscr{X}(n)$ by the rule $\langle x' | e_n'' \rangle = \langle e_n | x' \rangle$ for all $x' \in \mathscr{X}'(n)$. Put $G(n) = e_n''$ for all

$n \in \mathbb{N}$ and $G(\infty) = 0 \in \mathscr{X}'(\infty)'$. It is clear that the set $\mathscr{D} = \{H_y : y \in \ell^1\}$ is stalkwise dense in $\mathscr{X}'$. By applying [8, Theorem 1.4.9] (with $\mathscr{V} = \mathscr{D}$), we obtain the containment $G \in \operatorname{Hom}(\mathscr{X}', \mathscr{R})$. Furthermore, $|||G||| \notin C(Q)$ and, hence, in view of 3.3.2, the CBB $\mathscr{X}'$ has no dual bundle, i.e., $\mathscr{X}''$ does not exist.

3.4.11. *Proposition.* *The following are true:*

(1) *Suppose that a trivial* CBB *with stalk X has the second dual bundle. Then the latter is the trivial* CBB *with stalk X''.*

(2) *If a trivial* CBB *over a topological space Q with infinite-dimensional stalk has the second dual bundle, then Q is functionally discrete.*

(3) *Let X be an infinite-dimensional Banach space with separable dual. Then existence of the second dual for the bundle X_Q is equivalent to functional discreteness of Q.*

(4) *For every Banach space X, there exists a nondiscrete normal topological space Q such that the* CBB *X_Q has the second dual bundle.*

(5) *If a topological space Q is not functionally discrete then, for every Banach space X, the following are equivalent:*
 (a) *$(X_Q)''$ exists;*
 (b) *$(X'')_Q = (X_Q)''$;*
 (c) *$(X_Q)'$ exists and $C(Q, X'') = \operatorname{Hom}\left((X_Q)', \mathscr{R}\right)$;*
 (d) *X is finite-dimensional.*

◁ Assertions (1), (2), and (5) are simple consequences of Proposition 3.4.2.

A proof of assertion (4) can be obtained by a simple modification of the proof of Corollary 3.3.15 with Q a nondiscrete normal topological space such that the constant CBBs X_Q and $(X')_Q$ both have dual bundles.

Prove assertion (3). Necessity holds due to (2). Proceeding with sufficiency, observe first that the space X is itself separable. From 3.4.2 (6) it follows that the dual $(X_Q)'$ exists and, in view of 3.4.2 (3), the latter coincides with $(X')_Q$. Applying 3.4.2 (6) again, we complete the proof. ▷

3.4.12. In contrast to the situation described in Proposition 3.4.10, existence of $\mathscr{X}'$ in the following case implies existence of $\mathscr{X}''$.

Proposition. *If a* CBB *with Hilbert stalks over a topological space Q has the dual bundle then it has the second dual bundle. Moreover, the bundles $\mathscr{X}$, $\mathscr{X}'$, and $\mathscr{X}''$ are pairwise isometric.*

◁ Obviously, if two CBBs are isometric and one of them has the dual bundle then the other has the dual bundle too and these duals are isometric. This fact and Proposition 3.4.3 imply the claim. ▷

3.4.13. Let Q be a topological space and let $\mathscr{X}$ be a CBB over Q which has the dual bundle. The mapping $\imath$ that associates with every point $q \in Q$ the operator

$$\imath(q) : x \in \mathscr{X}(q) \mapsto x''|_{\mathscr{X}'(q)}$$

is called the *double prime mapping* for $\mathscr{X}$. (Here $x \mapsto x''$ is the canonical embedding into the second dual.)

Proposition. *Let Q be a topological space and let $\mathscr{X}$ be a CBB over Q which has the dual bundle. Suppose that $\mathrm{Hom}(\mathscr{X}, \mathscr{R})$ norms $\mathscr{X}$ and let $\imath$ be the double prime mapping for $\mathscr{X}$.*

(1) *For every point $q \in Q$, the operator $\imath(q)$ is an isometric embedding of $\mathscr{X}(q)$ into $\mathscr{X}'(q)'$.*

(2) *Assume that $\mathscr{X}$ has the second dual bundle. Then the mapping $\imath$ is an isometric embedding of $\mathscr{X}$ into $\mathscr{X}''$.*

$\lhd$ (1): For $q \in Q$ and $x \in \mathscr{X}(q)$, we have

$$\begin{aligned}
\|x''|_{\mathscr{X}'(q)}\| &= \sup\big\{\langle x'|x''\rangle : x' \in \mathscr{X}'(q),\ \|x'\| \leqslant 1\big\} \\
&= \sup\big\{\langle x|x'\rangle : x' \in \mathscr{X}'(q),\ \|x'\| \leqslant 1\big\} \\
&= \sup\big\{\langle x|v(q)\rangle : v \in C(Q, \mathscr{X}'),\ \|v(q)\| \leqslant 1\big\} \\
&= \sup\big\{\langle x|v(q)\rangle : v \in C(Q, \mathscr{X}'),\ |\!|\!|v|\!|\!| \leqslant 1\big\} \\
&= \sup\big\{\langle x|H\rangle : H \in \mathrm{Hom}(\mathscr{X}, \mathscr{R}),\ \|H(q)\| \leqslant 1\big\} \\
&= \|x\| \quad \text{(cf. [8, 1.3.9]).}
\end{aligned}$$

(2): In view of (1), the mapping $u \mapsto \imath \otimes u$ embeds the space $C(Q, \mathscr{X})$ into $\mathrm{Hom}(\mathscr{X}', \mathscr{R}) = C(Q, \mathscr{X}'')$ with pointwise norm preserved. It remains to employ [8, Theorem 1.4.4]. $\rhd$

3.4.14. ***Proposition.*** *Let $\mathscr{X}$ be a CBB with constant finite dimension over a completely regular topological space. Then the bundle $\mathscr{X}''$ exists, $\mathrm{Hom}(\mathscr{X}, \mathscr{R})$ norms $\mathscr{X}$, and the double prime mapping for $\mathscr{X}$ performs an isometry of $\mathscr{X}$ onto $\mathscr{X}''$.*

$\lhd$ By assertion 3.4.2 (1), in the situation under consideration, the dual bundle $\mathscr{X}'$ exists and $\dim \mathscr{X}' = \dim \mathscr{X}$. The same assertion implies that $\mathscr{X}''$ exists and the equality $\dim \mathscr{X}'' = \dim \mathscr{X}'$ holds. Hence, for every point q, the stalks $\mathscr{X}(q)$ and $\mathscr{X}''(q)$ have the same finite dimension. It remains to apply Proposition 3.4.13 (2) and [8, Theorem 1.4.12]. $\rhd$

3.4.15. Let Q be a topological space and let $\mathscr{X}$ be a CBB over Q which has the second dual bundle. In the following cases, the double prime mapping for $\mathscr{X}$ is an isometry of $\mathscr{X}$ onto $\mathscr{X}''$:

(1) $\mathscr{X}$ is a trivial CBB with reflexive stalk;

(2) $\mathscr{X}$ has constant finite dimension and the topological space Q is completely regular;

(3) $\mathscr{X}$ is a CBB with Hilbert stalks;

(4) $\mathscr{X}$ is an ample CBB over an extremally disconnected compact Hausdorff space Q and all stalks of $\mathscr{X}$ at nonisolated points are reflexive.

$\triangleleft$ Assertions (1)–(4) are easy from 3.4.11 (1), 3.4.14, 3.4.12, and [8, 2.3.5 (1), 2.3.7]. $\triangleright$

Observe that conditions (2) and (4) imply existence of $\mathscr{X}''$ without additional assumptions.

3.5. Weakly Continuous Sections

In this section, we introduce and study the notion of a weakly continuous section of a Banach bundle.

Since weakly continuous sections are closely connected with homomorphisms of the dual bundle (which are known to have locally bounded pointwise norms), the problem is natural of finding conditions that guarantee local boundedness for weakly continuous sections. Subsections 3.5.3–3.5.5 are devoted to this subject.

In 3.5.6–3.5.12, we study the question of continuity of weakly continuous sections for various classes of Banach bundles.

The remaining part of this section (3.5.13–3.5.18) is devoted to finding conditions for coincidence of the space of weakly continuous sections of a trivial Banach bundle and the space of weakly continuous vector valued functions acting into the corresponding stalk.

3.5.1. Let $\mathscr{X}$ be a CBB over a topological space Q and let $D \subset Q$.

DEFINITION. A section u over D of a bundle $\mathscr{X}$ is called *weakly continuous* if $\langle u|H\rangle \in C(D)$ for all $H \in \operatorname{Hom}(\mathscr{X}, \mathscr{R})$. The totality of all these sections is denoted by $C_w(D, \mathscr{X})$.

If $\mathscr{X}$ has the dual bundle then $\operatorname{Hom}(\mathscr{X}, \mathscr{R}) = C(Q, \mathscr{X}')$ and, in this case, weak continuity of a section u is equivalent to continuity of the functions $\langle u|u'\rangle$ for all $u' \in C(Q, \mathscr{X}')$.

It is clear that $C_w(D, \mathscr{X})$ is a vector subspace of the space of all sections over D of the bundle $\mathscr{X}$ and includes $C(D, \mathscr{X})$ as a vector subspace.

Note that a weakly continuous section need not be continuous. Indeed, considering the CBB $\mathscr{X}$ constructed in 3.4.9 and putting $u(n) = e_n$, $n \in \mathbb{N}$, and $u(\infty) = 0$, we obtain a weakly continuous (see Remark 3.4.10) but, obviously, discontinuous section of $\mathscr{X}$.

3.5.2. Lemma. *Let X be a Banach space and let Q be a topological space. Suppose that $D \subset Q$ and a sequence $(q_n) \subset D$ converges to a point $q \in D$.*

(1) *If Q is completely regular and $u \in C_w(D, X_Q)$ then the sequence $(u(q_n))$ w-w^*-converges to $u(q)$.*

(2) *For every $H \in \mathrm{Hom}(X_Q, \mathscr{R})$, the sequence $(H(q_n))$ is weakly* convergent to $H(q)$.*

(3) *If Q is a completely regular Fréchet–Urysohn space and the points q_n are pairwise distinct and distinct from q then, for every w-w^*-vanishing sequence $(x_n) \subset X$, there exists a section $u \in C_w(D, X_Q)$ taking the values $u(q_n) = x_n$ for all $n \in \mathbb{N}$ and $u(q) = 0$.*

(4) *If $u \in C_w(D, X)$ then the sequence $(u(q_n))$ converges weakly to $u(q)$.*

$\triangleleft$ (1): As is easily seen, we do not restrict generality by assuming that the points q_n are pairwise distinct and distinct from q. From 3.2.6 (4) it follows that, for every sequence $(x'_n) \subset X'$ convergent weakly* to an element $x' \in X'$, there exists a homomorphism $H \in \mathrm{Hom}(X_Q, \mathscr{R})$ taking the values $H(q_n) = x'_n$ for all $n \in \mathbb{N}$ and $H(q) = x'$. Hence, $\langle u(q_n)|x'_n\rangle = \langle u|H\rangle(q_n) \to \langle u|H\rangle(q) = \langle u(q)|x'\rangle$.

Assertions (2) and (4) are evident.

(3): Let (W_n) and (f_n) be sequences of open subsets of Q and of continuous functions from Q into $[0, 1]$ presented in Lemma 3.2.5. Then the section u over D defined by the formula

$$u(p) = \begin{cases} f_n(p)x_n, & p \in D \cap W_n, \\ 0, & p \in D \setminus \bigcup_{n\in\mathbb{N}} W_n \end{cases}$$

is weakly continuous. Indeed, consider an arbitrary $H \in \mathrm{Hom}(X_Q, \mathscr{R})$. The function $\langle u|H\rangle$ is continuous on each set $D \cap \mathrm{cl}\, W_n$, since $\mathrm{cl}\, W_n$ is included in

$$Q \setminus \mathrm{cl} \bigcup_{k \neq n} W_k,$$

the latter difference is open, and $\langle u|H\rangle$ and $\langle x_n|H\rangle f_n$ coincide on the intersection of D and the difference.

Assume that the function $\langle u|H\rangle$ is discontinuous at some point

$$p \in \left(\mathrm{cl} \bigcup_{n\in\mathbb{N}} W_n\right) \setminus \bigcup_{n\in\mathbb{N}} \mathrm{cl}\, W_n.$$

Then there exist an $\varepsilon > 0$, a sequence $(p_m) \subset D$, and a strictly increasing sequence $(n_m) \subset \mathbb{N}$ such that p belongs to $\mathrm{cl}\{p_m : m \in \mathbb{N}\}$, $p_m \in W_{n_m}$, and $|\langle u|H\rangle(p_m)| > \varepsilon$ for all $m \in \mathbb{N}$. Since Q is a Fréchet–Urysohn space, we can extract a subsequence (p_{m_k}) convergent to p. It is easy to verify that the sequence $(u(p_m))$ is w-w^*-vanishing; therefore, the subsequence $u(p_{m_k})$ of $(u(p_m))$ is w-w^*-vanishing too. At the same time, by (2), the sequence $(H(p_{m_k}))$ converges weakly* to $H(p)$. Consequently, $\varepsilon < |\langle u|H\rangle(p_{m_k})| \to |\langle u|H\rangle(p)| = 0$. The assumption that $\langle u|H\rangle$ is discontinuous at p yields a contradiction. It remains to observe that the function $\langle u|H\rangle$ vanishes on the set $Q\setminus \mathrm{cl}\bigcup_{n\in\mathbb{N}} W_n$. $\triangleright$

3.5.3. EXAMPLE. There exist a Fréchet–Urysohn space Q, a Banach space X, and a section $u \in C_w(Q, X_Q)$ that is not locally bounded.

$\triangleleft$ Consider the space Q constructed in Example 3.2.11.

As follows from Corollary 3.1.7 (2), the space ℓ^∞ contains a sequence (x_n) which is w-w^*-vanishing and does not converge in norm. Without loss of generality, we may assume that $\|x_n\| \geqslant 1$ for all $n \in \mathbb{N}$ (this may be fulfilled by extracting a subsequence and multiplying the latter by an appropriate constant elementwise). Put $u((m,n)) := mx_n$ for every $(m,n) \in \mathbb{N}\times\mathbb{N}$ and put $u(\infty) := 0 \in \ell^\infty$. Obviously, the section u is not locally bounded. Show that $H \otimes u \in C(Q)$ for an arbitrary homomorphism $H \in \mathrm{Hom}\,((\ell^\infty)_Q, \mathscr{R})$. By Lemma 3.5.2 (2), for every m, the sequence $(H((m,n)))_{n\in\mathbb{N}}$ is weakly* convergent, whence $(H\otimes u)((m,n)) = m\langle x_n|H((m,n))\rangle \to 0$ as $n \to \infty$. The latter relation implies continuity of the function $H \otimes u$ (see the description (1) of the elements of $C(Q)$ in Example 3.2.11). $\triangleright$

3.5.4. ***Proposition.*** *Let $\mathscr{X}$ be a CBB over a topological space Q. Suppose that* $\mathrm{Hom}(\mathscr{X}, \mathscr{R})$ *norms $\mathscr{X}$ and the space Q satisfies one of the following conditions:*

(a) *Q is first-countable and completely regular;*

(b) *Q is locally pseudocompact. Then every weakly continuous global section of $\mathscr{X}$ is locally bounded.*

$\triangleleft$ First suppose that Q satisfies condition (a). Assume that there is a weakly continuous and not locally bounded global section u of $\mathscr{X}$. In this case, the pointwise norm $|\!|\!| u |\!|\!|$ is unbounded on every neighborhood about some point $q \in Q$. By Dupré's Theorem (see [8, 1.3.5]), we may find a bounded continuous global section taking the value $u(q)$ at q and, next, subtract this section from u; therefore, we may assume that $|\!|\!| u |\!|\!|(q) = 0$.

Since Q is first-countable, there is a sequence $(q_n) \subset Q$ such that $|\!|\!| u |\!|\!|(q_n) > n^2$, $q_i \neq q_j$ for $i \neq j$, and $q_n \to q$. Using the hypotheses, for every number $n \in \mathbb{N}$, take a homomorphism $H_n \in \mathrm{Hom}(\mathscr{X}, \mathscr{R})$ satisfying the relations $\langle u|H_n\rangle(q_n) = \|u(q_n)\|$ and $|\!|\!| H_n |\!|\!| \leqslant 2$.

Corollary 3.2.6 (2) implies existence of an $H \in \operatorname{Hom}(\mathscr{X},\mathscr{R})$ such that $H(q) = 0$ and $H(q_n) = \frac{1}{n}H_n(q_n)$ for all $n \in \mathbb{N}$. On the other hand,

$$\langle u|H\rangle(q_n) = \frac{1}{n}\langle u|H_n\rangle(q_n) = \frac{1}{n}\|u(q_n)\| > n,$$

which contradicts weak continuity of u, since $q_n \to q$ and $\langle u|H\rangle(q) = 0$.

Now suppose that Q satisfies condition (b). Denote by $\operatorname{Hom}^b(\mathscr{X},\mathscr{R})$ the space of all bounded homomorphisms from $\mathscr{X}$ into $\mathscr{R}$. Fix an arbitrary weakly continuous section u of $\mathscr{X}$ and, for every point $q \in Q$, define a linear functional $T_q : \operatorname{Hom}^b(\mathscr{X},\mathscr{R}) \to \mathbb{R}$ by the formula $T_q(H) = \langle u(q)|H(q)\rangle$. Endowing the space $\operatorname{Hom}^b(\mathscr{X},\mathscr{R})$ with the uniform norm and considering an arbitrary pseudocompact subset $U \subset Q$, we conclude that $\|T_q\| \leqslant \|u(q)\|$; moreover,

$$\sup_{q\in U} \|T_q(H)\| = \sup_{q\in U} \bigl|\langle u|H\rangle(q)\bigr| < \infty$$

for all $H \in \operatorname{Hom}^b(\mathscr{X},\mathscr{R})$. By [8, 1.4.11], $\operatorname{Hom}^b(\mathscr{X},\mathscr{R})$ is a Banach space. Therefore, $\sup_{q\in U}\|T_q\| < \infty$ in view of the uniform boundedness principle. It remains to employ the relations

$$\|u(q)\| = \sup\bigl\{\bigl|\langle u(q)|H(q)\rangle\bigr| : H \in \operatorname{Hom}(\mathscr{X},\mathscr{R}),\ |\!|\!|H|\!|\!| \leqslant 1\bigr\} = \|T_q\|. \ \rhd$$

Observe that, in the last proposition, conditions (a) and (b) are essential even if the CBB $\mathscr{X}$ is trivial (see 3.5.3).

3.5.5. Corollary. *Let X be a Banach space and let Q be a topological space satisfying (a) or (b) of 3.5.4. Then every weakly global continuous section of X_Q is locally bounded.*

$\lhd$ The claim follows immediately from 3.5.4 and 3.4.4 (3). $\rhd$

3.5.6. REMARK. By the definition of continuity for sections (see [8, 1.1.2]), if $\mathscr{U}$ is a vector space of sections over $D \subset Q$ of a CBB $\mathscr{X}$ over a topological space Q and all elements of $\mathscr{U}$ have continuous pointwise norms, then the inclusion $C(D,\mathscr{X}) \subset \mathscr{U}$ implies the equality $C(D,\mathscr{X}) = \mathscr{U}$.

Proposition. *Let $\mathscr{X}$ be a CBB over a topological space Q.*

(1) *Suppose that $\mathscr{X}$ has the dual bundle and let $\imath$ be the double prime mapping for $\mathscr{X}$. For every subset $D \subset Q$, the mapping $u \mapsto \imath \otimes u$ performs a linear embedding of the space of locally bounded sections $u \in C_w(D,\mathscr{X})$ into $\operatorname{Hom}_D(\mathscr{X}',\mathscr{R})$. If, in addition, $\operatorname{Hom}(\mathscr{X},\mathscr{R})$ norms $\mathscr{X}$ then the embedding preserves the pointwise norm.*

(2) *Suppose that $\mathscr{X}$ has the second dual bundle and* $\mathrm{Hom}(\mathscr{X},\mathscr{R})$ *norms $\mathscr{X}$. If a section $u \in C_w(Q,\mathscr{X})$ is locally bounded then $u \in C(Q,\mathscr{X})$.*

$\lhd$ (1): The containment $\imath \otimes u \in \mathrm{Hom}_D(\mathscr{X}',\mathscr{R})$ holds in view of [8, Theorem 1.4.9]. Furthermore, if $\mathrm{Hom}(\mathscr{X},\mathscr{R})$ norms $\mathscr{X}$ then the equality $|\!|\!|\imath \otimes u|\!|\!| = |\!|\!|u|\!|\!|$ follows from 3.4.13 (1).

(2): Let a section $u \in C_w(Q,\mathscr{X})$ be locally bounded. Then, in view of assertion (1), the containment $\imath \otimes u \in \mathrm{Hom}(\mathscr{X}',\mathscr{R})$ holds which, together with the equality $\mathrm{Hom}(\mathscr{X}',\mathscr{R}) = C(Q,\mathscr{X}'')$, yields continuity of the pointwise norm of the homomorphism $\imath \otimes u$. Since, due to (1), the functions $|\!|\!|\imath \otimes u|\!|\!|$ and $|\!|\!|u|\!|\!|$ coincide, the latter function is continuous too. Therefore, the vector space $\mathscr{U}$ of locally bounded sections $u \in C_w(Q,\mathscr{X})$ consists of sections with continuous pointwise norms and contains $C(Q,\mathscr{X})$. The above Remark allows us to conclude that $\mathscr{U} = C(Q,\mathscr{X})$. $\rhd$

3.5.7. Corollary. *Let $\mathscr{X}$ be a* CBB *with constant finite dimension over a completely regular topological space Q. For every subset $D \subset Q$, the equality $C_w(D,\mathscr{X}) = C(D,\mathscr{X})$ holds.*

$\lhd$ The claim may be derived from Theorem 3.2.12, Proposition 3.5.6 (1), and Remark 3.5.6. $\rhd$

3.5.8. Corollary. *Suppose that a topological space Q and a* CBB *$\mathscr{X}$ over Q satisfy the conditions of Proposition* 3.5.4. *Then existence of $\mathscr{X}''$ implies continuity of all weakly continuous sections of $\mathscr{X}$.*

$\lhd$ This claims follows from Propositions 3.5.4 and 3.5.6 (2). $\rhd$

3.5.9. Proposition. *Let $\mathscr{X}$ be a* CBB *with Hilbert stalks over an arbitrary topological space. If a global section of $\mathscr{X}$ is locally bounded and weakly continuous then it is continuous.*

$\lhd$ Let Q be a topological space and let $\mathscr{X}$ be a CBB with Hilbert stalks over Q. Fix a locally bounded section $v \in C_w(Q,\mathscr{X})$ and use the mapping h of Lemma 3.4.3 which asserts that

$$h[C(Q,\mathscr{X})] \subset \mathrm{Hom}(\mathscr{X},\mathscr{R}).$$

Thus, the relations $\langle c|h(u)\rangle = \langle u|h(c)\rangle \in C(Q)$ are valid for all $c \in C(Q,\mathscr{X})$. By [8, 1.4.4] these relations imply $h(u) \in \mathrm{Hom}(\mathscr{X},\mathscr{R})$. Therefore,

$$|\!|\!|u|\!|\!|^2 = \langle u|h(u)\rangle \in C(Q).$$

Finally, since

$$|\!|\!|u - c|\!|\!|^2 = |\!|\!|u|\!|\!|^2 - 2\langle c|h(u)\rangle + |\!|\!|u|\!|\!|^2 \in C(Q)$$

for every $c \in C(Q,\mathscr{X})$, the section u is continuous. $\rhd$

3.5.10. Corollary. *Let $\mathscr{X}$ be a CBB with Hilbert stalks over a topological space Q satisfying* (a) *or* (b) *of 3.5.4. Then $C_w(Q, \mathscr{X}) = C(Q, \mathscr{X})$.*

$\lhd$ The claim follows immediately from Propositions 3.5.4 and 3.5.9, and Lemma 3.4.3. $\rhd$

3.5.11. Lemma. *Suppose that a CBB $\mathscr{X}$ over a topological space Q has the dual bundle. For arbitrary sections $u \in C(Q, \mathscr{X})$ and $v \in C_w(Q, \mathscr{X}')$, the real-valued function $\langle u|v\rangle$ is continuous.*

$\lhd$ Let $\imath$ be the double-prime mapping for $\mathscr{X}$. Then $\imath \otimes u$ is an element of $\mathrm{Hom}(\mathscr{X}', \mathscr{R})$ according to Proposition 3.5.6 (1). Consequently, $\langle u|v\rangle = \langle v \mid \imath \otimes u\rangle \in C(Q)$. $\rhd$

Proposition. *Suppose that a CBB $\mathscr{X}$ over a topological space Q has the dual bundle.*

(1) *If $v \in C_w(Q, \mathscr{X}')$ is locally bounded then $v \in C(Q, \mathscr{X}')$.*

(2) *If Q satisfies* (a) *or* (b) *of 3.5.4 then $C_w(Q, \mathscr{X}') = C(Q, \mathscr{X}')$.*

$\lhd$ (1): Let $v \in C_w(Q, \mathscr{X}')$ be a locally bounded section. In view of the above lemma, $\langle u|v\rangle \in C(Q)$ for all $u \in C(Q, \mathscr{X})$. Consequently, $v \in \mathrm{Hom}(\mathscr{X}, \mathscr{R})$ due to [8, Theorem 1.4.9] and local boundedness of v. It remains to recall that $\mathrm{Hom}(\mathscr{X}, \mathscr{R}) = C(Q, \mathscr{X}')$.

(2): It suffices to prove the inclusion $C_w(Q, \mathscr{X}') \subset C(Q, \mathscr{X}')$. Suppose that $v \in C_w(Q, \mathscr{X}')$. In view of the above lemma, $\langle u|v\rangle \in C(Q)$ for all $u \in C(Q, \mathscr{X})$. If Q satisfies 3.5.4 (a) then $v \in \mathrm{Hom}(\mathscr{X}, \mathscr{R})$ due to Theorem 3.2.10; if Q satisfies 3.5.4 (b) then $v \in \mathrm{Hom}(\mathscr{X}, \mathscr{R})$ due to [8, Theorem 1.4.7]. Therefore, in both cases, $v \in \mathrm{Hom}(\mathscr{X}, \mathscr{R}) = C(Q, \mathscr{X}')$. $\rhd$

3.5.12. Theorem. *Let X be a Banach space and let Q be a completely regular Fréchet–Urysohn space.*

(1) *If X possesses the* WS *property then $C_w(D, X_Q) = C(D, X_Q)$ for all subsets $D \subset Q$.*

(2) *If $C_w(D, X_Q) = C(D, X_Q)$ for some subset $D \subset Q$ which contains one of its limit points (in particular, if $D = Q$ and Q is nondiscrete), then X possesses the* WS *property.*

For instance, if Q is nondiscrete then the equality $C_w(Q, X_Q) = C(Q, X_Q)$ is equivalent to the fact that X possesses the WS *property.*

$\lhd$ (1): Suppose that the inclusion $C_w(D, X_Q) \supset C(D, X_Q)$ is strict for a subset $D \subset Q$ and show that X does not possess the WS property. Consider a section $u \in C_w(D, X_Q)$ discontinuous at a point $q \in D$. We may assume that $u(q) = 0$, since, otherwise, we can subtract from u the constant section taking the value $u(q)$.

Since Q is a Fréchet–Urysohn space, we may find a sequence $(q_n) \subset D$ convergent to q such that $|\!|\!|u|\!|\!|(q_n) > \varepsilon > 0$ for all $n \in \mathbb{N}$. By Lemma 3.5.2 (1), the sequence $\big(u(q_n)\big)$ is w-w^*-convergent to $u(q) = 0$. Consequently, X does not possess the WS property.

(2): Suppose that X does not enjoy the WS property and establish the inequality $C_w(D, X_Q) \neq C(D, X_Q)$ for every subset $D \subset Q$ that contains one of its limit point. Let $q \in D$ be a limit point of D. Since Q is a Fréchet–Urysohn space, there is a sequence $(q_n) \subset D\backslash\{q\}$ convergent to q. Without loss of generality, we may assume that $q_i \neq q_j$ whenever $i \neq j$. Since X does not possess the WS property, we may take a sequence $(x_n) \subset X$ which is w-w^*-vanishing and does not vanish in norm. By Lemma 3.5.2 (3), there is a section $u \in C_w(D, X_Q)$ taking the values $u(q_n) = x_n$ for all $n \in \mathbb{N}$ and $u(q) = 0$. It is clear that $u \notin C(D, X_Q)$. $\triangleright$

3.5.13. *Proposition.* *For every infinite-dimensional Banach space X, there exists a normal topological space Q such that the inclusion $C_w(Q, X_Q) \subset C_w(Q, X)$ is strict.*

$\triangleleft$ Let $(x_\alpha)_{\alpha\in\aleph}$ and $(x'_\alpha)_{\alpha\in\aleph}$ be the nets existent by Lemma 3.1.4. Put $Q = \aleph^\bullet$ (see 3.1.11) and consider vector valued functions $u : Q \to X$ and $H : Q \to X'$ satisfying the equalities $u(\alpha) = x_\alpha$, $H(\alpha) = x'_\alpha$ for all $\alpha \in \aleph$, $u(\infty) = 0$, and $H(\infty) = 0$.

In view of Remark 3.1.11 (2), the function u is weakly continuous and, in addition, $H \in C(Q, X')$. In particular, $H \in \operatorname{Hom}\big(X_Q, \mathscr{R}\big)$. Furthermore, $\langle u|H\rangle \equiv 1$ on $\aleph$ and $\langle u|H\rangle(\infty) = 0$, whence $u \notin C_w(Q, X_Q)$. $\triangleright$

3.5.14. *Corollary.* *Let X be a Banach space and let Q be an arbitrary topological space. The equality $C_w(Q, X_Q) = C_w(Q, X)$ holds for every topological space Q if and only if X is finite-dimensional.*

Observe that, in case X is finite-dimensional, we have

$$C(Q, X_Q) = C_w(Q, X_Q) = C_w(Q, X) = C(Q, X).$$

3.5.15. *Theorem.* *Let X be a Banach space and let Q be an arbitrary topological space.*

(1) *If Q is a Fréchet–Urysohn space and X possesses the* DP* *property, then $C_w(D, X_Q) = C_w(D, X)$ for every subset $D \subset Q$.*

(2) *Let a subset $D \subset Q$ be such that $C(Q)$ contains a function which is not locally constant on D. If $C_w(D, X_Q) = C_w(D, X)$ then X possesses the* DP* *property.*

In particular, if Q is a nondiscrete completely regular Fréchet–Urysohn space then the equality $C_w(Q, X_Q) = C_w(Q, X)$ is equivalent to the fact that X possesses the DP* *property.*

$\lhd$ (1): Suppose $C_w(D, X_Q) \neq C_w(D, X)$ for some subset $D \subset Q$. Show that X does not possess the DP* property. Take a vector valued function $u \in C_w(D, X) \setminus C_w(D, X_Q)$ and consider a homomorphism $H \in \operatorname{Hom}(X_Q, \mathscr{R})$ such that the function $\langle u|H \rangle$ is discontinuous at some point $q \in D$. Then the function $\langle u - u_q \mid H - H_q \rangle$ is discontinuous at q, where u_q and H_q are constant functions with values $u(q)$ and $H(q)$. (This is so due to the fact that the functions $\langle u|H_q \rangle$, $\langle u_q|H \rangle$, and $\langle u_q|H_q \rangle$ are continuous.) Since Q is a Fréchet–Urysohn space, there is a sequence $(q_n) \subset D \setminus \{q\}$ which converges to q and satisfies the condition $|\langle u(q_n) - u(q) \mid H(q_n) - H(q) \rangle| > \varepsilon$ for some $\varepsilon > 0$ and all $n \in \mathbb{N}$. Furthermore, in view of 3.5.2 (2), (4), the sequence $\big(u(q_n) - u(q)\big)$ is weakly vanishing and the sequence $\big(H(q_n) - H(q)\big)$ is weakly* vanishing. Consequently, X does not possess the DP* property.

(2): Suppose that X does not possess the DP* property. Consider a weakly null sequence $(x_n) \subset X$ and a weakly* null sequence $(x'_n) \subset X'$ such that $\langle x_n|x'_n \rangle$ does not vanish. By passing to a subsequence and multiplying all elements of one of them by $\pm\delta$ for a suitable $\delta \in \mathbb{R}$, we may achieve validity of the inequalities $\langle x_n|x'_n \rangle \geqslant 1$ for all $n \in \mathbb{N}$. We additionally require that $\langle x_{n+1}|x'_n \rangle + \langle x_n|x'_{n+1} \rangle \geqslant 0$ for all $n \in \mathbb{N}$, which in turn can be fulfilled by pairwise multiplication of the elements x_2 and x'_2, x_3 and x'_3, etc. by ± 1. Let vector valued functions $u : [0,1] \to X$ and $u' : [0,1] \to X'$ satisfy the equalities $u(0) = 0$, $u'(0) = 0$,

$$u\big(\lambda \tfrac{1}{n+1} + (1-\lambda)\tfrac{1}{n}\big) = \lambda x_{n+1} + (1-\lambda) x_n,$$
$$u'\big(\lambda \tfrac{1}{n+1} + (1-\lambda)\tfrac{1}{n}\big) = \lambda x'_{n+1} + (1-\lambda) x'_n$$

for all $\lambda \in [0,1]$ and $n \in \mathbb{N}$. By Lemma 3.1.12, the function u is weakly continuous and u' is weakly* continuous. Consider the function $\langle u|u' \rangle : [0,1] \to \mathbb{R}$. Given arbitrary $n \in \mathbb{N}$ and $0 \leqslant \lambda \leqslant 1$, we have

$$\begin{aligned}
\langle u|u' \rangle\big(\lambda \tfrac{1}{n+1} + (1-\lambda)\tfrac{1}{n}\big) &= \big\langle \lambda x_{n+1} + (1-\lambda) x_n \;\big|\; \lambda x'_{n+1} + (1-\lambda) x'_n \big\rangle \\
&= \lambda^2 \langle x_{n+1}|x'_{n+1} \rangle + (1-\lambda)^2 \langle x_n|x'_n \rangle \\
&\quad + \lambda(1-\lambda)\big(\langle x_{n+1}|x'_n \rangle + \langle x_n|x'_{n+1} \rangle\big) \\
&\geqslant \lambda^2 + (1-\lambda)^2 + 0 \\
&= 2(\lambda - 1/2)^2 + 1/2 \\
&\geqslant 1/2.
\end{aligned}$$

Thus, $\langle u|u' \rangle(0) = 0$ and, in addition, $\langle u|u' \rangle \geqslant 1/2$ on $(0,1]$. Next, take a continuous function $g \in C(Q)$ such that the restriction $g|_D$ is not constant on any neighborhood about a point $q \in D$. Without loss of generality, we may assume that $g : Q \to [0,1]$ and $g(q) = 0$ (see the proof of 3.1.13). As is easily seen, $u \circ g|_D \in C_w(D, X)$ and $u' \circ g \in \operatorname{Hom}(X_Q, \mathscr{R})$. It is clear that the function $\big\langle (u \circ g)|_D \;\big|\; u' \circ g \big\rangle = \langle u|u' \rangle \circ g|_D$

vanishes at q and, in addition, the image of the function on each neighborhood about q intersects the interval $[1/2, \infty)$. Consequently, $(u \circ g)|_D \notin C_w(D, X_Q)$.

The last assertion of the theorem follows from (1) and (2) and 3.1.10 (3). ▷

3.5.16. Corollary. *Let X be a Banach space and let Q be a topological space that is not functionally discrete. In each of the following cases, the inclusion $C_w(Q, X_Q) \subset C_w(Q, X)$ is strict:*

(1) *X is infinite-dimensional and reflexive;*

(2) *X is separable and does not possess the Schur property;*

(3) *X is a Banach space which does not possess the Schur property and satisfies one of the conditions* 3.1.6 (3), (5), *or* (6).

◁ In view of assertion (2) of Theorem 3.5.15, it suffices to show that, in each of the cases under consideration, X does not possess the DP* property. In cases (2) and (3) the latter is provided by Lemma 3.1.7 (3) and, in case (1), we can employ the Josefson–Niessenzweig Theorem [4, XII] according to which there exists a weakly* null sequence of norm-one vectors in X''. ▷

3.5.17. Proposition. *Let X be a Banach space and let Q be a functionally discrete topological space. If X' includes a countable total subspace then $C(Q, X) = C(Q, X_Q) = C_w(Q, X_Q) = C_w(Q, X)$.*

◁ The claim follows from Lemma 3.1.14, since the relations

$$C(Q, X) = C(Q, X_Q) \subset C_w(Q, X_Q) \subset C_w(Q, X)$$

are always true. ▷

3.5.18. Corollary. *Let Q be a topological space and let X be a separable Banach space that does not possess the Schur property. The equality $C_w(Q, X_Q) = C_w(Q, X)$ holds if and only if Q is functionally discrete.*

◁ Necessity follows from 3.1.6 (2), Lemma 3.1.7 (3), and Theorem 3.5.15 (2); sufficiency is justified by Proposition 3.5.17. ▷

References

1. Aliprantis C. D. and Burkinshaw O., Positive Operators, Academic Press, New York (1985).
2. Arkhangel′skiĭ A. V. and Ponomarëv V. I., Fundamentals of General Topology: Problems and Exercises, Reidel, Dordrecht etc. (1984).
3. Burden C. W. and Mulvey C. J., "Banach spaces in categories of sheaves," in: Applications of Sheaves, Springer-Verlag, Berlin, 1979, pp. 169–196 (Lecture Notes in Math.; **753**).

4. Diestel J., Sequences and Series in Banach Spaces,Springer-Verlag, Berlin etc. (1984) (Graduate Texts in Mathematics, **92**).
5. Engelking R., General Topology, Springer-Verlag, Berlin etc. (1985).
6. Fourman M. P., Mulvey C. J., and Scott D. S. (eds.), Applications of Sheaves, Springer-Verlag, Berlin (1979) (Lecture Notes in Math.; **753**).
7. Gierz G., Bundles of Topological Vector Spaces and Their Duality, Springer-Verlag, Berlin (1982) (Lecture Notes in Math.; **955**).
8. Gutman A. E., "Banach bundles in the theory of lattice normed spaces. I. Continuous Banach bundles," Siberian Adv. Math., **3**, No. 3, 1–55 (1993).
9. Gutman A. E., "Banach bundles in the theory of lattice normed spaces. II. Measurable Banach bundles," Siberian Adv. Math., **3**, No. 4, 8–40 (1993).
10. Gutman A. E., "Banach bundles in the theory of lattice normed spaces. III. Approximating sets and bounded operators," Siberian Adv. Math., **4**, No. 2, 54–75 (1994).
11. Gutman A. E., "Locally one-dimensional K-spaces and σ-distributive Boolean algebras," Siberian Adv. Math., **5**, No. 2, 99–121 (1995).
12. Gutman A. E., "Banach bundles in the theory of lattice normed spaces," in: Linear Operators Coordinated with Order [in Russian], Trudy Inst. Mat. (Novosibirsk), Novosibirsk, 1995, **29**, pp. 63–211.
13. Hofmann K. H., "Representations of algebras by continuous sections," Bull. Amer. Math. Soc., **78**, 291–373 (1972).
14. Hofmann K. H. and Keimel K., "Sheaf theoretical concepts in analysis: bundles and sheaves of Banach spaces, Banach $C(X)$-modules," in: Applications of Sheaves, Springer-Verlag, Berlin, 1979, pp. 414–441.
15. Kitchen J. W. and Robbins D. A., Gelfand Representation of Banach Modules, Polish Scientific Publishers, Warszawa (1982) (Dissertationes Mathematicae, **203**).
16. Koptev A. V., "A reflexivity criterion for stalks of a Banach bundle," Siberian Math. J., **36**, No. 4, 735–739 (1995) (Translated from Russian: Sibirsk. Mat. Zh. 1995. **36**, No. 4, 851–857).
17. Kusraev A. G., Vector Duality and Its Applications [in Russian], Nauka, Novosibirsk (1985).
18. Kutateladze S. S., Fundamentals of Functional Analysis, Kluwer Academic Publishers, Dordrecht (1996).
19. Schochetman I. E., "Kernels and integral operators for continuous sums of Banach spaces," Mem. Amer. Math. Soc., **14**, No. 202, 1–120 (1978).
20. Vulikh B. Z., Introduction to the Theory of Partially Ordered Spaces, Wolters-Noordhoff, Groningen, Netherlands (1967).
21. Wnuk W., Banach Lattices with Properties of the Schur Type. A Survey, Dipartimento di Matematica dell'Università di Bari, Bari (1993).

CHAPTER 4

Infinitesimals in Vector Lattices

BY

È. Yu. Emel'yanov

S.S. Kutateladze (ed.), Nonstandard Analysis and Vector Lattices, 161–230.

Infinitesimal analysis has lavishly contributed to various areas of mathematics since 1961 when A. Robinson published his famous paper [23]. The complete list of applications is very huge even as regards functional analysis. We give below some of them pertinent to the theory of vector lattices.

The first natural question is as follows: Where does infinitesimal analysis apply effectively? Clearly the methods of infinitesimal analysis are not a panacea and so they must fail in solving some problems. Furthermore, what new possibilities are open up by infinitesimal analysis when it applies? The book [17] offers a partial answer to this question. Roughly speaking, the nonstandard methods prove fruitful whenever the problem under consideration deals with such concepts as compactness or ultrafilter. On the other hand, these methods are often inapplicable to purely algebraic problems.

The main topic of further research is the theory of vector lattices and operators in them. We hope to convince the reader that vector lattice theory is a natural area for applying infinitesimal methods.

Vector lattices with some norm or other specific structure were studied by many authors (see, for example, [5, 11], and [18]) in the context of infinitesimal analysis. Our aim is to further pave the infinitesimal approach to vector lattices. In the sequel we proceed in the wake of the articles [6–10] with due modification, considering only real vector lattices.

The structure of this chapter is as follows. Section 4.0 is an introduction to Robinsonian infinitesimal analysis and its applications to normed spaces. In the end of the section, we make a short introduction to the theory of lattice normed spaces (for more information on this topic we refer the reader to the recent papers [13–16]).

Sections 4.1 and 4.2 deal mainly with an infinitesimal approach to representing vector lattices. It turns out that infinitesimal analysis provides new more natural and subtle possibilities for representing vector lattices as function spaces. We show below that construction of a representing topological space for a vector lattice is possible on using members of the lattice (more precisely, of a nonstandard enlargement of it) rather than ultrafilters or prime ideals (the latter inherent in Robinson's construction).

In Sections 4.3–4.7, we deal with infinitesimal interpretations of the basic concepts of the theory of vector lattices. We also consider the extension problem for a $*$-invariant homomorphism over a vector lattice or a Boolean algebra. Some types of elements of a nonstandard enlargement of a vector lattice are defined: limited or finite elements, (r)- and (o)-infinitesimals, prenearstandard and nearstandard elements, etc. We obtain some easily applicable nonstandard criteria for a vector lattice to be Archimedean, Dedekind complete, atomic, etc., and present a nonstandard construction of a Dedekind completion of an Archimedean vector lattice.

Using the concepts of Sections 4.3–4.7, we follow the Luxemburg scheme in Sections 4.8–4.11 for defining and studying two nonstandard hulls of a vector lattice: the order regular hulls. An elementary theory of these hulls is given in Sections 4.8 and 4.9. In Section 4.10 we introduce and study the concept of the nonstandard hull of a lattice normed space and that of the space associated with the order hull of a decomposable lattice normed space.

Section 4.11 discusses a nonstandard construction of an order completion of a decomposable lattice normed space. The scheme rests on embedding such a space into the associated Banach–Kantorovich space.

Throughout the sequel, we use the terminology and notations regarding vector lattices and operators from the books [1, 21, 24], and [28]. Lattice normed spaces are dealt with only in Sections 4.10 and 4.11. Therein we appeal to the terminology and notations of [13–16]. In the current chapter, we use some (standard and nonstandard) results on Boolean algebras and measure spaces which can be found in [2, 4], and [25]. For Robinsonian nonstandard analysis and its applications we refer the reader to [2, 12, 20] and [27]. Other explanations are to be made on the route.

4.0. Preliminaries

We start with a brief introduction to Robinsonian infinitesimal analysis and recall several facts of it without proofs. The level of formal requirements is chosen so as to avoid overloading the text with unnecessary details. We will mostly follow the books [2, 12] and [20]. We then recall some well-known results on applications of infinitesimal analysis to the theory of normed spaces and operators in them. For more applications we refer the reader to [2, 12] and [27]. In the end of the Section, we touch the theory of lattice normed spaces. Our exposition of this part rests on [13–16]. The concept of a quotient of a lattice normed space and Proposition 4.0.14 are new (cf. [9, Lemma 0.5.7]).

4.0.1. Let S be a set. A *superstructure* over S is the set $V(S) := \bigcup_n V_n(S)$, with $V_n(S)$ defined by recursion:

$$V_1(S) := S,$$
$$V_{n+1}(S) := V_n(S) \cup \{X : X \subseteq V_n(S)\}.$$

Superstructures are fragments of the von Neumann universe, providing a basis for various mathematical theories in dependence on the choice of the basic set. For example, superstructures over the reals serve the needs of calculus. When working with the superstructure over some set S, we suppose throughout that $\mathbb{R} \subseteq S$.

We need some formal language L. The alphabet of L contains

(1) variables: small and capital letters with possible indices;

(2) the symbols $=$ and $\in$ for equality and membership;

(3) symbols for propositional connectives and quantifiers;

(4) auxiliary symbols.

Atomic formulas of the language L are expressions of the form $x = y$ or $x \in y$. Arbitrary formulas are obtained from atomic formulas by applying propositional connectives and bounded quantifiers for sets (i.e. the prefixes $\forall x \in y$ and $\exists x \in y$).

Given an arbitrary set S on which some (partial) operations and relations are defined, we introduce some language $L_{V(S)}$ of the superstructure $V(S)$. To make the presentation easier, we construct the language $L_{V(S)}$ in a simple particular case.

Let $S = E \cup \mathbb{N}$, with E a lattice. The set E is equipped with the operations $\wedge$, $\vee$, and the relation $\widehat{\leq}$; the set $\mathbb{N}$ of naturals is equipped with the operations $+$, $\cdot$, and the relation $\leq$. In this case, the language $L_{V(S)}$ is obtained from L by enriching the alphabet with the symbols $\wedge$, $\vee$, $+$, and $\cdot$ for operations and $\widehat{\leq}$ and $\leq$ for predicates. The list of atomic formulas extends to include the expressions of the form $t_1 \vee t_2 = t_3$, $t_1 \cdot t_2 = t_3$, $t_1 \leq t_2$, etc., where t_1, t_2, and t_3 are arbitrary terms of L. Every formula of the language $L_{V(S)}$ is naturally interpreted in the superstructure $V(S)$. For example, the formula $\Psi(x, y, z) := x + y \leq z$ is true for a triple (a, b, c) of elements of $V(S)$ if and only if $a, b, c \in \mathbb{N}$ and $a + b \leq c$. It is clear that interpretation of arbitrary formulas of $L_{V(S)}$ involves no difficulty either.

In what follows, we choose the basic set S depending on the context. This set will be assumed to contain various objects: real and complex numbers, vector spaces, vector lattices, etc. Our presentation is structured so that it may be translated into the formal language $L_{V(S)}$ if need be. Throughout the article, the word "interpretation" means the natural interpretation in the respective superstructure.

4.0.2. Let S be a set equipped with some operations and relations (not necessarily defined everywhere). Then there exist an enlargement ${}^*S$ of S and embedding $* : V(S) \hookrightarrow V({}^*S)$ satisfying the following principles:

Extension Principle. *The set ${}^*S$ is a proper enlargement of S. Moreover, ${}^*S$ is equipped with the same set of operations and relations as S. In addition, ${}^*x = x$ for every element $x \in S$.*

Transfer Principle. *Let $\psi(x_1, x_2, \dots, x_n)$ be a formula of $L_{V(S)}$, and let $A_1, A_2, \dots, A_n$ be elements of the superstructure $V(S)$. Then the assertion*

$$\psi(A_1, A_2, \dots, A_n)$$

about elements of $V(S)$ is true if and only if the assertion

$$\psi({}^*A_1, {}^*A_2, \dots, {}^*A_n)$$

about elements of $V(^*S)$ is true.

Construction of the enlargement $^*S$ and embedding $* : V(S) \hookrightarrow V(^*S)$ with the required properties can be found, for example, in [2]. For convenience, we suppose that $*$ is the identical embedding and so $V(S) \subseteq V(^*S)$.

DEFINITION 1. The superstructure $V(^*S)$ is called a *nonstandard enlargement* of $V(S)$ if the embedding $V(S) \subseteq V(^*S)$ satisfies the transfer and extension principles.

Dealing with some superstructure in the sequel, we will not specify the basic set over which it is constructed. This set is chosen to be sufficiently substantive. We will denote the superstructure under consideration by M.

DEFINITION 2. Let $^*M$ be a nonstandard enlargement of a superstructure M. An element $x \in {}^*M$ is called:

(1) *standard* if $x = {}^*B$ for some $B \in M$;

(2) *internal* if $x \in {}^*B$ for some $B \in M$;

(3) *external* if $x \notin {}^*B$ for every $B \in M$.

Note that every standard set is internal and every element of an internal set is internal too. The following is easy from the transfer principle:

Internal Definition Principle. *Let $\psi(x, x_1, x_2, \dots, x_n)$ be a formula of the language L_M, and let $A, A_1, A_2, \dots, A_n$ be internal sets. Then the set $\{x \in A : \psi(x, A_1, A_2, \dots, A_n)\}$ is internal too.*

It is well known that a nonstandard enlargement $^*M$ of the superstructure M may be chosen so that the following principle holds:

General Saturation Principle. *For each family $\{X_\gamma\}_{\gamma\in\Gamma}$ of internal sets which has standard cardinality (i.e., card$(\Gamma) <$ card(M)) and enjoys the finite intersection property, the condition $\bigcap_{\gamma\in\Gamma} X_\gamma \neq \varnothing$ is valid.*

In the sequel, we deal only with nonstandard enlargements satisfying the general saturation principle. These nonstandard enlargements are called *polysaturated.*

4.0.3. Let X be an element of a superstructure M. We denote by $\mathscr{F}(X)$ the family of all finite subsets of X. Recall that elements of $^*\mathscr{F}(X)$ are exactly the subsets $A \subseteq {}^*X$ for which there exist an internal function f and element $\nu \in {}^*\mathbb{N}$ such that $\mathrm{dom}(f) = \{1, \dots, \nu\}$ and $\mathrm{im}(f) = A$. These subsets of $^*X$ are called *hyperfinite* and denoted, for example, by $\{x_n\}_{n=1}^{\nu}$ in analogy with finite families.

Lemma. *Let $X \in M$ be an infinite set and $\nu \in {}^*\mathbb{N} \setminus \mathbb{N}$. Then there exists an internal function f such that $\mathrm{dom}(f) = \{1, \dots, \nu\}$ and $X \subseteq \mathrm{im}(f) \subseteq {}^*X$. In other words, there is a hyperfinite set $\{x_n\}_{n=1}^{\nu}$ with $X \subseteq \{x_n\}_{n=1}^{\nu} \subseteq {}^*X$.*

◁ Let Ψ be the set of all functions ψ such that $\mathrm{dom}(\psi) \subseteq \mathbb{N}$ and $\mathrm{im}(\psi) \subseteq X$. Given $\pi \in \mathscr{F}(X)$, we assign

$$A_\pi := \{\varphi \in {}^*\Psi : \mathrm{dom}(\varphi) = \{1, \dots, \nu\} \ \&\ \pi \subseteq \mathrm{im}(\varphi)\}.$$

Since X is infinite, every set A_π is not empty. By the internal definition principle, the sets A_π are internal. They form a family with the finite intersection property. Since $\mathrm{card}(\mathscr{F}(X)) < \mathrm{card}(M)$, by the general saturation principle we have $f \in \bigcap_{\pi \in \mathscr{F}(X)} A_\pi$ for some $f \in {}^*\Psi$. It is easy to see that f is an internal function with $\mathrm{dom}(f) = \{1, \dots, \nu\}$ and $X \subseteq \mathrm{im}(f) \subseteq {}^*X$. ▷

4.0.4. Lemma. *Suppose that $X \in M$ and $X \subseteq M$. Then, for every $\nu \in {}^*\mathbb{N} \backslash \mathbb{N}$, we have $\mathrm{card}(X) < \mathrm{card}(\nu)$.*

◁ Consider the set $\mathscr{P}(X)$ of all subsets of X. Since $\mathscr{P}(X) \in M$, by the preceding lemma, $\mathscr{P}(X) \subseteq \{x_n\}_{n=1}^{\nu}$, where $\{x_n\}_{n=1}^{\nu}$ is a hyperfinite subset of ${}^*\mathscr{P}(X)$. Then $\mathrm{card}(X) < \mathrm{card}(\mathscr{P}(X)) \leq \mathrm{card}(\nu)$. ▷

4.0.5. Let Θ be a directed set which is an element of the basic superstructure M. Denote the set $\{\xi \in {}^*\Theta : (\forall \tau \in \Theta)\ \xi \geq \tau\}$ by ${}^a\Theta$. Elements of ${}^a\Theta$ are called *(infinitely) remote.*

Lemma. *For every directed set $\Theta \in M$, there is at least one remote element $\alpha \in {}^a\Theta$.*

◁ In the case when Θ is a finite set there is nothing to prove. So, we assume that Θ is infinite. By Lemma 4.0.3, there is a hyperfinite set A such that $\Theta \subseteq A \subseteq {}^*\Theta$. Since ${}^*\Theta$ is an internal directed set, there is an element $\alpha \in {}^*\Theta$ satisfying $\alpha \geq \tau$ for all $\tau \in A$. It is clear that $\alpha \in {}^a\Theta$. ▷

4.0.6. The main tool for applying nonstandard analysis to normed spaces is the following simple construction discovered by W. A. J. Luxemburg [20]. Let X be a normed space. Consider the two external subspaces

$$\mathrm{Fin}({}^*X) := \{\varkappa \in {}^*X : (\exists r \in \mathbb{R}) \|\varkappa\| \leq r\},$$
$$\mu({}^*X) := \{\varkappa \in {}^*X : (\exists r \in \mathbb{R})(\forall n \in \mathbb{N})\ \|n\varkappa\| \leq r\}$$

in ${}^*X$. Elements of $\mathrm{Fin}({}^*X)$ are called *(norm) limited* (or *finite in norm*) and elements of $\mu({}^*X)$ are called *infinitesimal.* Obviously, $\mu({}^*X)$ is a subspace of $\mathrm{Fin}({}^*X)$. Thus we can take the quotient space

$$\ddot{X} := \mathrm{Fin}({}^*X)/\mu({}^*X)$$

under the norm $\|[\varkappa]\| := \mathrm{st}(\|\varkappa\|)$. In the wake of W. A. J. Luxemburg, we call $\ddot{X}$ the *nonstandard hull* of X. Define the mapping $\widehat{\mu}_X : X \to \ddot{X}$ as

$$\widehat{\mu}_X(x) := [x] \quad (x \in X).$$

It is easy that $\widehat{\mu}_X$ is an embedding of X into $\ddot{X}$. The following is well known:

Proposition. *For every normed space X, the quotient $\ddot{X}$ is a Banach space and the map $\widehat{\mu}_X$ is onto if and only if X is finite-dimensional.*

4.0.7. A nonstandard construction of a norm completion of a normed space lies very closely to the construction of the nonstandard hull of the space under study. Let X be a normed space. Consider the external subspace

$$\mathrm{pns}({}^*X) := \{\varkappa \in {}^*X : (\forall n \in \mathbb{N})(\exists y \in X)\ n\rho(\varkappa - y) \leq 1\}$$

of ${}^*X$.

Proposition. *The quotient normed space $pns({}^*X)/\mu({}^*X)$ is a norm completion of X under the embedding $\widehat{\mu}_X$.*

4.0.8. Let A be a subset in a normed space X. We have a simple and useful criterion for boundedness of A.

Proposition. *The following are equivalent:*

(1) *A is a norm bounded set;*

(2) *${}^*A \subseteq Fin({}^*X)$.*

4.0.9. Let X and Y be normed spaces and let $T : X \to Y$ be a linear operator. The next well-known proposition is immediate from 4.0.8.

Proposition. *The following are equivalent:*

(1) *T is a bounded operator;*

(2) *${}^*T(Fin({}^*X)) \subseteq Fin({}^*Y)$;*

(3) *${}^*T(\mu({}^*X)) \subseteq \mu({}^*Y)$;*

(4) *${}^*T(\mu({}^*X)) \subseteq Fin({}^*Y)$.*

Thus the operator $\ddot{T} : \ddot{X} \to \ddot{Y}$, acting as $\ddot{T}([x]) := [Tx]$ for all $x \in \mathrm{Fin}({}^*X)$, is well defined and bounded together with T. This operator is the *nonstandard hull* of T.

Now we briefly present necessary facts from the theory of lattice normed spaces and dominated operators. Our exposition follows [13–16].

4.0.10. A lattice normed space is a triple (X, p, E) with X a vector space, E a vector lattice, and p a mapping $X \to E_+$ such that

(1) $p(x) = 0 \Leftrightarrow x = 0$;

(2) $p(\lambda x) = |\lambda| p(x)$ $(\lambda \in \mathbb{R},\ x \in X)$;

(3) $p(x+y) \leq p(x) + p(y)$ $(x, y \in X)$.

The mapping p is called an *E-valued norm* on X. The lattice norm p is called *decomposable* (*(d)-decomposable*) if, for all $e_1, e_2 \in E_+$ (for all disjoint $e_1, e_2 \in E_+$) and every $x \in X$, the condition $p(x) = e_1 + e_2$ implies existence of $x_1, x_2 \in X$ such that $x_1 + x_2 = x$ and $p(x_k) = e_k$ for $k = 1, 2$. A lattice normed space with decomposable ((d)-decomposable) norm is called *decomposable* (*(d)-decomposable*).

We say that a sequence (x_n) in (X, p, E) *(r)-converges* to $x \in X$ if there exist a sequence $(\varepsilon_n) \subseteq \mathbb{R}$, $\varepsilon_n \downarrow 0$, and an element $u \in E$ such that $p(x_n - x) \leq \varepsilon_n u$ for all $n \in \mathbb{N}$. By definition, a sequence $(x_n) \subseteq X$ is *(r)-Cauchy* if there is a sequence $(\varepsilon_n) \subseteq \mathbb{R}$, $\varepsilon_n \downarrow 0$, such that $p(x_k - x_m) \leq \varepsilon_n u$ for all $k, m, n \in \mathbb{N}$ such that $k, m \geq n$. A lattice normed space X is called *(r)-complete* if every (r)-Cauchy sequence in X (r)-converges to some element of X. The following assertion is a consequence of [13, 1.1.3] and the Freudenthal Spectral Theorem.

Proposition. *A (d)-decomposable (r)-complete lattice normed space is decomposable.*

4.0.11. A net $(x_\alpha)_{\alpha \in \mathrm{A}}$ in a lattice normed space X is called *(o)-convergent* to $x \in X$ if there exists a decreasing net $(e_\gamma)_{\gamma \in \Gamma}$ in E such that $e_\gamma \downarrow 0$ and, for every $\gamma \in \Gamma$, there is a subscript $\alpha(\gamma)$ for which $a(x_\alpha - x) \leq e_\gamma$ whenever $\alpha \geq \alpha(\gamma)$. A net $(x_\alpha)_{\alpha \in \mathrm{A}}$ is called *(o)-Cauchy* if the net $(x_\alpha - x_\beta)_{(\alpha,\beta) \in \mathrm{A} \times \mathrm{A}}$ (o)-converges to zero. A lattice normed space is called *(o)-complete* if every (o)-Cauchy net in it (o)-converges to an element of the space. A decomposable (o)-complete lattice normed space is said to be a *Banach–Kantorovich space.*

Consider a decomposable lattice normed space (X, p, E). Let $(\pi_\xi)_{\xi \in \Xi}$ be some partition of unity in the Boolean algebra $\mathscr{B}(E)$ and let $(x_\xi)_{\xi \in \Xi}$ be a family of elements in X. If there exists an $x \in X$ satisfying the condition

$$\pi_\xi \circ p(x - x_\xi) = 0 \quad (\xi \in \Xi),$$

then such an element x is uniquely determined. It is called the *mixing* of (x_ξ) by (π_ξ) and denoted by $\mathrm{mix}(\pi_\xi x_\xi)_{\xi \in \Xi}$ or simply $\mathrm{mix}(\pi_\xi x_\xi)$. A lattice normed space (X, p, E) is said to be *(d)-complete* if the mixing $\mathrm{mix}(\pi_\xi x_\xi) \in X$ exists for every partition of unity $(\pi_\xi) \subseteq \mathscr{B}(E)$ and every norm bounded family $(x_\xi) \subseteq X$.

Proposition [13, Theorem 1.3.2]. *A decomposable lattice normed space is (o)-complete if and only if it is (r)- and (d)-complete.*

4.0.12. Let (X, p, E) be a decomposable lattice normed space whose norm lattice E is Dedekind complete. Then there is a unique lattice normed space (X', p', E) to within isometric isomorphism with the following properties (see [13, Theorem 1.3.8]).

(1) (X', p', E) is a Banach–Kantorovich space;

(2) there exists a linear embedding $\imath : X \to X'$ such that $p'(\imath(x)) = p(x)$ for all $x \in X$;

(3) X' is the least (o)-complete lattice normed subspace of X' that contains $\imath(X)$.

The space (X', p', E) is an (o)-*completion* of the lattice normed space (X, p, E).

Consider some properties of lattice normed spaces connected with decomposability and (r)-completeness.

4.0.13. *Lemma.* *Let (X, p, E) be a decomposable lattice normed space and let I be an ideal in E. Then, for arbitrary elements $x, y \in X$, $q \in E_+$, and $\eta \in I$ satisfying the condition $p(x - y) \le q + \eta$, there is an element $y' \in X$ such that*

(1) $p(x - y') \le q$;

(2) $p(y - y') \in I$.

$\triangleleft$ Clearly, $p(x - y) \le q + \eta_+$. By the Riesz decomposition property, there are elements $a_1, a_2 \in X$ such that

$$0 \le a_1 \le q, \ 0 \le a_2 \le \eta_+, \ a_1 + a_2 = p(x - y).$$

Since p is decomposable, we may find elements $z_1, z_2 \in X$ with

$$p(z_1) = a_1, \ p(z_2) = a_2, \ z_1 + z_2 = x - y.$$

It is easy to see that the element $y' := y - z_2$ satisfies (1) and (2). $\triangleright$

4.0.14. Let (X, p, E) be an arbitrary lattice normed space. Take an ideal I in E and consider the quotient space X' of X by the subspace $X(I) := \{x \in X : p(x) \in I\}$. Given $x \in X$ $(e \in E)$, assign $[x] := x + X(I)$ (respectively, $[e] := e + I \in E/I$). Define the mapping $p' : X' \to E/I$ by the rule

$$p'([x]) := [p(x)] \quad (x \in X).$$

It is easy that the mapping p' is defined correctly and presents an E/I-valued norm on the space X'. The so-obtained lattice normed space $(X', p', E/I)$ is called the *quotient space of X by the ideal I of the norm lattice E.*

Proposition. *Let (X, p, E) be a decomposable (r)-complete lattice normed space, and let I be an ideal of the norm lattice E. Then the quotient space $(X', p', E/I)$ is decomposable and (r)-complete.*

◁ Verify that the norm p' is decomposable. Let $p'([x]) = [e_1] + [e_2]$, where $[e_1], [e_2] \in (E/I)_+$. We may assume that $e_1, e_2 \geq 0$. There is an element $\eta \in E$ such that $p(x) = e_1 + e_2 + \eta$. By Lemma 4.0.13, there is an element $x' \in X$ such that $p(x - x') \leq e_1 + e_2$ and $p(x') \in I$. In particular,

$$p(x - x') = e_1 + e_2 + \eta' \tag{1}$$

for some $\eta' \in I$. Applying the Riesz decomposition property to the inequality $p(x - x') \leq e_1 + e_2$, we find elements $e_1', e_2' \in E$ for which

$$p(x - x') = e_1' + e_2', \quad 0 \leq e_k' \leq e_k \quad (k = 1, 2). \tag{2}$$

Then $[e_k'] = [e_k]$ $(k = 1, 2)$. Indeed, supposing that for instance $[e_1 - e_1'] > 0$, in view of (1) and (2) we obtain a contradiction:

$$\begin{aligned} 0 &\leq [e_2 - e_2'] \\ &= [p(x - x') - e_1 - \eta' - e_2'] = [e_1' - e_1 - \eta'] \\ &= [e_1' - e_1] < 0. \end{aligned}$$

Since the norm p is assumed decomposable, it follows from (2) that there exist $y_1, y_2 \in X$ such that $x - x' = y_1 + y_2$ and $p(y_k) = e_k'$ $(k = 1, 2)$. Then $p'([y_k]) = [p(y_k)] = [e_k'] = [e_k]$ for $k = 1, 2$, and

$$[y_1] + [y_2] = [x - x'] = [x],$$

as required.

We now verify that X' is (r)-complete. Take some (r)-Cauchy sequence $(x_i) \subseteq X'$. There exist $e \in E$ and $(x_{i(n)}) \subseteq (x_i)$, for which

$$p'(x_{i(k)} - x_{i(m)}) \leq 2^{-n}[e] \tag{3}$$

whenever k, m, and n are such that $k, m \geq n$. Choose elements $\varkappa_n \in X$ so that $x_{i(n)} = [\varkappa_n]$. Then, by (3), there are $\eta_{k,m} \in I$ with

$$p(\varkappa_k - \varkappa_m) \leq 2^{-n}e + \eta_{k,m} \tag{4}$$

for all $k, m, n \in \mathbb{N}$ such that $k, m \geq n$. By induction, construct a sequence $(\varkappa_n') \subseteq X$ that satisfies the conditions

$$p(\varkappa_n' - \varkappa_{n+1}') \leq 2^{-n}e; \tag{5}$$

$$p(\varkappa_{n+1} - \varkappa'_{n+1}) \in I \tag{6}$$

for all $n \in \mathbb{N}$. Assign $\varkappa'_1 := \varkappa_1$ and assume that the elements $\varkappa'_j$ are already defined for $j \leq n$. From (4) it follows that

$$p(\varkappa'_n - \varkappa_{n+1}) \leq p(\varkappa_n - \varkappa_{n+1}) - p(\varkappa_n - \varkappa'_{n+1}) \leq 2^{-n}e + \eta_{n,n+1} + p(\varkappa_n - \varkappa'_n).$$

Since by the induction hypothesis $p(\varkappa_n - \varkappa'_n) \in I$, we may apply Lemma 4.0.13 to the elements

$$\varkappa'_n, \varkappa_{n+1} \in X, \quad 2^{-n}e \in E, \quad \eta := \eta_{n,n+1} + p(\varkappa_n - \varkappa'_n).$$

In result, we come to an element $\varkappa'_{n+1}$ satisfying conditions (5) and (6). It follows from (5) that the sequence $(\varkappa'_n)$ is (r)-Cauchy in X. Consequently, it (r)-converges to some element $\varkappa_0 \in X$. Then, as is easy to see, the sequence $(x_{i(n)}) = ([\varkappa_n])$ (r)-converges in the norm p'' to $[\varkappa_0] \in X'$. Since the initial sequence (x_i) is (r)-Cauchy in the norm p', we obtain $x_i \xrightarrow{(r)} [\varkappa_0]$. ▷

4.0.15. Assume that E and F are some Dedekind complete vector lattices, while (X, a, E) and (Y, b, F) are decomposable LNSs. A linear operator $T : E \to F$ is called *dominated* if there exists an order bounded linear operator $S : E \to F$ such that

$$|Tx| \leq U(|x|) \quad (x \in X).$$

The operator S is called a *dominant* of T. The least dominant of an operator T is denoted by $|T|$.

By $M(X, Y)$ we will denote the vector space of all dominated operators from an LNS X into an LNS Y. The mapping $T \mapsto |T|$ $(T \in M(E, F))$ satisfies all axioms of an E-valued norm from 4.0.10. Consequently, $M(E, F)$ is also an LNS with norm lattice the Dedekind complete vector lattice $L_b(E, F)$ (see, for example, [28, Theorem 83.4]) of all order bounded linear operators from E to F.

4.1. Saturated Sets of Indivisibles

Here we deal with lattices with zero and present some elementary facts about them, standard and nonstandard. We prove that a nonstandard enlargement of a lattice with zero contains a saturating family of indivisible elements.

4.1.1. Let L be an ordered set whose order is denoted by $\geq$. We write $x > y$ whenever $x \geq y$ and $x \neq y$. The set L is called a *lattice* if every two-element subset $\{x, y\}$ in L has the least upper bound $x \vee y := \sup\{x, y\}$ and the greatest lower bound $x \wedge y := \inf\{x, y\}$. If a lattice contains the smallest (largest) element then this element is called *zero* (*unity*) and denoted by 0 (respectively, 1). We always assume that each lattice under consideration has some zero. Elements x and y of a lattice are *disjoint* if $x \wedge y = 0$. A lattice is called *distributive* if every triple x, y, z of its elements satisfies $x \wedge (y \vee z) = (x \wedge y) \vee (x \wedge z)$ and $x \vee (y \wedge z) = (x \vee y) \wedge (x \vee z)$. A distributive lattice L with zero 0 and unity 1 is called a *Boolean algebra* if every element $x \in L$ possesses the *complement*, i.e., a (unique) element $x' \in L$ such that $x \wedge x' = 0$ and $x \vee x' = 1$.

Let L be a lattice. An element $e \in L$ is said to be a *weak* (*order*) *unity* if $e \wedge x > 0$ for every $x \in L$, $x > 0$. We call $y \in L$ a *pseudocomplement* of an element $x \in L$ if $x \wedge y = 0$ and $x \vee y$ is a weak unity. The lattice L is called a *pseudocomplemented lattice* if, for each $x \in L$, there is at least one pseudocomplement in L. An example of a pseudocomplemented lattice is a Boolean algebra. A less trivial example is the lattice of all nonnegative continuous functions on an arbitrary metric space. Also, we will use the following notation: Given an element $\varkappa$ of a nonstandard enlargement ${}^*L$, we denote by $U(\varkappa) := \{x \in E : x \geq \varkappa\}$ the set of standard upper bounds of $\varkappa$ and by $L(\varkappa) := \{y \in E : \varkappa \geq y\}$, the set of standard lower bounds of $\varkappa$.

4.1.2. Let L be a distributive lattice. An *ideal* of the lattice L is a nonempty set $I \subseteq L$ such that $x, y \in I$ implies $x \vee y \in I$, and $z \in I$ if $z \leq v$ for some $v \in I$. An ideal P of L is called *prime* if, for every $x, y \in L$, the condition $x \wedge y = 0$ implies that either $x \in P$ or $y \in P$. A prime ideal P is called *minimal* if, for every prime ideal $P_1 \subseteq L$, the condition $P_1 \subseteq P$ implies $P_1 = P$. Every subset $S \subseteq L$ such that $0 \notin S$ and $x, y \in S$ implies $x \wedge y \in S$ is called a *lower sublattice*. A lower sublattice S is called *maximal* or an *ultrafilter* if for every lower sublattice $S_1 \subseteq L$ the condition $S \subseteq S_1$ implies $S_1 = S$. The following assertion is well-known [21, Theorem 5.4].

Lemma. *Let L be a distributive lattice and let P be some prime ideal in L. Then $L \setminus P$ is a lower sublattice of L. Furthermore, $L \setminus P$ is an ultrafilter if and only if P is a minimal prime ideal.*

4.1.3. Consider a nonstandard enlargement ${}^*L$ of a lattice L. By the transfer principle, ${}^*L$ has the same zero element 0 as the initial lattice L. We give two important definitions.

DEFINITION 1. An element $\varkappa \in {}^*L$ is called *indivisible* if $\varkappa > 0$ and, for every $x \in L$, either $x \geq \varkappa$ or $x \wedge \varkappa = 0$.

DEFINITION 2. A subset Λ of the lattice ${}^*L$ is called *saturating* if Λ is internal and, for every $x \in L$, $x > 0$, there is some $a \in \Lambda$ such that $x \geq a > 0$.

The following simple lemma is a key for many of our further results.

Lemma. *A nonstandard enlargement of an arbitrary lattice contains a hyperfinite saturating set of disjoint indivisible elements.*

$\triangleleft$ Let L be a lattice. We denote by $\mathscr{F}$ the family of all finite subsets of $L \setminus \{0\}$ and put

$$\begin{aligned}\mathscr{L}_\pi := \{X \in {}^*\mathscr{F} : (\forall y \in \pi)(\forall x \in X)(y \geq x \text{ or } y \wedge x = 0)\\ \& \ (\forall y \in \pi)(\exists x \in X)(y \geq x)\\ \& \ (\forall x \in X)(\forall z \in X)(x \neq z \ \to \ x \wedge z = 0)\}\end{aligned}$$

for all $\pi \in \mathscr{F}$. Show that the sets $\mathscr{L}_\pi$ are nonempty. Take an arbitrary $\pi \in \mathscr{F}$. For every element $x \in \pi$, there is a set B_x such that $x \in B_x \subseteq \pi$, $\inf B_x > 0$, and $\inf(B_x \cup \{y\}) = 0$ for all $y \in \pi \setminus B_x$. It is easy to verify that the set $\{\inf B_x : x \in \pi\}$ belongs to $\mathscr{L}_\pi$. The fact that the sets $\mathscr{L}_\pi$ are nonempty and the condition $\mathscr{L}_\pi \cap \mathscr{L}_\gamma = \mathscr{L}_{\pi \cup \gamma}$ implies that the family $\{\mathscr{L}_\pi\}_{\pi \in \mathscr{F}}$ enjoys the finite intersection property. All elements of this family are internal sets by construction. Thus, by the general saturation principle, there is a $\Lambda \in \bigcap_{\pi \in \mathscr{F}} \mathscr{L}_\pi$. It is easy to see that Λ is a desired saturating family of disjoint indivisible elements. $\triangleright$

4.1.4. Let L be a lattice. By Lemma 4.1.3, there exists a saturating set of indivisible elements in the lattice ${}^*L$. Let Λ be such a set. Denote $\Lambda^x := \{\varkappa \in \Lambda : x \geq \varkappa\}$ for all $x \in L$. It is easy to see that, in the case when L possesses a weak unity e, the family $\{\Lambda^x : x \in L\}$ is an open base for a topology τ on Λ. This topology is called *canonical* on Λ.

Theorem. *Let Λ be a saturating set of indivisible elements in a nonstandard enlargement of a lattice with weak unity and let τ be the canonical topology on Λ. Then (Λ, τ) is a compact space.*

$\triangleleft$ Assume that (Λ, τ) is not compact. In this case we may extract a subset $\{\Lambda^x\}_{x \in X}$ from the base $\{\Lambda^x\}_{x \in L}$ of the topology τ such that $\Lambda = \bigcup_{x \in X} \Lambda^x$ and, for every finite subset π of X, the following holds:

$$A_\pi := \Lambda \setminus \bigcup_{x \in \pi} \Lambda^x \neq \varnothing.$$

It is easy to verify that $\{A_\pi : \pi \in \mathscr{P}_{\text{fin}}(X)\}$ is a family of internal sets with the finite intersection property. By the general saturation principle, there is an element $\varkappa \in \bigcap \{A_\pi : \pi \in \mathscr{P}_{\text{fin}}(X)\}$. Then $\varkappa \in \Lambda \setminus \bigcup_{x \in X} \Lambda^x$, a contradiction with the fact that $\{\Lambda^x\}_{x \in X}$ is an open covering of Λ. Thus, (Λ, τ) is a compact space. $\triangleright$

4.1.5. We introduce some equivalence relation $\sim$ in the lattice ${}^*L$ by putting $\varkappa_1 \sim \varkappa_2$ whenever the inequalities $x \geq \varkappa_1$ and $x \geq \varkappa_2$ are equivalent for all $x \in L$. Suppose that a lattice L possesses a weak unity e. Take a saturating set Λ of indivisible elements in ${}^*L$ (such a set exists by Lemma 4.1.3). Let τ be the canonical topology on Λ. The topological space (Λ, τ) is compact by Theorem 4.1.4. Its quotient space by the equivalence $\sim$ is a compact T_0-space. We denote this quotient space by $\ddot{\Lambda}$. It is clear that the sets of the type

$$\ddot{\Lambda}^x := \{[\varkappa] \in \ddot{\Lambda} : x \geq \varkappa\} \quad (x \in L)$$

form an open base for the quotient topology (throughout, we denote by $[\varkappa]$ the coset containing an element $\varkappa \in \Lambda$).

Take a Boolean algebra B as L and consider a saturating set Υ of indivisible elements in ${}^*B$. It is easy to see that $\ddot{\Upsilon}$ is a totally disconnected compact Hausdorff space, while the mapping associating with each element $b \in B$ the subset $\ddot{\Upsilon}^b$ of the space $\ddot{\Upsilon}$ is a Boolean isomorphism of B onto the algebra $\operatorname{clop}(\ddot{\Upsilon})$ of clopen (closed and open) subsets of the compact Hausdorff space $\ddot{\Upsilon}$. Thus we obtained the following

Theorem. *Let B be a Boolean algebra and let Υ be a saturating set of indivisible elements in a nonstandard enlargement of B. Then the corresponding topological space $\ddot{\Upsilon}$ is the Stone space of B.*

4.1.6. The following theorem describes connection between the properties of a lattice and the corresponding topological space.

Theorem. *Let L be a distributive lattice with weak unity. Then the following are equivalent:*

(1) *L is a pseudocomplemented lattice;*

(2) *The topological space $\ddot{\Lambda}$ is totally disconnected for every saturating set Λ of indivisible elements in ${}^*L$;*

(3) *The topological space $\ddot{\Lambda}$ satisfies the T_1-separation axiom for every saturating set Λ of indivisible elements in ${}^*L$;*

(4) *The set $\{x \in L : x \wedge \varkappa = 0\}$ is a minimal prime ideal in L for every indivisible element $\varkappa \in {}^*L$.*

$\lhd$ (1)$\to$(2): It is easy to see that if condition (1) holds, then the base $\{\ddot{\Lambda}^x\}_{x \in L}$ for the topology of $\ddot{\Lambda}$ consists of clopen sets. Indeed, given an $x \in L$, we have $\ddot{\Lambda}^x \cup \ddot{\Lambda}^y = \ddot{\Lambda}$ and $\ddot{\Lambda}^x \cap \ddot{\Lambda}^y = \varnothing$, where y is some pseudocomplement of x.

(2)$\to$(3): Obvious.

(3)→(4): Let condition (3) be satisfied. Take some indivisible element $\varkappa \in {}^*L$ and consider the set $I_\varkappa := \{x \in L : x \wedge \varkappa = 0\}$. It is easy to see that $I_\varkappa$ is a prime ideal in the lattice L. Indeed, by distributivity, it follows from $x, y \in I_\varkappa$ that $(x \vee y) \wedge \varkappa = (x \wedge \varkappa) \vee (x \wedge \varkappa) = 0$, and consequently $x \vee y \in I_\varkappa$. If $x \wedge y \in I_\varkappa$ then either $x \in I_\varkappa$ or $y \in I_\varkappa$ (otherwise, since the element $\varkappa$ is indivisible, we would have $x \geq \varkappa$ and $y \geq \varkappa$). It remains to verify that the ideal $I_\varkappa$ is minimal.

Take an arbitrary prime ideal $P \subseteq I_\varkappa$. Assume that $y \in I_\varkappa \backslash P$ for some element $x \in L$. Then $x \wedge y > 0$ for every $x \in L \backslash I_\varkappa$. Indeed, otherwise it would be valid that $x \wedge y = 0$, and hence we would have either $x \in P$ or $y \in P$, which is impossible. By Lemma 4.1.3, there exists a saturating set of indivisible elements in the lattice ${}^*L$. Let Λ' be such a set. Assign $\Lambda := \Lambda' \cup \{\varkappa\}$. Then Λ is also a saturating set of indivisible elements, and $\varkappa \in \Lambda$. As was mentioned above, $x \wedge y > 0$ for every $x \in L \setminus I_\varkappa$. Using this observation, it is easy to show that $\{\Lambda^{x \wedge y}\}_{x \in L \backslash I_\varkappa}$ is a system of internal sets with the finite intersection property. Applying the general saturation principle, we find an element $\delta \in \Lambda$ such that

$$\delta \in \cap \{\Lambda^{x \wedge y} : x \in L \setminus I_\varkappa\}.$$

The indivisible element δ satisfies the condition $\delta \leq y$. At the same time, $\varkappa \wedge y = 0$ because $y \in I_\varkappa$. Consequently, $\delta \not\geq \varkappa$. By condition (3), the topological space $\bar{\Lambda}$ satisfies the T_1-separation axiom. Therefore, there is $z \in L$ for which $[\varkappa] \in \bar{\Lambda}^z$ and $[\delta] \notin \bar{\Lambda}^z$. Then the relations $\varkappa \leq z$ and $z \wedge \delta = 0$ are valid. The first of them implies $z \in L \backslash I_\varkappa$, which contradicts the second relation. The obtained contradiction shows that $I_\varkappa \setminus P = \varnothing$. Since the choice of the prime ideal P satisfying the condition $P \subseteq I_\varkappa$ was arbitrary, $I_\varkappa$ is a minimal prime ideal.

(4)→(1): Suppose now that condition (4) is satisfied. Show that every element of the lattice L has a pseudocomplement. Take an arbitrary $a \in L$. By Lemma 4.1.3, there exists some saturating set Λ of indivisible elements in ${}^*L$. Consider the topological space (Λ, τ), where τ is the canonical topology in Λ. Let $\varkappa \in \Lambda \setminus \Lambda^a$. By hypothesis, the set

$$I_\varkappa := \{x \in L : x \wedge \varkappa = 0\}$$

is a minimal prime ideal of L. By the choice of $\varkappa$, we have $a \wedge \varkappa = 0$, and so $a \notin L \setminus I_\varkappa$. Since $L \setminus I_\varkappa$ is an ultrafilter of the lattice L by Lemma 4.1.2, there exists an element $y(\varkappa) \in L \setminus I_\varkappa$ such that $y(\varkappa) \wedge a = 0$. In other words, $\varkappa \in \Lambda^{y(\varkappa)}$ and $\Lambda^a \cap \Lambda^{y(\varkappa)} = 0$. The family $\{\Lambda^{y(\varkappa)}\}_{\varkappa \in \Lambda \backslash \Lambda^a}$ is an open covering of the closed set $\Lambda \setminus \Lambda^a$ in the space (Λ, τ). By Theorem 4.1.4, it contains a finite subcovering $\{\Lambda^{y(\varkappa_k)}\}_{k=1}^n$. The element $b := \bigvee_{k=1}^n y(\varkappa_k)$ satisfies the conditions

$$\Lambda^b \cap \Lambda^a = \varnothing, \quad \Lambda^b \cup \Lambda^a = \Lambda$$

and, consequently, it is the desired pseudocomplement of a. ▷

4.1.7. Let L be a distributive lattice. If L is pseudocomplemented then, by Theorem 4.1.6, for every indivisible element $\varkappa \in {}^*L$, there exists a respective minimal prime ideal $I_\varkappa := \{x \in L : x \wedge \varkappa = 0\}$. The converse is true in a more general setting:

Lemma. *Let I be a minimal prime ideal in a distributive lattice L. Then there is an indivisible element $\varkappa \in {}^*L$ such that $I = \{x \in L : x \wedge \varkappa = 0\}$.*

$\lhd$ Observe that the subset $U := L \setminus I$ in L is directed downwards. By Lemma 4.0.5, there exists a remote element $\varkappa \in {}^*U$. An easy check shows that $\varkappa$ is an indivisible element in the lattice ${}^*L$ and $I = \{x \in L : x \wedge \varkappa = 0\}$. $\rhd$

4.1.8. Let L be a distributive lattice. Denote by $\mathscr{M}$ the set of all minimal prime ideals in L. The set $\mathscr{M}$ is equipped with the canonical topology generated by the open base of all sets of the form

$$\mathscr{M}^u := \{P \in \mathscr{M} : u \notin P\} \quad (u \in L)$$

(see, for example, [21, Section 7]).

Theorem. *Let L be a pseudocomplemented distributive lattice. Then, for every saturating set Λ of indivisible elements of ${}^*L$, the mapping φ_Λ, defined by the rule*

$$\varphi_\Lambda([\varkappa]) := \{x \in L : x \wedge \varkappa = 0\} \quad ([\varkappa] \in \ddot{\Lambda}),$$

is a homeomorphism of the topological space $\ddot{\Lambda}$ onto $\mathscr{M}$.

$\lhd$ Let Λ be a saturating set of indivisible elements in the lattice ${}^*L$. By Theorem 4.1.6, the mapping φ_Λ ranges in the space $\mathscr{M}$. The mapping φ_Λ is injective. Indeed, take arbitrary elements $\varkappa_1, \varkappa_2 \in \Lambda$ such that $\varkappa_1 \not\sim \varkappa_2$. Then

$$U(\varkappa_1) \neq U(\varkappa_2), \varphi_\Lambda([\varkappa_1]) \neq \varphi_\Lambda([\varkappa_2]),$$

since

$$\varphi_\Lambda([\varkappa_1]) = L \setminus U(\varkappa)$$

for all $\varkappa \in \Lambda$. Show that $\varphi_\Lambda(\ddot{\Lambda}) = \mathscr{M}$. Take an arbitrary $P \in \mathscr{M}$. As is easy to verify, $\{\Lambda^x\}_{x \in L \setminus P}$ is a system of internal sets with the finite intersection property. Thus, by the general saturation principle we may find $\varkappa \in \bigcap_{x \in L \setminus P} \Lambda^x$. Then

$$\begin{gathered}\varphi_\Lambda([\varkappa]) = \{x \in L : x \wedge \varkappa = 0\} \\ = \{x \in L : \varkappa \notin \Lambda^x\} = L \setminus \{x \in L : \varkappa \in \Lambda^x\} \\ \subseteq L \setminus (L \setminus P) = P\end{gathered}$$

and so $\varphi_\Lambda([\varkappa]) = P$, because the ideal P is minimal. It remains to verify that φ_Λ is a homeomorphism. This readily follows on observing that

$$\varphi_\Lambda(\bar{\Lambda}^x) = \{\varphi_\Lambda([\varkappa]) : \varkappa \leq x\} = \{P \in \mathscr{M} : x \notin P\} = \mathscr{M}^x. \ \triangleright$$

4.1.9. Let Λ_1 and Λ_2 be saturating sets of indivisible elements in a distributive pseudocomplemented lattice L. Then, by the preceding theorem, the mapping $\psi := \varphi_{\Lambda_2}^{-1} \circ \varphi_{\Lambda_1}$ is a homeomorphism of the topological space $\bar{\Lambda}_1$ onto $\bar{\Lambda}_2$. Note that this homeomorphism can be defined explicitly as follows:

$$\psi([\varkappa_1]) := \{\varkappa_2 \in \Lambda_2 : U(\varkappa_1) \geq \varkappa_2\}$$

for every element $\varkappa_1 \in \Lambda_1$. Thus, the topological space $\bar{\Lambda}$ is uniquely determined to within a homeomorphism by the distributive pseudocomplemented lattice L and does not depend on the choice of a saturating set Λ of indivisible elements.

4.2. Representation of Archimedean Vector Lattices

In this section, we prove some nonstandard variant of the representation theorem for Archimedean vector lattices. We then give nonstandard proofs for the Brothers Kreĭn–Kakutani and Ogasawara–Vulikh representation theorems.

Throughout this section we suppose that E is an Archimedean vector lattice. The positive cone E_+ of E is a distributive lattice with zero. Therefore, by Lemma 4.1.3, there exists a saturating set of indivisible elements in ${}^*E_+$. Here we fix such a set and denote it by Λ up to the end of this section.

4.2.1. Let $e \in E$ and $\varkappa \in \Lambda$ be such that $e \geq \varkappa$. Given $f \in E$, define the element $f^\wedge(\varkappa)$ of $\overline{\mathbb{R}}$ as

$$f^\wedge(\varkappa) := \inf\{\lambda \in \mathbb{R} : (\lambda e - f)_+ \geq \varkappa\}. \tag{1}$$

Granted $f \in E$, let $\mathscr{D}(f)$ stand for the subset $\{\varkappa \in \Lambda : |f^\wedge(\varkappa)| < \infty\}$ of Λ. We establish some properties of the mapping $f \mapsto f^\wedge(\varkappa)$.

Lemma. *For every $f, g \in E$ and every $\alpha \in \mathbb{R}$, the following hold:*

(1) $f^\wedge(\varkappa) = \sup\{\lambda \in \mathbb{R} : (\lambda e - f)_- \geq \varkappa\}$;

(2) $(\alpha f)^\wedge(\varkappa) = \alpha f^\wedge(\varkappa)$;

(3) $(f + g)^\wedge(\varkappa) = f^\wedge(\varkappa) + g^\wedge(\varkappa)$ *for all* $\varkappa \in \mathscr{D}(f) \cap \mathscr{D}(g)$;

(4) $(f \wedge g)^{\wedge}(\varkappa) = \min\{f^{\wedge}(\varkappa),\, g^{\wedge}(\varkappa)\}$
and $(f \vee g)^{\wedge}(\varkappa) = \max\{f^{\wedge}(\varkappa),\, g^{\wedge}(\varkappa)\}$.

$\lhd$ (1): Denote the right side of the equality under proof by $f^{\sim}(\varkappa)$. We consider only the case in which both $f^{\wedge}(\varkappa)$ and $f^{\sim}(\varkappa)$ are finite. Let $\alpha > f^{\wedge}(\varkappa)$. Then $(\alpha e - f)_+ \geq \varkappa$, and so $(\alpha e - f)_- \wedge \varkappa = 0$. Consequently, $\alpha \geq f^{\sim}(\varkappa)$, which implies $f^{\wedge}(\varkappa) \geq f^{\sim}(\varkappa)$, since the choice of the number $\alpha > f^{\wedge}(\varkappa)$ is arbitrary. Conversely, suppose that $\alpha > f^{\sim}(\varkappa)$. Then $(\alpha e - f)_- \not\geq \varkappa$, and hence $(\alpha e - f)_- \wedge \varkappa = 0$, because $\varkappa$ is an indivisible element. At the same time, since $e \geq \varkappa$, we have

$$((\alpha + 1/n)e - f)_+ + (\alpha e - f)_- \geq (1/n)e \geq \varkappa$$

for all natural n. Hence

$$((\alpha + 1/n)e - f)_+ \geq \varkappa \text{ and } \alpha + 1/n \geq f^{\wedge}(\varkappa)$$

for every $\alpha > f^{\sim}(\varkappa)$ and $n \in \mathbb{N}$, which is possible only if $f^{\sim}(\varkappa) \geq f^{\wedge}(\varkappa)$.

(2): Omitting easy verification of the relation $(\alpha f)^{\wedge}(\varkappa) = \alpha f^{\wedge}(\varkappa)$ with $0 \leq \alpha < \infty$, we show only that $(-f)^{\wedge}(\varkappa) = -f^{\wedge}(\varkappa)$. Indeed, the required condition follows from the equalities

$$(-f)^{\wedge}(\varkappa) = \inf\{\lambda : (\lambda e - f)_+ \geq \varkappa\}$$
$$= -\sup\{\beta : (-\beta e + f)_+ \geq \varkappa\} = -\sup\{\beta : (\beta e - f)_- \geq \varkappa\} = -f^{\wedge}(\varkappa).$$

The last equality is valid by assertion (1) proven above.

(3): Let $\varkappa \in \mathscr{D}(f) \cap \mathscr{D}(g)$. Observe that the conditions $(\lambda e - f)_+ \geq \varkappa$ and $(\beta e - g)_+ \geq \varkappa$ imply

$$((\lambda + \beta)e - (f + g))_+ = ((\lambda e - f) + (\beta e - g)_+)$$
$$\geq ((\lambda e - f) \wedge (\beta e - g))_+ = (\lambda e - f)_+ \wedge (\beta e - g)_+ \geq \varkappa.$$

The following inequality is easy from this remark:

$$f^{\wedge}(\varkappa) + g^{\wedge}(\varkappa) = \inf\{\lambda : (\lambda e - f)_+ \geq \varkappa\} + \inf\{\beta : (\beta e - g)_+ \geq \varkappa\}$$
$$\geq \inf\{\gamma : (\gamma e - (f + g))_+ \geq \varkappa\} = (f + g)^{\wedge}(\varkappa).$$

Replacing f by $-f$ and g by $-g$ and applying (2), we obtain the reverse inequality. Thus we have $f^{\wedge}(\varkappa) + g^{\wedge}(\varkappa) = (f + g)^{\wedge}(\varkappa)$, as required.

(4): It suffices to prove that $(f \wedge g)^{\wedge}(\varkappa) = \min\{f^{\wedge}(\varkappa), g^{\wedge}(\varkappa)\}$. Since

$$(\lambda e - (f \wedge g))_- = l((f \wedge g) - \lambda e)_+ = ((f - \lambda e) \wedge (g - \lambda e))_+$$
$$= (f - \lambda e)_+ \wedge (g - \lambda e)_+ = (\lambda e - f)_- \wedge (\lambda e - g)_-,$$

the condition $(\lambda e - (f \wedge g))_- \geq \varkappa$ is valid if and only if $(\lambda e - f)_- \geq \varkappa$ and $(\lambda e - g)_- \geq \varkappa$. The required assertion follows now from (1). $\rhd$

4.2.2. Let $\varkappa \in \Lambda$ and $e \in E$ be such that $e \geq \varkappa$. Consider the mapping $h_\varkappa : E \to \overline{\mathbb{R}}$ assigning to each $f \in E$ the element $f^\wedge(\varkappa)$ defined by (1). By the preceding lemma, the restriction of the mapping $h_\varkappa$ onto the vector sublattice $E_\varkappa := \{x \in E : |h_\varkappa(x)| < \infty\}$ of the lattice E is an $\mathbb{R}$-valued Riesz homomorphism on $E_\varkappa$.

Let h be an arbitrary $\mathbb{R}$-valued Riesz homomorphism on E. By the general saturation principle, there exists an element $\varkappa \in \Lambda$ satisfying the condition $\varkappa \leq x$ for every $x \in E_+$ whenever $h(x) > 0$. Take an arbitrary $e \in E_+$ with $h(e) = 1$. Clearly, $h = h_\varkappa$. In other words, each real-valued Riesz homomorphism on the vector lattice E may be represented as $h_\varkappa$.

4.2.3. Take some maximal family $(e_\sigma)_{\sigma\in S}$ of disjoint nonzero positive elements in a vector lattice E. Put

$$^0\Lambda := \{\varkappa \in \Lambda : (\exists \sigma \in S)\ \varkappa \leq e_\sigma\} \tag{2}$$

and consider the family of subsets

$$^0\Lambda^x := \{\varkappa \in \Lambda : \varkappa \leq x\} \quad (x \in E_S)$$

of the set $^0\Lambda$, where E_S is the union of the order intervals $I_\sigma = [0, e_\sigma]$ for all $\sigma \in S$. It is easy to see that $\{^0\Lambda^x\}_{x\in E_S}$ is a base for some topology τ on $^0\Lambda$. Throughout this section, we denote by $(^0\Lambda, \tau)$ the corresponding topological space. It is easy to verify that the condition

$$f^\wedge(\varkappa) := \inf\{\lambda \in \mathbb{R} : (\lambda e_{\sigma(\varkappa)} - f)_+ \geq \varkappa\} \tag{3}$$

soundly defines an $\overline{\mathbb{R}}$-valued function on $(^0\Lambda, \tau)$. We now state the main result of the section:

Theorem. *Under the above assumptions, the function $f^\wedge$ belongs to $C_\infty(^0\Lambda)$ for every $f \in E$. Moreover, the mapping assigning to each element $f \in E$ the function $f^\wedge$ is a Riesz isomorphism of the vector lattice E onto the vector sublattice $f^\wedge(E)$ of the space $C_\infty(^0\Lambda)$.*

$\triangleleft$ We show that the functions defined by (3) are continuous in the topology of $^0\Lambda$. Take an arbitrary $f \in E$. It suffices to establish the continuity of the function $f^\wedge$ on the subspaces $^0\Lambda^{e_\sigma}$ $(\sigma \in S)$ of $^0\Lambda$. Fix some $\sigma \in S$ and let $e := e_\sigma$. Consider the sets

$$P_\lambda := \{\varkappa \in {}^0\Lambda^e : (\lambda e - f)_+ \geq \varkappa\},$$
$$N_\lambda := \{\varkappa \in {}^0\Lambda^e : (\lambda e - f)_- \geq \varkappa\}$$

for all $\lambda \in \mathbb{R}$. Then $\{P_\lambda\}_{\lambda\in\mathbb{R}}$ is an increasing family of open subsets of ${}^0\Lambda^e$, while $\{N_\lambda\}_{\lambda\in\mathbb{R}}$ is a decreasing family; moreover, $P_\lambda \cap N_\lambda = \varnothing$ for all λ. In addition, since

$$(se - f)_+ + (te - f)_+ \geq (s - t)e$$

for arbitrary $s, t \in \mathbb{R}$, $s > t$, we have $P_s \cup N_t = {}^0\Lambda^e$. Hence,

$$\mathrm{cl}P_t \subseteq {}^0\Lambda^e \setminus N_t \subseteq P_s = \mathrm{int}P_s \quad (s > t). \tag{4}$$

By the definition of $f^\wedge$, the following holds for all $\varkappa \in {}^0\Lambda^e$:

$$f^\wedge(\varkappa) = \inf\{\lambda \in \mathbb{R} : \varkappa \in P_\lambda\}. \tag{5}$$

It is easy to verify that conditions (4) and (5) imply continuity of $f^\wedge$ on ${}^0\Lambda^e$, as required.

Now we show that functions of the form $f^\wedge$ are finite on dense subsets of the space ${}^0\Lambda$. As in the proof of continuity, we confine exposition to considering the functions on the subspaces ${}^0\Lambda^e$, where e is some element of the family $(e_\sigma)_{\sigma\in S}$. Thus, we must prove that the set $\mathscr{D}(f)$ is dense in ${}^0\Lambda^e$ for every $f \in E$. Take an arbitrary $f \in E$ (we may assume that $f \geq 0$) and suppose that an element $u \in E$, $0 < u \leq e$, satisfies $f^\wedge(\varkappa) = \infty$ $(\varkappa \in {}^0\Lambda^u)$. Then $u = 0$. Indeed, the condition $(\lambda e - f)_+ \geq \varkappa$ fails for all $\varkappa \in {}^0\Lambda^e$ and $\varkappa \leq u$. Since elements of the set ${}^0\Lambda^e$ are indivisible; therefore,

$$\varkappa \wedge (\lambda e - f)_+ = 0 \text{ for all } \varkappa \in {}^0\Lambda^u, \quad \lambda \in \mathbb{R}. \tag{6}$$

The set Λ is saturating, and so (6) implies that the elements u and $(\lambda e - f)_+$ of the lattice E are disjoint for all $\lambda \in \mathbb{R}$. Consequently,

$$u \wedge (e - (1/n)f)_+ = 0 \quad (n \in \mathbb{N}).$$

It follows that

$$u \wedge e = u \wedge \sup_E\{(e - (1/n)f)_+ : n \in \mathbb{N}\} = 0,$$

because the lattice E is Archimedean. At the same time, $u \leq e$. Hence, $u = 0$. Thus, the set $\{f^\wedge = \infty\}$ does not contain any nonempty open subset of the space Λ.

By Lemma 4.2.1, the mapping $f \mapsto f^\wedge$ is a Riesz homomorphism of the vector lattice E onto the vector sublattice $f^\wedge(E)$ of the space $C_\infty({}^0\Lambda)$.

To complete the proof of the theorem, it remains to establish that this mapping is injective. To this end, it suffices to verify that the conditions $f \in E_+$ and $f^\wedge = 0$

imply $f = 0$. Let an element $f \in E_+$ satisfy $f^\wedge(\varkappa) = 0$ for all $\varkappa \in {}^0\Lambda$. Choose an arbitrary $\sigma \in S$. Then

$$\inf\{\lambda : (\lambda e_\sigma - f)_+ \geq \varkappa\} = 0 \quad (\varkappa \leq e_\sigma),$$

and so $(f - (1/n)e_\sigma)_+ \wedge \varkappa = 0$ for all $n \in \mathbb{N}$ and $\varkappa \leq e_\sigma$. Since the set Λ is saturating, it follows that

$$(f - (1/n)e_\sigma)_+ \wedge e_\sigma = 0 \quad (n \in \mathbb{N}).$$

The vector lattice E is Archimedean. Therefore, the last relation implies

$$\begin{aligned} e_\sigma \wedge f &= e_\sigma \wedge \sup\{(f - (1/n)e_\sigma)_+ : n \in \mathbb{N}\} \\ &= \sup\{(f - (1/n)e_\sigma)_+ \wedge e_\sigma : n \in \mathbb{N}\} = 0. \end{aligned}$$

Thus, since the choice of $\sigma \in S$ was arbitrary, the element f is disjoint from every element of the family $(e_\sigma)_{\sigma \in S}$, which is possible (since this family is maximal) only if $f = 0$. So, the mapping $f \mapsto f^\wedge$ is injective. The proof of the theorem is complete. $\triangleright$

4.2.4. Define some equivalence relation $\mathscr{R}$ on Λ as follows: $\varkappa_1 \mathscr{R} \varkappa_2$ means that $f^\wedge(\varkappa_1) = f^\wedge(\varkappa_2)$ for every $f \in E$. By Theorem 4.1.4, Λ is a compact topological space. It follows immediately that the quotient space $\Lambda_{\mathscr{R}}$ of Λ by $\mathscr{R}$ is compact too. This quotient space is Hausdorff by construction. Given $\varkappa \in \Lambda$, denote by $\langle \varkappa \rangle$ the coset of $\varkappa$ in the space $\Lambda_{\mathscr{R}}$. It is easy to see that the formula

$$\varphi(f)(\langle \varkappa \rangle) := f^\wedge(\varkappa) \quad (f \in E,\ \varkappa \in \Lambda) \tag{7}$$

soundly defines the mapping $\varphi : E \to C_\infty(\Lambda_{\mathscr{R}})$, where $C_\infty(\Lambda_{\mathscr{R}})$ is the space of extended continuous functions on the compact Hausdorff space $\Lambda_{\mathscr{R}}$. The following lemma is a consequence of Theorem 4.2.3 and the definition of φ.

Lemma. *The mapping φ is a Riesz isomorphism of the vector lattice E onto the vector sublattice $\varphi(E)$ of $C_\infty(\Lambda_{\mathscr{R}})$. Furthermore, $\varphi(E)$ separates points of $\Lambda_{\mathscr{R}}$, and φ maps the element e to the identically one function.*

4.2.5. Let E be a relatively uniformly complete Archimedean vector lattice with a strong unity e. Then, by the preceding lemma, E is Riesz isomorphic to the vector sublattice $\varphi(E)$ of the space $C(\Lambda_{\mathscr{R}})$ of continuous functions on the compact Hausdorff space $\Lambda_{\mathscr{R}}$; moreover, $\varphi(E)$ separates points of $\Lambda_{\mathscr{R}}$ and contains all constant functions. Since E is relatively uniformly complete, the sublattice $\varphi(E)$ is uniformly closed in $C(\Lambda_{\mathscr{R}})$. Applying the Stone theorem, we obtain $\varphi(E) = C(\Lambda_{\mathscr{R}})$. Thus, we have

Theorem (S. Kakutani; M. G. Kreĭn and S. G. Kreĭn). *For every relatively uniformly complete Archimedean vector lattice E with a strong unity e, there exists a compact Hausdorff space Q such that E is Riesz isomorphic to the vector lattice $C(Q)$. Moreover, such an isomorphism may be constructed so as to send the element e to the identically one function.*

4.2.6. We also give a sketch of a nonstandard proof of the Ogasawara–Vulikh Theorem.

Theorem (T. Ogasawara; B. Z. Vulikh). *For every Dedekind complete vector lattice E with unity e, there is an extremally disconnected compact Hausdorfl space Q such that E is Riesz isomorphic to an order dense ideal E' of the Dedekind complete vector lattice $C_\infty(Q)$. Moreover, some isomorphism may be constructed so that $C(Q) \subseteq E'$ and the identically one function corresponds to e.*

$\lhd$ Let E be a Dedekind complete vector lattice with unity e. Take $\Lambda_{\mathscr{R}}$ as the compact Hausdorff space Q. We first verify that $\Lambda_{\mathscr{R}}$ is extremally disconnected. It suffices to show that the closure of the union of every family of sets in some base for the topology on $\Lambda_{\mathscr{R}}$ is open. Consider the base $\{\Lambda_{\mathscr{R}}^x\}_{x\in E_+}$ of the topology of the space $\Lambda_{\mathscr{R}}$ constituted by the sets

$$\Lambda_{\mathscr{R}}^x := \{\langle\varkappa\rangle \in \Lambda_{\mathscr{R}} : \varkappa \le x\}.$$

Take an arbitrary family $\{\Lambda_{\mathscr{R}}^x\}_{x\in A}$ of sets in this base. The closure of the union $\bigcup_{x\in A}\Lambda_{\mathscr{R}}^x$ is open, because it coincides with the set $\Lambda_{\mathscr{R}}^y$ where y is the band projection of e onto the band generated by the set A. The space $C_\infty(\Lambda_{\mathscr{R}})$ of extended continuous functions on the extremally disconnected compact Hausdorff space $\Lambda_{\mathscr{R}}$ is a Dedekind complete vector lattice (see [21, Theorem 47.4]). By Lemma 4.2.1, the mapping φ defined by (7) is a Riesz isomorphism of E onto the point-separating vector sublattice $\varphi(E)$ of $C_\infty(\Lambda_{\mathscr{R}})$; furthermore,

$$\varphi(e)[x] = 1 \text{ for all } x \in \Lambda_{\mathscr{R}}.$$

To complete the proof, it remains to show that $\varphi(E)$ is an order-dense ideal in $C_\infty(\Lambda_{\mathscr{R}})$. The vector lattice E is Dedekind complete. So, it is relatively uniform complete. According to the result of the preceding subsection, $\varphi(E_e) = C(\Lambda_{\mathscr{R}})$, where E_e is the principal ideal in E generated by the element e. Thus, the vector lattice $\varphi(E)$ contains the order-dense ideal $C(\Lambda_{\mathscr{R}})$ of $C_\infty(\Lambda_{\mathscr{R}})$. Therefore, to be an order-dense ideal in $C_\infty(\Lambda_{\mathscr{R}})$, together with each element $\varphi(x) \ge 0$ the set $\varphi(E)$ must contain all elements $f \in C_\infty(\Lambda_{\mathscr{R}})$ with $0 \le f \le \varphi(x)$. Take an arbitrary $x \in E_+$ and let a function $f \in C_\infty(\Lambda_{\mathscr{R}})$ be such that $0 \le f \le \varphi(x)$. Consider the elements

$$f_n := f \wedge n\varphi(e) \quad (n \in \mathbb{N})$$

of the space $C_\infty(\Lambda_{\mathscr{R}})$. It is clear that $f_n \in C(\Lambda_{\mathscr{R}})$. Consequently, there are $y_n \in E$ with $f_n = \varphi(y_n)$. Since φ is a Riesz isomorphism, $y_n \uparrow \leq x$. By Dedekind completeness of E, there exists $y = \sup_E\{y_n : n \in \mathbb{N}\}$. Obviously, we obtain $f = \varphi(y)$. ▷

4.2.7. In conclusion, we show that in the case when E is a lattice with the principal projection property, the equivalence relation $\mathscr{R}$ can be described in a simpler manner.

Lemma. *Suppose that a vector lattice E has the principal projection property. Then for every $\varkappa_1, \varkappa_2 \in \Lambda$ we have $\varkappa_1 \mathscr{R} \varkappa_2$ if and only if $\varkappa_1$ and $\varkappa_2$ have the same standard upper bounds in E.*

◁ Assume that the elements $\varkappa_1$ and $\varkappa_2$ have the same sets of standard upper bounds. Then $\varkappa_1 \mathscr{R} \varkappa_2$ follows immediately from the definition of $\mathscr{R}$. Conversely, assume that the sets $\{f \in E : f \geq \varkappa_1\}$ and $\{f \in E : f \geq \varkappa_2\}$ are distinct. For instance, take $x \in E_+$ so that $x \geq \varkappa_1$ and $x \ngeq \varkappa_2$. Then $x \wedge \varkappa = 0$ because the element $\varkappa$ is indivisible. Consider the band projection $pr_x(e)$ of the unity e of the lattice E onto the principal band generated by x. It is easy to see that $\widehat{y}(\varkappa_1) = 1$ and $\widehat{y}(\varkappa_2) = 0$. Consequently, $(\varkappa_1, \varkappa_2) \notin \mathscr{R}$. ▷

4.3. Order, Relative Uniform Convergence, and the Archimedes Principle

We now introduce some types of infinitesimal elements in a nonstandard enlargement of a vector lattice and use them for a nonstandard description of various kinds of convergence. Also we obtain a nonstandard criterion for a vector lattice to be Archimedean.

4.3.1. Let E be a vector lattice. Given $\varkappa \in {}^*E$, we consider the set $U(\varkappa) := \{x \in E : x \geq \varkappa\}$ of standard upper bounds of $\varkappa$ and the set $L(\varkappa) := \{y \in E : \varkappa \geq y\}$ of standard lower bounds of $\varkappa$. Define the following external subsets of some nonstandard enlargement ${}^*E$ of the vector lattice E:

$$\mathrm{fin}({}^*E) := \{\varkappa \in {}^*E : U(|\varkappa|) \neq \varnothing\},$$
$$o\text{-pns}({}^*E) := \{\varkappa \in {}^*E : \inf_E (U(\varkappa) - L(\varkappa)) = 0\},$$
$$\eta({}^*E) := \{\varkappa \in {}^*E : \inf_E U(|\varkappa|) = 0\},$$
$$\lambda({}^*E) := \{\varkappa \in {}^*E : (\exists y \in E)(\forall n \in \mathbb{N})\ |n\varkappa| \leq y\}.$$

It is easy to see that $\mathrm{fin}({}^*E)$, o-pns$({}^*E)$, $\eta({}^*E)$, and $\lambda({}^*E)$ are vector lattices with respect to the lattice operations, addition, and multiplication by scalars in $\mathbb{R}$

which are inherited from the standard vector lattice ${}^*E$. The elements of $\mathrm{fin}({}^*E)$ we call *finite* or *limited;* the elements of o-$\mathrm{pns}({}^*E)$ we call (o)-*prenearstandard;* the elements of $\eta({}^*E)$ we call (o)-*infinitesimal;* the elements of $\lambda({}^*E)$ we call (r)-*infinitesimal.* The elements of $E+\eta({}^*E)$ (of $E+\lambda({}^*E)$) we call (o)-*nearstandard* (respectively, (r)-*nearstandard*). Note some simple properties:

(1) E is a vector sublattice in o-$\mathrm{pns}({}^*E)$, while o-$\mathrm{pns}({}^*E)$ is a vector sublattice in $\mathrm{fin}({}^*E)$;

(2) $\eta({}^*E)$ is an ideal both in $\mathrm{fin}({}^*E)$ and o-$\mathrm{pns}({}^*E)$;

(3) $E\cap\eta({}^*E)=\{0\}$;

(4) $\lambda({}^*E)$ is an ideal in $\mathrm{fin}({}^*E)$.

4.3.2. There are simple nonstandard conditions for a monotone net to be order convergent.

Let $(x_\alpha)_{\alpha\in\Xi}$ be a decreasing or increasing net in a vector lattice E. Then the following are equivalent:

(1) *the net (x_α) converges to 0;*

(2) *$x_\beta\in\eta({}^*E)$ for all $\beta\in{}^a\Xi$;*

(3) *$x_\beta\in\eta({}^*E)$ for some $\beta\in{}^a\Xi$.*

$\lhd$ We consider only the case of a decreasing net.

(1)→(2): Assume that $x_\alpha\downarrow 0$. Then every remote element $\beta\in{}^a\Xi$ satisfies $x_\alpha\geq x_\beta\geq 0$ for all $\alpha\in\Xi$. Consequently, $\inf_E U(|x_\beta|)=0$ and $x_\beta\in\eta({}^*E)$.

(2)→(3): This is an immediate consequence of Lemma 4.0.5.

(3)→(1): Take an element $\beta\in{}^a\Xi$ for which $x_\beta\in\eta({}^*E)$. Since $x_\alpha\downarrow$, we have $(x_\beta)_-\geq(x_\alpha)_-\geq 0$ for all $\alpha\in\Xi$. In view of $x_\beta\in\eta({}^*E)$, we have $(x_\alpha)_-=0$ for all $\alpha\in\Xi$. Consequently, $x_\alpha\downarrow\geq 0$. Let $y\in E_+$ be an arbitrary element such that $x_\alpha\downarrow\geq y$. By the transfer principle, each $\alpha\in{}^*\Xi$ satisfies $x_\alpha\geq y$. In particular, $x_\beta\geq y$. This is possible only if $y=0$. Hence $x_\alpha\downarrow 0$. $\rhd$

It is easy to see that (1)→(2) and (2)→(3) are true for an arbitrary (not necessarily monotone) net $(x_\alpha)_{\alpha\in\Xi}\subseteq E$. But the implication (3)→(1) may be false without the monotonicity condition. Indeed, let $E:=L_1[0,1]$. For each $n\in\mathbb{N}$, we take elements $f_1^n,f_2^n,\dots,f_{2^n}^n\in E$ such that f_k^n is the equivalence class containing the characteristic function of the interval $[\frac{k-1}{2^n},\frac{k}{2^n}]$. Arrange these elements in the sequence

$$f_1^1,f_2^1,\ \ f_1^2,f_2^2,f_3^2,f_4^2,\ \ \dots,\ \ f_1^n,f_2^n,\dots,f_{2^n}^n,\ \ \dots\,.$$

Obviously, (2) and (3) hold for this sequence, but it does not converge in order to any element of E.

4.3.3. We now give nonstandard conditions under which a monotone sequence converges relatively uniformly.

Let (x_n) be a decreasing or increasing sequence of elements in a vector lattice E. Then the following are equivalent:

(1) $x_n \xrightarrow{(r)} 0$;

(2) $x_\nu \in \lambda({}^*E)$ *for every* $\nu \in {}^*\mathbb{N} \setminus \mathbb{N}$;

(3) $x_\nu \in \lambda({}^*E)$ *for some* $\nu \in {}^*\mathbb{N} \setminus \mathbb{N}$.

$\triangleleft$ We verify only (2)→(1) in the case of a decreasing sequence. Let $x_\nu \in \lambda({}^*E)$ for some $\nu \in {}^*\mathbb{N}\setminus\mathbb{N}$. It is clear that $x_n \geq 0$ for all $n \in \mathbb{N}$. Assume that the condition $x_n \xrightarrow{(r)} 0$ is false. Then, for every $d \in E$, there is a number $n(d) \in \mathbb{N}$ such that $n(d) \cdot x_k \nleq d$ for all $k \in \mathbb{N}$. By the transfer principle, $n(d) \cdot x_k \nleq d$ for all $k \in {}^*\mathbb{N}$. In particular, $n(d) \cdot x_\nu \nleq d$, contradicting to $x_\nu \in \lambda({}^*E)$. So, $x_n \xrightarrow{(r)} 0$. $\triangleright$

As in 4.3.2, observé that (1)→(2) and (2)→(3) hold for every sequence $(x_n) \subseteq E$. At the same time, the implication (3)→(1) can be false without the monotonicity condition. This may be checked by considering the example in 4.3.2. Indeed, it is easy to see that the constructed sequence satisfies conditions (2) and (3) but not (1).

4.3.4. We now give nonstandard conditions for a vector lattice to be Archimedean. We start with proving one auxiliary assertion.

Lemma. *Let u be an element of a vector lattice E and let $\nu \in {}^*\mathbb{N} \setminus \mathbb{N}$. Then either $u = 0$ or $\nu u \notin$ o-pns$({}^*E)$.*

$\triangleleft$ Let $u \neq 0$. Take arbitrary $x \in L(|\nu u|)$ and $y \in U(|\nu u|)$. Then $x \leq |\nu u| \leq y$. By the transfer principle, we find $m \in \mathbb{N}$ such that $x \leq |mu| \leq y$. Comparing this inequality with the previous, we obtain $|u| \leq |\nu u| - |mu| \leq y - x$. Since the choice of the elements x in $L(|\nu u|)$ and y in $U(|\nu u|)$ is arbitrary, we have $U(|\nu u|) - L(|\nu u|) \geq |u| > 0$. Thus, $|\nu u|$ and, consequently, νu do not belong to o-pns$({}^*E)$. $\triangleright$

4.3.5. *Theorem.* *For every vector lattice E, the following are equivalent:*

(1) *E is an Archimedean vector lattice;*

(2) $\lambda({}^*E) \cap E = \{0\}$;

(3) $\lambda({}^*E) \subseteq \eta({}^*E)$;

(4) $\lambda({}^*E) \subseteq$ *o-pns*$({}^*E)$;

(5) *the set o-pns$({}^*E)$ is a relatively uniformly closed vector sublattice of* fin$({}^*E)$;

(6) *$\eta(^*E)$ is a relatively uniformly closed ideal of* $\operatorname{fin}(^*E)$.

$\lhd$ We will prove the theorem by the following scheme:

$$(1) \to (2) \to (3) \to (4) \to (1) \quad \text{and} \quad (1) \to (5) \to (6) \to (1).$$

(1)$\to$(2): This is obvious.

(2)$\to$(3): Let $\varkappa \in \lambda(^*E) \setminus \eta(^*E)$. Then there exists a $y \in E$ such that $|n\varkappa| \leq y$ for all $n \in \mathbb{N}$, and therefore, $(1/n)y \in U(|\varkappa|)$ for all $n \in \mathbb{N}$. Since $\varkappa \notin \eta(^*E)$, there exists a $z \in E$ satisfying $0 < z \leq U(|\varkappa|)$. In particular, $0 < z \leq (1/n)y$ for all $n \in \mathbb{N}$. So, $0 \neq z \in \lambda(^*E) \cap E$, which contradicts (2).

(3)$\to$(4): This is true since $\eta(^*E) \subseteq o\text{-pns}(^*E)$.

(4)$\to$(1): Take arbitrary elements $u, v \in E$ such that $0 \leq nu \leq v$ $(n \in \mathbb{N})$ and let $\nu \in {}^*\mathbb{N} \setminus \mathbb{N}$. Then it is obvious that $\nu u \in \lambda(^*E)$, and, by hypothesis, $\nu u \in o\text{-pns}(^*E)$. Hence, by Lemma 4.3.4, we have $u = 0$. So, E is Archimedean.

(1)$\to$(5): Take a sequence $(\varkappa_n)$ of elements in the vector lattice o-pns$(^*E)$ which converges relatively uniformly to some element $\varkappa \in \operatorname{fin}(^*E)$. Show that $\varkappa$ belongs to o-pns$(^*E)$. We may suppose that $(\varkappa_n)$ converges e-uniformly to $\varkappa$ for some $e \in E$. Then there is a sequence $\varepsilon_n \downarrow 0$ of real numbers such that $|\varkappa_k - \varkappa| \leq \varepsilon_n e$ for all natural $k \geq n$. For every $n \in \mathbb{N}$, we have $\varkappa_n - \varepsilon_n e \leq \varkappa \leq \varkappa_n + \varepsilon_n e$, and so

$$L(\varkappa_n - \varepsilon_n e) \leq \varkappa \leq U(\varkappa_n + \varepsilon_n e). \tag{1}$$

Given $n \in \mathbb{N}$, assign

$$\mathscr{E}_n := U(\varkappa_n + \varepsilon_n e) - L(\varkappa_n - \varepsilon_n e).$$

The inclusion $(\varkappa_n) \subseteq o\text{-pns}(^*E)$ implies, by (1), that

$$\inf_E \mathscr{E}_n = 2\varepsilon_n e \quad (n \in \mathbb{N}). \tag{2}$$

Since E is Archimedean, it follows from (2) that $\inf_E \bigcup_{n=1}^{\infty} \mathscr{E}_n = 0$, and hence $\inf_E(U(\varkappa) - L(\varkappa)) = 0$. We have used the inclusion $\bigcup_{n=1}^{\infty} \mathscr{E}_n \subseteq U(\varkappa) - L(\varkappa)$ which ensues from (1). Thus, $\varkappa \in o\text{-pns}(^*E)$. Consequently, o-pns$(^*E)$ is relatively uniformly closed in $\operatorname{fin}(^*E)$.

(5)$\to$(6): Let $\varkappa_n \in \eta(^*E)$ and $\varkappa_n \xrightarrow{(r)} \varkappa \in \operatorname{fin}(^*E)$. Then $\varkappa \in o\text{-pns}(^*E)$, by hypothesis. Check that $\varkappa \in \eta(^*E)$. We may assume that $\varkappa_n \xrightarrow{(r)} \varkappa$ d-uniformly for some $d \in E$. This means that $|\varkappa_n - \varkappa| \leq \varepsilon_n d$ for all $n \in \mathbb{N}$ and some appropriate sequence $(\varepsilon_n) \subseteq \mathbb{R}$, $\varepsilon_n \downarrow 0$. Assume that $\varkappa \notin \eta(^*E)$. Take an arbitrary element

$a \in E$ satisfying $U(|\varkappa|) \geq a \geq 0$. For each $n \in \mathbb{N}$, choose an arbitrary $u_n \in U(|\varkappa_n|)$. It is obvious that $u_n + \varepsilon_n d \geq |\varkappa_n| + \varepsilon_n d \geq |\varkappa|$. We thus have

$$U(|\varkappa_n|) + \varepsilon_n d \subseteq U(|\varkappa|) \text{ and } U(|\varkappa_n|) + \varepsilon_n d \geq a.$$

Now it follows from $\inf_E U(|\varkappa_n|) = 0$ that $\varepsilon_n d \geq a$. This inequality is true for all natural n and the sequence (ε_n) decreasing to zero. Therefore, $d \geq ka$ for every $k \in \mathbb{N}$. Applying the transfer principle, we obtain $d \geq ka$ for all $k \in {}^*\mathbb{N}$. Take some $\nu \in {}^*\mathbb{N} \setminus \mathbb{N}$. It is easy to see that the sequence $(ka)_{k=1}^{\infty}$ of elements of E converges d-uniformly to the element νa of the vector lattice $\operatorname{fin}({}^*E)$. Since, by hypothesis, $o\text{-pns}({}^*E)$ is a relatively uniformly closed vector sublattice of $\operatorname{fin}({}^*E)$, we have $\nu a \in o\text{-pns}({}^*E)$. By Lemma 4.3.4, this implies $a = 0$. Thus, $\inf_E U(|\varkappa|) = 0$, and so $\varkappa \in \eta({}^*E)$, as required.

(6)→(1): Take arbitrary $u, v \in E$ satisfying $0 \leq nu \leq v$ for all $n \in \mathbb{N}$. Show that $u = 0$. Let $\nu \in {}^*\mathbb{N} \setminus \mathbb{N}$. It is easy to see that a sequence (x_n) with $x_n = 0$ for all $n \in \mathbb{N}$ converges v-uniformly to an element νu. By hypothesis, the ideal $\eta({}^*E)$ is relatively uniformly closed in $\operatorname{fin}({}^*E)$, so $\nu u \in \eta({}^*E)$. Therefore, by Lemma 4.3.4, we have $u = 0$. ▷

4.3.6. Theorem. *For a vector lattice E the following are equivalent:*

(1) *E is an order separable Archimedean vector lattice in which order convergence and relative uniform convergence coincide for any sequence;*

(2) $\eta({}^*E) = \lambda({}^*E)$.

◁ (1)→(2): E satisfies the inclusion $\lambda({}^*E) \subseteq \eta({}^*E)$ by Theorem 4.3.5. Prove the reverse inclusion. Take an arbitrary $\varkappa \in \eta({}^*E)$. Then $\inf_E U(|\varkappa|) = 0$. Since $U := U(|\varkappa|)$ is a downwards-directed set such that $U \downarrow 0$ and since E is an order separable vector lattice, there is a sequence $(u_n) \subseteq U$ with $u_n \downarrow 0$. The condition (1) implies $u_n \xrightarrow{(r)} 0$. Then, by 4.3.3, $u_\nu \in \lambda({}^*E)$ for all $\nu \in {}^*\mathbb{N} \setminus \mathbb{N}$. Consequently, $\varkappa \in \lambda({}^*E)$, because $|\varkappa| \leq u_n$ for all $n \in {}^*\mathbb{N}$.

(2)→(1): E is Archimedean by Theorem 4.3.5. Show that order convergence and relative uniform convergence coincide for every sequence in E. It is sufficient to prove that $u_n \downarrow 0$ implies $u_n \xrightarrow{(r)} 0$. Take an arbitrary sequence $(u_n) \subseteq E$ such that $u_n \downarrow 0$. Then, by 4.3.2, $u_\nu \in \eta({}^*E)$ for all $\nu \in {}^*\mathbb{N} \setminus \mathbb{N}$. Hence $u_\nu \in \lambda({}^*E)$ for all $\nu \in {}^*\mathbb{N} \setminus \mathbb{N}$. Now we see from 4.3.3 that $u_n \xrightarrow{(r)} 0$.

It remains to verify that the vector lattice E is order separable. Take an arbitrary net $(x_\xi)_{\xi \in \Xi} \subseteq E$ such that $x_\xi \downarrow 0$. By Lemma 4.1.5, there exists remote element τ in the standard directed set ${}^*\Xi$. Then, by 4.3.2, we have $x_\tau \in \eta({}^*E)$, and so $x_\tau \in \lambda(^*E)$. In this case, there is $d \in E$ satisfying $nx_\tau \leq d$ for all $n \in \mathbb{N}$. Assume that (x_ξ) does not contain a subsequence convergent relatively uniformly to zero. Then there is a number $n_0 \in \mathbb{N}$ such that $n_0 x_\xi \not\leq d$ for all $\xi \in \Xi$. By the transfer principle, $n_0 x_\xi \not\leq d$ for every $\xi \in {}^*\Xi$. The contradiction with $n_0 x_\tau \leq d$ ensures existence of a subsequence $(x_{\xi_n}) \subseteq (x_\xi)$ with $x_{\xi_n} \xrightarrow{(r)} 0$. Since E is Archimedean, we have $x_{\xi_n} \xrightarrow{(o)} 0$. Thus, E is order separable. $\triangleright$

4.4. Conditional Completion and Atomicity

In this section, we give a nonstandard construction of a Dedekind completion of an Archimedean vector lattice. Also, we give an infinitesimal interpretation for the property of a vector lattice to be atomic.

4.4.1. Let E be a vector lattice. Consider the quotient vector lattice $\widehat{E} := o\text{-pns}(^*E)/\eta(^*E)$ and denote by $\widehat{\eta}$ the mapping $x \mapsto [x]$, where $x \in E$ and $[x] \in \widehat{E}$ is the coset containing x.

Theorem. *For every Archimedean vector lattice E, the following hold:*

(1) *$\widehat{E}$ is Dedekind complete;*

(2) *$\widehat{\eta}$ is a Riesz isomorphism of the vector lattice E into the vector lattice $\widehat{E}$;*

(3) *for every $x \in E$*

$$x = \sup_{\widehat{E}}\{y \in \widehat{\eta}(E) : y \leq x\} = \inf_{\widehat{E}}\{y \in \widehat{\eta}(E) : y \geq x\}.$$

In other words, the vector lattice $\widehat{E}$ is a Dedekind completion of E.

We prove the theorem in four steps:

Step 1. *Let E be an arbitrary vector lattice. Then, for every $0 < x \in \widehat{E}$, there exists an element $e \in E$ such that $0 < \widehat{\eta}(e) \leqslant x$.*

$\triangleleft$ Take an element $x \in \widehat{E}$, $x > 0$. Let $\varkappa \in o\text{-pns}(^*E)$ be such that $\varkappa > 0$ and $x = [\varkappa]$. Then there exists an $e \in E$ for which $0 < e \leqslant \varkappa$. Indeed, in the other case, $\sup_E L(\varkappa) = 0$. Consequently, $\inf_E U(\varkappa) = 0$, since $\varkappa \in o\text{-pns}(^*E)$. This contradicts $[\varkappa] = x \neq 0$. So, e is a sought element. $\triangleright$

Step 2. *Given $x \in \widehat{E}$, assign*

$$\widehat{\mathscr{U}}(x) := \{y \in \widehat{\eta}(E) : y \geqslant x\}; \quad \widehat{\mathscr{L}}(x) := \{z \in \widehat{\eta}(E) : x \geqslant z\}.$$

Then we have

$$x = \inf_{\widehat{E}} \widehat{\mathscr{U}}(x) = \sup_{\widehat{E}} \widehat{\mathscr{L}}(x).$$

$\lhd$ Let $\varkappa \in o\text{-pns}({}^*E)$ and let $x = [\varkappa]$. To prove the claim it suffices to check that every element $y \in \widehat{E}$ satisfying $\widehat{\mathscr{L}}(x) \leqslant y \leqslant \widehat{\mathscr{U}}(x)$ is equal to x. Take an arbitrary $y \in \widehat{E}$ such that $\widehat{\mathscr{L}}(x) \leqslant y \leqslant \widehat{\mathscr{U}}(x)$. Assume $|x - y| > 0$. Since the $\widehat{\eta}(E)$ is minorant in $\widehat{E}$ by Step 1, there exists an $e \in E$ satisfying

$$\widehat{\mathscr{U}}(x) - \widehat{\mathscr{L}}(x) \geqslant |x - y| \geqslant \widehat{\eta}(e) > 0. \tag{3}$$

It is easy to see that

$$\widehat{\eta}(U(\varkappa)) \subseteq \widehat{\mathscr{U}}(x) \text{ and } \widehat{\eta}(L(\varkappa)) \subseteq \widehat{\mathscr{L}}(x).$$

Now, the inequality (3) implies

$$\widehat{\eta}(U(\varkappa) - L(\varkappa)) \geqslant \widehat{\eta}(e) > 0,$$

and consequently $U(\varkappa) - L(\varkappa) \geqslant e > 0$. We obtained a contradiction to $\varkappa \in o\text{-pns}({}^*E)$. Hence, $|x - y| = 0$, and $y = x$. $\rhd$

Step 3. *Given an Archimedean vector lattice E, every nonempty subset D in $\widehat{\eta}(E)$ bounded above has a least upper bound in $\widehat{E}$.*

$\lhd$ Let $D \subseteq \widehat{\eta}(E)$ is nonempty and bounded above. Then the subset $\mathscr{D} := \widehat{\eta}^{-1}(D)$ of E is bounded above in E. Denote by $U(\mathscr{D})$ the set of all upper bounds of $\mathscr{D}$ in E. Since E is Archimedean, we have

$$\inf_E (U(\mathscr{D}) - \mathscr{D}) = 0. \tag{4}$$

Applying the general saturation principle, find an element $\delta \in {}^*E$ such that

$$\mathscr{D} \leqslant \delta \leqslant U(\mathscr{D}). \tag{5}$$

From (4) and (5), it follows that $\inf_E (U(\delta) - L(\delta)) = 0$. Hence $\delta \in o\text{-pns}({}^*E)$. The element $[\delta] \in \widehat{E}$ is an upper bound of the set $D = \widehat{\eta}(\mathscr{D})$. Show that $[\delta] = \sup_{\widehat{E}} D$. Let $y \in \widehat{E}$ be some upper bound of D such that $[\delta] \geqslant y$. By (5) we have

$$0 \leqslant [\delta] - y \leqslant \widehat{\eta}(U(\mathscr{D}) - \mathscr{D}). \tag{6}$$

According to Step 1, the vector lattice $\widehat{\eta}(E)$ is minorant in $\widehat{E}$. Therefore, (4) and (6) imply $y = [\delta]$. Thus, $[\delta] = \sup_{\widehat{E}} D$. $\rhd$

Step 4. PROOF OF THE THEOREM.

◁ Assertion (2) is obvious. Assertion (3) is valid in view of Step 2. It is interesting to note that (2) and (3) hold in an arbitrary vector lattice. Verify the condition (1). Take an arbitrary nonempty subset A in $\widehat{E}$ bounded above. Denote

$$\mathscr{A} := \{x \in E : (\exists a \in A)\widehat{\eta}(x) \leqslant a\}.$$

According to Step 3, the set $\widehat{\eta}(\mathscr{A})$ has a least upper bound in $\widehat{E}$. Assign $a := \sup_{\widehat{E}} \widehat{\eta}(\mathscr{A})$. It is easy to see that $a = \sup_{\widehat{E}} A$. Thus, every nonempty subset in $\widehat{E}$ bounded above has a least upper bound, as required. ▷

4.4.2. The preceding theorem easily implies the following assertion to which we prefer to give a simpler and more direct proof:

Theorem. *For every Archimedean vector lattice E, the following are equivalent:*

(1) *The vector lattice E is Dedekind complete;*

(2) *o-pns$(^*E) = E + \eta(^*E)$.*

◁ (1)→(2): Obviously, $E + \eta(^*E) \subseteq o\text{-pns}(^*E)$. Show the reverse inclusion. Take an arbitrary $\varkappa \in o\text{-pns}(^*E)$. Then $\varkappa \in \operatorname{fin}(^*E)$. So, $U(\varkappa)$ is nonempty. Therefore, $L(\varkappa)$ is bounded above. Since E is Dedekind complete, $L(\varkappa)$ has a least upper bound. Assign $a := \sup_E L(\varkappa)$. It is easy to see that $L(\varkappa) \le a \le U(\varkappa)$. Hence $|\varkappa - a| \le U(\varkappa) - L(\varkappa)$. Since $\varkappa \in o\text{-pns}(^*E)$, the last inequality implies that $\inf_E U(|\varkappa - a|) = 0$. We have $\varkappa = a + (\varkappa - a)$ with $a \in E$ and $\varkappa - a \in \eta(^*E)$. Consequently $\varkappa \in E + \eta(^*E)$.

(2)→(1): It suffices to show that every net $(u_\xi)_{\xi\in\Xi} \subseteq E$ such that $u_\xi\uparrow \le d \in E$ is order convergent. Assume that $u_\xi\uparrow \le d \in E$. It is well known (see, for example, [21, Theorem 22.5]) that the following condition holds in an Archimedean vector lattice E:

$$\inf_E\{y - u_\xi : \xi \in \Xi,\ y \in E,\ u_\xi\uparrow \le y\} = 0. \tag{7}$$

Fix a remote element $\tau \in {}^a\Xi$. It is easy to see that $\{y \in E : u_\xi\uparrow \le y\} = U(u_\tau)$. Moreover, $(u_\xi) \subseteq L(u_\tau)$. Thus, (7) implies $\inf_E\{U(u_\tau) - L(u_\tau)\} = 0$ or, in other words, $u_\tau \in o\text{-pns}(^*E)$. Then $u_\tau \in E + \eta(^*E)$ by (2). Let $u \in E$ be such that $u_\tau - u \in \eta(^*E)$. Thus, 4.3.2 implies that the net (u_ξ) converges in order to u. ▷

4.4.3. Now we consider the property of a vector lattice to be atomic. Recall that a vector lattice E is *atomic* if E is Archimedean and for every $0 < x \in E$ there exists an atom $a \in E$ such that $0 < a \le x$. Also, we recall that, for every atom a in an Archimedean vector lattice and for each element $0 \le x \le a$, there is a real α such that $x = \alpha a$. We start with the following

Lemma. *Let E be an atomic vector lattice. Then* $\mathrm{fin}({}^*E) = o\text{-}\mathrm{pns}({}^*E)$.

$\lhd$ It suffices to verify that every element $\varkappa \in \mathrm{fin}({}^*E)$, $\varkappa \geq 0$, satisfies $\varkappa \in o\text{-}\mathrm{pns}({}^*E)$. Let $\varkappa$ be an arbitrary positive element of $\mathrm{fin}({}^*E)$. Assume that $U(\varkappa) - L(\varkappa) \geq x > 0$. Then, by hypothesis, there exists an atom $a \in E$ such that $U(\varkappa) - L(\varkappa) \geq a > 0$. Take an element $u \in U(\varkappa)$. Since E is an Archimedean vector lattice, there exists a number $n \in \mathbb{N}$ for which $na \not\leq u$. The element a is an atom; so we have $u \wedge na = \alpha a$ and $\varkappa \wedge na = \beta a$ for appropriate $\alpha, \beta \in {}^*[0, n]$. Assign

$$l' := \mathrm{st}(\beta - 1/3) \cdot a \text{ and } u' := u - \mathrm{st}(\alpha - \beta - 1/3) \cdot a,$$

where st is the taking of the standard part of a real. Then $u' \in U(\varkappa)$ and $l' \in L(\varkappa)$, but $u' - l' \not\geq a$; a contradiction. Consequently, $\inf_E(U(\varkappa) - L(\varkappa)) = 0$, and so $\varkappa \in o\text{-}\mathrm{pns}({}^*E)$. $\rhd$

4.4.4. The condition $\mathrm{fin}({}^*E) = o\text{-}\mathrm{pns}({}^*E)$ is not only necessary but also sufficient for a vector lattice E to be atomic. To prove this, we need to introduce the concept of *punch* of a positive element of a vector lattice.

Definition. Let E be a vector lattice and let $e \in E_+$. An element $\varkappa$ of a nonstandard enlargement ${}^*E$ of E is said to be an *e-punch* if

(1) $0 \leq \varkappa \leq e$;

(2) $\inf_E\{y \in E : y \geq \varkappa\} = e$;

(3) $\sup_E\{z \in E : \varkappa \geq z\} = 0$.

We recall that an element e of the vector lattice E is called *nonatomic* if $|e| \wedge a = 0$ for any atom $a \in E$. Below we will need the following easy

Remark. For every nonatomic element $e \in E$, $e > 0$, and every natural number n, there is a family $\{e_k\}_{k=1}^n \subseteq E$ of disjoint elements satisfying $0 < e_k \leq e$ for all $k = 1, \dots, n$.

Lemma. *Let E be an arbitrary Archimedean vector lattice. Then, for every $\nu \in {}^*\mathbb{N}$ and every nonatomic element $e \in E$, $e \geq 0$, there exists a family $\{e_k\}_{k=1}^{\nu} \subseteq {}^*E$ of disjoint e-punches.*

$\lhd$ Take an arbitrary $\nu \in {}^*\mathbb{N}$, and let $e \geq 0$ be some nonatomic element in E. Since the assertion of the lemma is obvious for $e = 0$, we suppose that $e > 0$. Denote by L the set of positive elements of the principal ideal E_e generated by e in E. It is obvious that L is a lattice with zero. By Lemma 4.1.3, there exists a hyperfinite saturating family $\{\varkappa_n\}_{n=1}^{\omega}$ of disjoint indivisible elements in a nonstandard enlargement ${}^*L$ of L. Clearly, every $n = 1, \dots, \omega$ satisfies $0 < \varkappa_n \leq e$. Applying the transfer principle and remark before the lemma under proof, we can easily see that, for each $n = 1, \dots, \omega$, there exists a hyperfinite family $\{\gamma_n^k\}_{k=1}^{\nu+1} \subseteq {}^*E$ such that

$$0 < \gamma_n^k \leq \varkappa_n \quad (k = 1, \dots, \nu + 1); \quad \gamma_n^k \wedge \gamma_n^p = 0 \quad (k \neq p).$$

Applying the transfer principle once again and using the fact that the vector lattice E is Archimedean, we find a hyperfinite family $\{\alpha_n^k\}_{n=1;\,k=1}^{\omega;\,\nu} \subseteq {}^*\mathbb{R}$ with the following properties:

(1) $0 < \alpha_n^k \gamma_n^k \leq \varkappa_n$ for all $n = 1, \dots, \omega$ and $k = 1, \dots, \nu$;

(2) the condition $\alpha > \alpha_n^k$ implies that

$$\alpha \gamma_n^k \not\leq \varkappa_n \text{ for all } n = 1, \dots, \omega, \ k = 1, \dots, \nu, \text{ and } \alpha \in {}^*\mathbb{R}.$$

Put $e_k := \bigvee_{n=1}^{\omega} \alpha_n^k \gamma_n^k$ for all $k = 1, \dots, \nu$. It is easy to verify that $\{e_k\}_{k=1}^{\nu}$ is the desired family of ν disjoint e-punches. $\triangleright$

4.4.5. Theorem. *For every vector lattice E, the following are equivalent:*

(1) *E is an atomic vector lattice;*

(2) $\mathrm{fin}({}^*E) = o\text{-}\mathrm{pns}({}^*E)$.

$\triangleleft$ (1)$\to$(2): This is established in Lemma 4.4.3.

(2)$\to$(1): Let $\mathrm{fin}({}^*E) = o\text{-}\mathrm{pns}({}^*E)$. In particular, $o\text{-}\mathrm{pns}({}^*E)$ is an (r)-closed vector sublattice of $\mathrm{fin}({}^*E)$. Therefore, E is Archimedean by Theorem 4.3.5. Verify that E is atomic. It is sufficient to show that E has no nonzero nonatomic elements. Take an arbitrary nonatomic element $e \in E$. We may suppose that $e \geq 0$. By Lemma 4.4.4, there exists an e-punch $\varkappa \in {}^*E$. The element $\varkappa$ satisfies $\inf_E(U(\varkappa) - L(\varkappa)) = e$. At the same time, $\varkappa$ is finite and, by hypothesis, $\varkappa$ is an (o)-prenearstandard element of ${}^*E$. So, $e = 0$. $\triangleright$

4.4.6. As an application of the last theorem, we establish some useful nonstandard criterion for a vector lattice to be atomic and Dedekind complete.

Theorem. *For every vector lattice E, the following are equivalent:*

(1) *E is a Dedekind complete atomic vector lattice;*

(2) $\mathrm{fin}({}^*E) = E + \eta({}^*E)$.

$\triangleleft$ Observe that $E + \eta({}^*E) \subseteq o\text{-}\mathrm{pns}({}^*E) \subseteq \mathrm{fin}({}^*E)$ and use Theorems 4.4.5 and 4.4.2. $\triangleright$

We point out that, in the proof of Theorem 4.4.5, we used only the fact that there exists one e-punch for a nonatomic element $e \in E_+$. We will need the Lemma 4.4.4 in full strength below, in the proof of a criterion for a vector lattice E to be isomorphic to the order hull of E.

4.5. Normed Vector Lattices

In this section, we consider normed vector lattices and study some of their infinitesimal interpretations. Throughout the section we assume (E, ρ) to be a normed vector lattice.

4.5.1. It is well known that, in a nonstandard enlargement of E, together with $\operatorname{fin}({}^*E)$, o-$\operatorname{pns}({}^*E)$, $\eta({}^*E)$, and $\lambda({}^*E)$, we may also consider the following subsets:

$$\operatorname{Fin}({}^*E) := \{\varkappa \in {}^*E : \rho(\varkappa) \in \operatorname{fin}({}^*\mathbb{R})\};$$
$$\operatorname{pns}({}^*E) := \{\varkappa \in {}^*E : (\forall n \in \mathbb{N})(\exists y \in E)\ n\rho(\varkappa - y) \leq 1\};$$
$$\mu({}^*E) := \{\varkappa \in {}^*E : \rho(\varkappa) \approx 0\}.$$

It is easy to see that these are vector lattices over $\mathbb{R}$ under the operations inherited from ${}^*E$. Furthermore, $\operatorname{pns}({}^*E)$ is a vector sublattice of $\operatorname{Fin}({}^*E)$, while $\mu({}^*E)$ is an ideal in $\operatorname{pns}({}^*E)$ as well as in $\operatorname{Fin}({}^*E)$.

4.5.2. Let E be a vector lattice. If there exists a strong unity $e \in E$ then we may introduce the Riesz norm $\|\cdot\|_e$ on E by the well known formula

$$\|x\|_e := \inf\{\lambda \in \mathbb{R} : |x| \leq \lambda e\} \quad (x \in E).$$

We prove the next

Theorem. *Let $(E, \|\cdot\|)$ be a normed vector lattice. Then the following are equivalent:*

(1) *E possesses a strong unity e, and the norm $\|\cdot\|_e$ is equivalent to $\|\cdot\|$;*

(2) *$\operatorname{Fin}({}^*E) = \operatorname{fin}({}^*E)$;*

(3) *$\mu({}^*E) \subseteq \operatorname{fin}({}^*E)$;*

(4) *$\mu({}^*E) = \lambda({}^*E)$;*

(5) *$\mu({}^*E) \subseteq \eta({}^*E)$;*

(6) *$\operatorname{Fin}({}^*E) = \operatorname{fin}({}^*E) + \mu({}^*E)$.*

$\lhd$ First of all, we prove the equivalence of conditions (2)–(5). To this end, it suffices to show that (2) $\to$ (3) $\to$ (4) $\to$ (5) $\to$ (3) and (4)$\to$(2). The implications (2)$\to$(3), (4)$\to$(5), and (5)$\to$(3) do not require checking.

(3)$\to$(4): Let $\mu({}^*E) \subseteq \operatorname{fin}({}^*E)$. To prove the implication, it is sufficient to verify the inclusion $\mu({}^*E) \subseteq \lambda({}^*E)$. Take an arbitrary $\varkappa \in \mu({}^*E)$. Then $\|\alpha\varkappa\| \approx 0$ with $\alpha = \|\varkappa\|^{-1/2}$, and consequently $\alpha\varkappa \in \mu({}^*E) \subseteq \operatorname{fin}({}^*E)$. Thus, there is an element $y \in E$ for which $|\alpha\varkappa| \leq y$. Then $|n\varkappa| \leq |\alpha\varkappa| \leq y$ for all $n \in \mathbb{N}$, and so $\varkappa \in \lambda(^*E)$.

(4)→(2): Let $\mu(^*E) = \lambda(^*E)$. It is obvious that $\mathrm{fin}(^*E) \subseteq \mathrm{Fin}(^*E)$. Assume that the inclusion is proper. Then there is a $\varkappa \in {}^*E$ such that $\|\varkappa\| = 1$ and $|\varkappa| \not\leq y$ for all $y \in E$. Consider the internal sets

$$A_y^n := \{r \in {}^*\mathbb{R}_+ : n \leq r \ \&\ |\varkappa| \not\leq ry\}$$

for $y \in E_+$ and $n \in \mathbb{N}$. Since $|\varkappa| \not\leq (n+1)y$ for every $y \in E_+$ and every $n \in \mathbb{N}$, we have $n+1 \in A_y^n$, and all these sets are nonempty. The family $\{A_y^n\}_{y \in E_+}^{n \in \mathbb{N}}$ possesses the finite intersection property since

$$A_{y \vee z}^{\max\{n,m\}} \subseteq A_y^n \cap A_z^m.$$

By the general saturation principle, there is some $r \in {}^*\mathbb{R}$ satisfying

$$r \in \cap\{A_y^n : y \in E_+,\ n \in \mathbb{N}\}.$$

Then r is an infinite positive number such that $|\varkappa| \not\leq ry$ for all $y \in E_+$. However, $(1/r)\varkappa \in \mu(^*E)$, since $\|(1/r)\varkappa\| = 1/r \approx 0$. By hypothesis, $\mu(^*E) = \lambda(^*E)$, therefore, $|(1/r)\varkappa| \leq z$ for some $z \in E_+$, and so $|\varkappa| \leq rz$, which contradicts $r \in A_z^1$. Thus, the equivalence of conditions (2)–(5) is established.

In order to finish the proof, we show that (1) → (2) → (6) → (1). The implications (1)→(2) and (2)→(6) are obviously true.

(6)→(1): Let $\mathrm{Fin}(^*E) = \mathrm{fin}(^*E) + \mu(^*E)$. First, we prove that the unit ball $B := \{x \in E : \|x\| \leq 1\}$ of the vector lattice E is order bounded. Assume the contrary. Take an arbitrary $x \in E_+$. There is a $y \in E_+$ such that $\|y\| = 1$ and $y \not\leq x$. Consider $z = y - y \wedge x$. Then $0 < z \leq y$, and so $0 < \|z\| \leq 1$. Show that

$$tz \wedge x \leq y \quad (t \in \mathbb{R}_+). \tag{8}$$

Let $t \in \mathbb{R}_+$. We represent x as $(x - x \wedge y) + (x \wedge y)$ and assign $u = tz$, $v = x - x \wedge y$, and $w = x \wedge y$. It is clear that $u, v, w \in E_+$ and $tz \wedge x = u \wedge (v + w)$. The easy relations

$$u \wedge (v + w) - u \wedge v \leq w,$$
$$u \wedge (v + w) - u \wedge v \leq u \wedge (v + w) \leq u$$

imply the inequality

$$u \wedge (v + w) \leq u \wedge v + u \wedge w. \tag{9}$$

The elements $u = tz$ and $v = x - x \wedge y$ are disjoint because

$$z \wedge v = (y - x \wedge y) \wedge (x - x \wedge y) = y \wedge x - x \wedge y = 0.$$

Hence, by (9), we have

$$tz \wedge x = u \wedge (v + w) \leq w \wedge v = tz \wedge x \wedge y \leq y.$$

Inequality (8) is proven. Consider the element $s = (2/\|z\|)z$ of the vector lattice E. It is clear that $\|s\| = 2$. Since $s \wedge x \leq y$, we have $\|s \wedge x\| \leq \|y\| = 1$. Consequently, the internal set

$$A_x := \{s \in {}^*E_+ : \|s\| = 2 \ \&\ s \wedge x \in B\}$$

is nonempty for all $x \in E_+$. Since $A_{x \vee y} \subseteq A_x \cap A_y \quad (x, y \in E_+)$, the family $\{A_x\}_{x \in E_+}$ possesses the finite intersection property. By the general saturation principle, there exists $y_0 \in {}^*E_+$ such that

$$y_0 \in \cap\{A_x : x \in E_+\}.$$

It is clear that $\|y_0\| = 2$. In particular, $y_0 \in \mathrm{Fin}({}^*E)$. By assumption (6), $y_0 \in \mathrm{fin}({}^*E) + \mu({}^*E)$. Therefore, there are elements $x_0 \in X_+$ and $h \in \mu({}^*E)$, for which $y_0 \leq x_0 + h$. Obviously, $\|y_0 \wedge x_0\| \approx \|y_0\|$. At the same time, $\|y_0\| = 2$ and $\|y_0 \wedge x_0\| \leq 1$, since $y_0 \in A_{x_0}$. The obtained contradiction shows that the unit ball of E is order bounded.

Choose an $e \in E$ so that $|x| \leq e$ for all $x \in B$. Then

$$|x| \leq \|x\|_e \cdot e \quad (x \in E).$$

This implies that e is a strong order unity of E. Moreover, $\|x\|_e \leq \|x\| \quad (x \in E)$. At the same time, $\|x\| \leq c\|x\|_e \quad (x \in E)$ for $c = \|e\|^{-1}$. Consequently, the norms $\|\cdot\|_e$ and $\|\cdot\|$ are equivalent. The implication (6)→(1) is established. The proof of the theorem is complete. ▷

4.5.3. Now we give a nonstandard condition for a norm to be order continuous.

Theorem. *The norm ρ of a normed vector lattice (E, ρ) is order continuous if and only if $\eta({}^*E) \subseteq \mu({}^*E)$.*

◁ Assume that the norm ρ is order continuous. Take an arbitrary $\varkappa \in \eta({}^*E)$. Since $U(|\varkappa|)$ is directed downwards and $\inf_E U(|\varkappa|) = 0$, order continuity of the norm ρ implies

$$\inf\{\rho(u) : u \in U(|\varkappa|)\} = 0.$$

Then $\rho(\varkappa) \approx 0$. Since the choice of $\varkappa \in \eta(^*E)$ is arbitrary, it follows that $\eta(^*E) \subseteq \mu(^*E)$.

Now, let $\eta(^*E) \subseteq \mu(^*E)$. Assume that ρ is not order continuous. In this case there are a net $(x_\xi)_{\xi \in \Theta} \subseteq E$, $x_\xi \downarrow 0$, and a number $0 < a \in \mathbb{R}$ such that $\rho(x_\xi) \geq a$ for all $\xi \in \Theta$. Take some remote element $\beta \in {}^a\Theta$. Then, by 4.3.2, $x_\beta \in \eta(^*E)$. Thus $\rho(x_\beta) \approx 0$. On the other hand, by the transfer principle, $\rho(x_\xi) \geq a$ for all $\xi \in {}^*\Theta$. The contradiction shows that the norm ρ is order continuous. $\rhd$

As an example of applying Theorem 4.5.2, we propose a nonstandard proof for the following well-known assertion:

Let a Banach lattice E have an order continuous norm. Then E is order separable and Dedekind complete. Moreover, order convergence in E coincides with relative uniform convergence.

$\lhd$ In view of 4.4.2 and 4.3.6, it suffices to verify the relations

$$o\text{-pns}(^*E) \subseteq E + \eta(^*E) \text{ and } \eta(^*E) \subseteq \lambda(^*E).$$

Let $\varkappa \in o\text{-pns}(^*E)$. Using the fact that the norm is order continuous, it is easy to see that $\varkappa \in \text{pns}(^*E)$. According to Proposition 4.0.7, the Banach lattice (E, ρ) satisfies $\text{pns}(^*E) = E + \mu(^*E)$. So, there is an $x \in E$ such that $\varkappa - x \in \mu(^*E)$. Obviously, $L(\varkappa) \leq x \leq U(\varkappa)$. Since $\varkappa \in o\text{-pns}(^*E)$, we have $\varkappa - x \in \eta(^*E)$. Thus, $\varkappa \in E + \eta(^*E)$.

Verify that $\eta(^*E) \subseteq \lambda(^*E)$. Let $\varkappa \in \eta(^*E)$. Then $U(|\varkappa|) \downarrow 0$. By order continuity of ρ, for every $n \in \mathbb{N}$, there is an $u_n \in U(|\varkappa|)$ with $\rho(u_n) \leq 2^{-n}$. The sum $u := \sum_{n=1}^{\infty} u_n$ exists in the Banach lattice E. Obviously, $|n\varkappa| \leq u$ for all $n \in \mathbb{N}$. Thus $\varkappa \in \lambda(^*E)$. $\rhd$

4.5.4. Concluding this section, we will establish a nonstandard criterion for a normed vector lattice to be finite-dimensional.

Theorem. *A normed vector lattice (E, ρ) is finite-dimensional if and only if $\eta(^*E) = \mu(^*E)$.*

$\lhd$ Necessity is obvious. To prove sufficiency, we let $\eta(^*E) = \mu(^*E)$. By Theorem 4.5.2, E possesses a strong unity e. Moreover, the norm $\|\cdot\|_e$ is equivalent to the initial norm ρ. According to Theorem 4.5.3, ρ is order continuous. Consequently, $\|\cdot\|_e$ is order continuous. Next, by Theorem 4.5.2, $\eta(^*E) = \lambda(^*E)$. Applying Theorem 4.3.6, we conclude that E is order separable.

Assume that $\dim E = \infty$. It is easy to see that in this case there exists an infinite disjoint order basis $A \subseteq E_+$ such that $a \in A$ implies $\|a\|_e = 1$. Since E is order separable, the set A is at most countable, because it is order bounded in E by some element e. Thus, we may suppose $A = \{a_n\}_{n=1}^{\infty}$. For each natural n, define the element

$$u_n := e - \left(n \sum_{k=1}^{n} a_k\right) \wedge e.$$

It is easy to see that $u_n \downarrow 0$. Since the norm $\|\cdot\|_e$ is order continuous, it follows that $\|u_n\|_e \to 0$. On the other hand, by the construction of the sequence (u_n) we have $\|u_n\|_e \geq \|a_{n+1}\|_e = 1$; a contradiction. Hence, $\dim E < \infty$. ▷

4.6. Linear Operators Between Vector Lattices

In this section, we establish nonstandard criteria for linear operators in vector lattices to be order continuous and order bounded. These criteria are similar to those in 4.0.9. Below, the symbols E and F denote some vector lattices and $T : E \to F$ is a linear operator.

4.6.1. We first prove one useful auxiliary assertion (see also 4.0.8):

Lemma. *For every nonempty subset D of a vector lattice E, the following are equivalent:*

(1) *D is order bounded;*

(2) *${}^*D \subseteq \mathrm{fin}({}^*E)$.*

◁ We need to prove only the implication (2)→(1). Let ${}^*D \subseteq \mathrm{fin}({}^*E)$. Assume that D is not contained in any order interval. Then, for every $u \in E_+$, there is $d_u \in D$ satisfying $(d_u - u)_+ > 0$. By the general saturation principle, there exists some $d \in {}^*D$ such that $(d-u)_+ > 0$ for all $u \in E_+$. Then d satisfies $d \in {}^*D \setminus \mathrm{fin}({}^*E)$. This contradiction with ${}^*D \subseteq \mathrm{fin}({}^*E)$ shows that D is order bounded. ▷

4.6.2. *Theorem.* *Let E and F be vector lattices, and let $T : E \to F$ be a linear operator. Then the following are equivalent:*

(1) *T is an order bounded operator;*

(2) *${}^*T(\mathrm{fin}({}^*E)) \subseteq \mathrm{fin}({}^*F)$;*

(3) *${}^*T(\lambda({}^*E)) \subseteq \lambda({}^*F)$;*

(4) *${}^*T(\lambda({}^*E)) \subseteq \mathrm{fin}({}^*F)$.*

◁ (1)→(2): Obvious.

(2)→(3): Let ${}^*T(\mathrm{fin}({}^*E)) \subseteq \mathrm{fin}({}^*F)$. Take an arbitrary $\varkappa \in \lambda({}^*E)$. Then, for some $d \in E$, the condition $|n\varkappa| \leq d$ holds for all $n \in \mathbb{N}$ simultaneously. It is easy to see that in this case there is a $\nu \in {}^*\mathbb{N} \setminus \mathbb{N}$ such that $|\nu\varkappa| \leq d$. Consequently, $\nu\varkappa \in \mathrm{fin}({}^*E)$ and, by hypothesis, $\nu {}^*T(\varkappa) = {}^*T(\nu\varkappa) \in \mathrm{fin}({}^*F)$. This implies

${}^*T\varkappa \in \lambda({}^*F)$. The latter means that condition (3) is valid, since the choice of the element $\varkappa \in \lambda({}^*E)$ was arbitrary.

(3)→(4): Obvious.

(4)→(2): Assume that ${}^*T\varkappa \notin \mathrm{fin}({}^*F)$ for some $\varkappa \in \mathrm{fin}({}^*E)$. For every $n \in \mathbb{N}$ and every $f \in F$, assign

$$A_{n,f} := \{k \in {}^*\mathbb{N} : k \geq n \ \& \ (|{}^*T(k^{-1}\varkappa)| - f)_+ > 0\}.$$

The sets $A_{n,f}$ are nonempty by the choice of $\varkappa$. By construction, they are internal, comprising a system with the finite intersection property since

$$A_{\max(n,p),\sup(f,g)} \subseteq A_{n,f} \cap A_{p,g}$$

for arbitrary $n, p \in \mathbb{N}$ and $f, g \in F$. Applying the general saturation principle, we find a $\nu \in \bigcap_{n,f} A_{n,f}$. It is obvious that $\nu \in {}^*\mathbb{N} \setminus \mathbb{N}$, and so $|\nu^{-1}\varkappa| \in \lambda({}^*E)$. By assumption, ${}^*T(\lambda({}^*E)) \subseteq \mathrm{fin}({}^*F)$. Therefore, there is a $y \in F$ such that $(|{}^*T(\nu^{-1}\varkappa)| - y)_+ = 0$, which is impossible since $\nu \in A_{1,y}$. The obtained contradiction means that ${}^*T(\mathrm{fin}({}^*E)) \subseteq \mathrm{fin}({}^*F)$.

(2)→(1): Take an arbitrary $u \in E_+$. By condition (2),

$${}^*(T([-u, u])) = {}^*T({}^*[-u, u]) \subseteq \mathrm{fin}({}^*F).$$

Hence, by Lemma 4.6.1, the set $T([-u, u])$ is order bounded. ▷

4.6.3. Before stating a nonstandard criterion for a linear operator to be order continuous, we find a connection between (o)-infinitesimal elements of an Archimedean vector lattice F and those of a Dedekind completion of F.

Lemma. *Let F be an Archimedean vector lattice and let F_1 be a Dedekind completion of F. Then $\eta({}^*F) = {}^*F \cap \eta({}^*F_1)$.*

◁ Given $\varkappa \in {}^*F_1$, assign

$$U_F(\varkappa) := \{x \in F : x \geq \varkappa\}, \quad U_{F_1}(\varkappa) := \{x \in F_1 : x \geq \varkappa\}.$$

Let $\varkappa \in \eta({}^*F)$. Then $\inf_F U_F(|\varkappa|) = 0$ and, since F_1 is a Dedekind completion of F, we have $\inf_{F_1} U_F(|\varkappa|) = 0$. Furthermore, $U_F(|\varkappa|) \subseteq U_{F_1}(|\varkappa|)$, which implies $\inf_{F_1} U_{F_1}(|\varkappa|) = 0$. Hence, $\varkappa \in \eta({}^*F_1)$. At the same time, $\varkappa \in {}^*F$. Consequently, $\varkappa \in {}^*F \cap \eta({}^*F_1)$. Conversely, let $\varkappa \in {}^*F \cap \eta({}^*F_1)$. Then $\inf_{F_1} U_{F_1}(|\varkappa|) = 0$. Since F_1 is a Dedekind completion of F, it is easy to verify that $\inf_{F_1} U_F(|\varkappa|) = 0$. This immediately implies $\inf_F U_F(|\varkappa|) = 0$. Thus, $\varkappa \in \eta({}^*F)$. ▷

4.6.4. Theorem. *Let E and F be Archimedean vector lattices, let F_1 be a Dedekind completion of F, and let $T : E \to F$ be a linear operator. Then the following are equivalent:*

(1) *T is an order continuous operator;*

(2) $^*T(\eta(^*E)) \subseteq \eta(^*F_1)$;

(3) $^*T(\eta(^*E)) \subseteq \eta(^*F)$.

$\lhd$ (1)$\to$(2): Let T be an order continuous operator in $L(E,F)$. Then it is easy to verify that T is an order continuous operator in $L(E,F_1)$. Since F_1 is Dedekind complete; $|T|$ is defined, presenting an order continuous operator from E into F_1.

In view of the inequality $|Tx| \le |T|(|x|)$ $(x \in E)$, to verify the required implication it is sufficient to show

$$^*|T|(\eta(^*E)) \subseteq \eta(^*F_1).$$

Take an arbitrary $\varkappa \in \eta(^*E)$. Then, since $|T|$ is order continuous, $\inf_E U(|\varkappa|) = 0$ implies $\inf_{F_1} |T|(U(|\varkappa|)) = 0$. But

$$|T|(U(|\varkappa|)) \subseteq U(|^*T(\varkappa)|),$$

so $\inf_{F_1} U(|^*T\varkappa|) = 0$ and, consequently, $^*T\varkappa \in \eta(^*F_1)$.

(2)$\to$(3): This follows readily from Lemma 4.6.3 since $^*T(^*E) \subseteq {}^*F$.

(3)$\to$(1): Let $^*T(\eta(^*E)) \subseteq \eta(^*F)$. Since E is Archimedean, $\lambda(^*E) \subseteq \eta(^*E)$ by Theorem 4.3.5. Consequently,

$$^*T(\lambda(^*E)) \subseteq \eta(^*F) \subseteq \operatorname{fin}(^*F).$$

By Theorem 4.6.2, this implies $T \in L_r(E,F)$. To verify order continuity, it remains to prove that $\inf_F |Tx_\xi| = 0$ for every net $x_\xi \downarrow 0$ in E. Take an arbitrary net $(x_\xi)_{\xi\in\Xi} \subseteq E$ such that $x_\xi \downarrow 0$. Assume that, for some element $f \in F$, $f > 0$, the condition $|Tx_\xi| \ge f$ holds for all $\xi \in \Xi$ simultaneously. Then, by the transfer principle, $|Tx_\xi| \ge f$ for all $\xi \in {}^*\Xi$. Let β be some remote element of the directed set $^*\Xi$ (such an element exists by Lemma 4.0.5). According to the criterion established in 4.3.2, we have $x_\beta \in \eta(^*F)$. Then, by condition (3), $^*Tx_\beta \in \eta(^*F)$, which contradicts $|^*Tx_\beta| \ge f$. Thus, $\inf |Tx_\xi| = 0$ for every net (x_ξ) decreasing to zero, and so the operator T is order continuous. $\rhd$

4.7. $*$-Invariant Homomorphisms

One of the important facts of nonstandard analysis is the assertion that each limited internal real number $\alpha \in \operatorname{fin}({}^*\mathbb{R})$ is infinitely close to a unique standard real number $\operatorname{st}(\alpha)$ called the *standard part* of α. The operation st of taking the standard part of a real is a Riesz homomorphism of the external vector space $\operatorname{fin}({}^*\mathbb{R})$ into $\mathbb{R}$ such that $\operatorname{st}(a) = a$ for all $a \in \mathbb{R}$ and $\operatorname{st}(\alpha_1) = \operatorname{st}(\alpha_2)$ whenever $\alpha_1 \approx \alpha_2$. This leads to the problem whether or not we may take the standard part of an element in a nonstandard enlargement of a vector lattice or a Boolean algebra. In other words: What conditions will guarantee existence of a Riesz or Boolean homomorphism keeping standard elements invariant and not distinguishing infinitely close elements? In this section, we discuss this question for nonstandard enlargements of vector lattices and Boolean algebras and establish that such an *invariant* homomorphism exists if and only if the vector lattice (Boolean algebra) in question is Dedekind complete (complete). In the end, we consider the structure of invariant homomorphisms on nonstandard enlargements of complete normed Boolean algebras and establish that, for atomless complete normed Boolean algebras, every invariant homomorphism is almost singular with respect to the measure, in the sense that the carrier of the homomorphism is contained in an internal set whose measure is arbitrarily small but nonzero. The considerations in this section rest on [10].

4.7.1. Let E be a vector lattice and let ${}^*E$ be a nonstandard enlargement of E. We may assume that E is a vector sublattice of ${}^*E$.

DEFINITION. A mapping $\psi : \operatorname{fin}({}^*E) \to E$ is called a *$*$-invariant Riesz homomorphism* if ψ is a Riesz homomorphism such that $\psi(x) = x$ for $x \in E$.

Henceforth we abbreviate a $*$-invariant Riesz homomorphism to a $*$-IRH. It is easy to see that the inequalities

$$\sup_E\{x \in E : x \leq \chi\} \leq \psi(\chi) \leq \inf_E\{y \in E : y \geq \chi\} \tag{10}$$

are valid for every $*$-IRH ψ and every $\chi \in \operatorname{fin}({}^*E)$ provided that the supremum and infimum exist on the right and left sides. In particular this implies

$$(x - \chi) \in \eta({}^*E) \to \psi(\chi) = x \quad (x \in E,\ \chi \in \operatorname{fin}({}^*E)). \tag{11}$$

Theorem. *Let E be a vector lattice. There exists a $*$-invariant Riesz homomorphism ψ on $\operatorname{fin}({}^*E)$ if and only if the vector lattice E is Dedekind complete. If E is atomic and Dedekind complete, then a $*$-IRH on $\operatorname{fin}({}^*E)$ is uniquely defined by*

$$\psi(\chi) = \sup_E\{x \in E : x \leq \chi\} = \inf_E\{y \in E : y \geq \chi\} \quad (\chi \in \operatorname{fin}({}^*E)). \tag{12}$$

◁ Let E be a Dedekind complete vector lattice. We apply the extension theorem of Bernau–Lipecki–Luxemburg–Schep (see, for example, [3, Theorem 2.1]) to the triple $(E, \mathrm{fin}({}^*E), E)$ and the identical Riesz homomorphism $\imath : E \to E$. Then we obtain the Riesz homomorphism $\psi : \mathrm{fin}({}^*E) \to E$ which extends $\imath$. Obviously, ψ is a $*$-IRH on $\mathrm{fin}({}^*E)$.

Suppose there is a $*$-IRH $\psi : \mathrm{fin}({}^*E) \to E$. Take an order bounded upwards-directed nonempty set $\mathscr{D} \subseteq E$. By the general saturation principle, in ${}^*\mathscr{D}$ there exists an element $\delta \in \mathrm{fin}({}^*E)$ satisfying $\delta \geq d$ for all $d \in \mathscr{D}$. Then, as is easy to see, $\psi(\delta) = \sup_E \mathscr{D}$. Since the set $\mathscr{D} \subseteq E$ was chosen arbitrarily, this implies that E is a Dedekind complete vector lattice.

Now let E be atomic and Dedekind complete. Take a $*$-IRH $\psi : \mathrm{fin}({}^*E) \to E$ and $\chi \in \mathrm{fin}({}^*E)$. By Theorem 4.4.6, $\mathrm{fin}({}^*E) = E + \eta({}^*E)$. Consequently, there exists a unique $x \in E$ obeying the condition $(x - \chi) \in \eta({}^*E)$ By (11), we obtain $\psi(\chi) = x$. Thus the $*$-IRH ψ is defined uniquely and satisfies (12). ▷

Note that uniqueness of a $*$-IRH on a Dedekind complete vector lattice E implies that E is atomic. For a proof of this assertion we refer the reader to [10, Theorem 2.1].

4.7.2. We now consider a similar problem for Boolean algebras. Let B be a Boolean algebra and let ${}^*B$ be a nonstandard enlargement of B. We assume that B is a Boolean subalgebra of ${}^*B$.

DEFINITION. A mapping $h : {}^*B \to B$ is a *$*$-invariant Boolean homomorphism* if h is a Boolean homomorphism such that $h(b) = b$ for all $b \in B$.

For brevity, a $*$-invariant Boolean homomorphism will be called a $*$-IBH henceforth. It is easy to see that every $*$-IBH h satisfies

$$\sup_B \{x \in B : x \leq \beta\} \leq h(\beta) \leq \inf_B \{y \in B : y \geq \beta\} \tag{13}$$

for all $\beta \in {}^*B$ provided that the supremum on the left side and the infimum on the right side both exist. This implies in particular that $h(\beta) = 0$ for every element $\beta \in {}^*B$ such that $\inf_B\{b \in B : b \geq \beta\} = 0$.

Theorem. *There exists a $*$-invariant Boolean homomorphism ${}^*B \to B$ if and only if the Boolean algebra B is complete. Moreover, a $*$-IBH $h : {}^*B \to B$ is defined uniquely if the complete Boolean algebra B is atomic. In this case*

$$h(\beta) = \sup_B \{x \in B : x \leq \beta\} = \inf_B \{y \in B : y \geq \beta\}$$

for all $\beta \in {}^*B$.

Before proving the theorem, we present one nonstandard characterization of an atomic complete Boolean algebra which is due to H. Conshor. For this we need some notations. Let B be a Boolean algebra. Given $\varkappa \in {}^*B$, consider the set $U(\varkappa) := \{x \in B : x \geq \varkappa\}$ of standard upper bounds of $\varkappa$ and the set $L(\varkappa) := \{y \in B : \varkappa \geq y\}$ of standard lower bounds of $\varkappa$. Define the external subsets of ${}^*B$:

$$o\text{-pns}({}^*B) := \{\varkappa \in {}^*B : \inf_B (U(\varkappa) - L(\varkappa)) = 0\},$$
$$\eta({}^*B) := \{\varkappa \in {}^*B : \inf_B U(|\varkappa|) = 0\}.$$

It can be shown that o-pns$({}^*B)$ is a Boolean subalgebra of ${}^*B$, and $\eta({}^*B)$ is an ideal in o-pns$({}^*B)$ and the quotient o-pns$({}^*B)/\eta({}^*B)$ is Boolean isomorphic to B (see [4, Theorem 4.1]). From this and from [4, Theorem 4.3] we have immediately the next

Lemma (H. Conshor). *For every Boolean algebra B, the following are equivalent:*

(1) *B is an atomic complete Boolean algebra;*

(2) ${}^*B = B + \eta({}^*B)$.

PROOF OF THE THEOREM:

◁ Let B be a complete Boolean algebra. We apply Sikorski's Extension Theorem (see, for example, [25, Theorem 33.1]) to the triple $(B, {}^*B, B)$ and the identical Boolean homomorphism $\imath : B \to B$. We then obtain a Boolean homomorphism $h : {}^*B \to B$ that extends $\imath$. Obviously, h is a $*$-IBH on ${}^*B$).

We now show that the existence of a $*$-IBH $h : {}^*B \to B$ implies completeness of B. Let $h : {}^*B \to B$ be a $*$-IBH. Take an upwards-directed nonempty set $\mathscr{D} \subseteq B$. By the general saturation principle, there exists an element δ in ${}^*\mathscr{D}$ satisfying $\delta > d$ for all $d \in \mathscr{D}$. Then, as is easy to verify, $h(\delta) = \sup_B \mathscr{D}$. Since the upwards-directed nonempty set $\mathscr{D} \subseteq B$ was chosen arbitrarily, this implies that B is a complete Boolean algebra.

Let E be atomic and Dedekind complete. Then the proof of uniqueness of a $*$-IBH $h : {}^*B \to B$ is similar to the proof of uniqueness of a $*$-IRH in 4.7.1. We must use the preceding lemma instead of Theorem 4.4.6 only. ▷

Note that uniqueness of a $*$-IBH on a complete Boolean algebra B implies that B is atomic. For a proof of this assertion we refer the reader to [10, Theorem 1.1].

4.7.3. Let B be a complete Boolean algebra. For convenience, given a family $(a_\tau) \subseteq B$, we denote $\sup_\tau a_\tau$ by $\bigoplus_\tau a_\tau$ whenever the elements a_τ are disjoint. A *partition* of an element $b \in B$ is a family $(b_\tau) \subseteq B$ such that $b = \bigoplus_\tau b_\tau$. Let $\mu : B \to \mathbb{R}_+$ be a mapping on B satisfying the following conditions:

(1) $\mu(b) > 0 \leftrightarrow b > 0$;

(2) The equality $\mu(\bigoplus_{n=1}^{\infty} a_n) = \sum_{n=1}^{\infty} \mu(a_n)$ is valid for every sequence $a_1, a_2, \dots$ of disjoint elements of B.

Recall that a mapping μ with the above properties is a *σ-additive measure* and the pair (B, μ) is a *complete normed Boolean algebra.*

Let (B, μ) be a complete normed Boolean algebra and let $h : {}^*B \to B$ be a $*$-IBH. Represent B as a direct sum of atomic and atomless components: $B = B_a \bigoplus B_c$. Then (B_a, μ) and (B_c, μ) are complete normed Boolean algebras. The restriction of h to ${}^*B_a$ is a $*$-invariant Boolean homomorphism preserving the measure μ in the sense that $\mu(h(\alpha)) = \mathrm{st}({}^*\mu(\alpha))$ for all $\alpha \in {}^*B_a$. The restriction of h to ${}^*B_c$ with respect to the measure μ has an opposite behavior. Namely, the following holds:

Theorem. *Let (B, μ) be an atomless complete normed Boolean algebra and let $h : {}^*B \to B$ be a $*$-invariant Boolean homomorphism. Then, for every real $\varepsilon > 0$, there exists $\chi_\varepsilon \in {}^*B$, ${}^*\mu(\chi_\varepsilon) < \varepsilon$, satisfying the condition $h(b) = h(b \wedge \chi_\varepsilon)$ for all $b \in {}^*B$.*

Before proving the theorem, we establish one elementary property of atomless complete normed Boolean algebras.

Lemma. *For every atomless complete normed Boolean algebra (B, μ) and every natural n, there exists a partition $(\chi_i)_{i=1}^n \subseteq B$ of $\mathbf{1}_{^*B}$ such that ${}^*\mu(\chi_i \wedge d) = \frac{1}{n}\mu(d)$ for all $d \in B$, $i = 1, \dots, n$.*

$\lhd$ Take an arbitrary hyperfinite partition $(e_k)_{k=1}^{\nu}$ of $\mathbf{1}_{^*B}$ in the Boolean algebra ${}^*B$ which is refined into each finite standard partition. Existence of such a partition is easy on using the general saturation principle. Since B is atomless and the measure μ is σ-additive, there exist partitions $e_k = \bigoplus_{i=1}^n e_i^k$ such that

$$ {}^*\mu(e_i^k) = \frac{1}{n} {}^*\mu(e_k) $$

for all $k = 1, \dots, \nu$ and $i = 1, \dots, n$. Put $\chi_i := \bigoplus_{k=1}^{\nu} e_i^k$. The family $(\chi_i)_{i=1}^n \subseteq {}^*B$ is a required partition of unity. $\rhd$

PROOF OF THE THEOREM:

$\lhd$ Take an $n \in \mathbb{N}$ such that $\frac{1}{n}\mu(\mathbf{1}) < \varepsilon$. According to the preceding lemma, there exists a partition $(\chi_i)_{i=1}^n \subseteq {}^*B$ of $\mathbf{1}_{^*B}$ satisfying the condition

$$ {}^*\mu(\chi_i \wedge d) = \frac{1}{n}\mu(d) $$

for arbitrary $d \in B$, $i = 1, \dots, n$. In particular,

$$ {}^*\mu(\chi_k \wedge h(\chi_m)) = \frac{1}{n}\mu(h(\chi_m)) $$

for $k, m \in 1, \dots, n$. Consider the element

$$\chi_\varepsilon := \bigoplus_{k=1}^{n} \chi_k \wedge h(\chi_k).$$

Then

$$^*\mu(\chi_\varepsilon) = \sum_{k=1}^{n} {^*\mu}(\chi_k \wedge h(\chi_k)) = \sum_{k=1}^{n} \frac{1}{n}\mu(h(\chi_k)) = \frac{1}{n}\sum_{k=1}^{n} \mu(h(\chi_k))$$

$$= \frac{1}{n}\mu\Big(\bigoplus_{k=1}^{n} h(\chi_k)\Big) = \frac{1}{n}\mu\Big(h\Big(\bigoplus_{k=1}^{n} \chi_k\Big)\Big) = \frac{1}{n}\mu(h(\mathbf{1}_{^*B})) = \frac{1}{n}\mu(\mathbf{1}_B) < \varepsilon.$$

At the same time

$$h(\mathbf{1}_{^*B} \setminus \chi_\varepsilon) = h\Big(\bigoplus_{m=1}^{n} \bigoplus_{k \neq m} \chi_k \wedge h(\chi_m)\Big) = \bigoplus_{m=1}^{n} \bigoplus_{k \neq m} h(\chi_k \wedge h(\chi_m)) = 0,$$

since

$$h(\chi_k \wedge h(\chi_m)) = h(\chi_k) \wedge h^2(\chi_m)$$
$$= h(\chi_k) \wedge h(\chi_m) = h(\chi_k \wedge \chi_m) = h(0) = 0$$

for $k \neq m$. Thus,

$$h(b) = h(b \wedge \mathbf{1}_{^*B}) = h((b \wedge \chi_\varepsilon) \oplus (b \wedge (\mathbf{1}_{^*B} \setminus \chi_\varepsilon)))$$
$$= h(b \wedge \chi_\varepsilon) \oplus (h(b) \wedge h(\mathbf{1}_{^*B} \setminus \chi_\varepsilon)) = h(b \wedge \chi_\varepsilon)$$

for all $b \in {^*B}$. ▷

4.7.4. Consider the real-valued mappings $\mathrm{st} \circ {^*\mu}$ and $\mu \circ h$ defined on a Boolean algebra $^*B$, where h is a $*$-IBH. Obviously, the mappings $\mathrm{st} \circ {^*\mu}$ and $\mu \circ h$ are finitely additive measures on $^*B$. Moreover, these mappings are σ-additive, since the condition $b = \bigoplus_{n=1}^{\infty} b_n$ imposed on elements of $^*B$ implies $b = \bigoplus_{n=1}^{m} b_n$ for some $m \in \mathbb{N}$. Thus, $\mathrm{st} \circ {^*\mu}$ and $\mu \circ h$ extend to σ-additive measures $\tilde{\mu}$ and $\tilde{\mu}_h$ on the σ-completion $^*B_\sigma$ of the Boolean algebra $^*B$. Observe that $\tilde{\mu}$ is the Loeb measure corresponding to the initial measure μ.

Theorem. *Let (B, μ) be an atomless complete normed Boolean algebra and let $h : {^*B} \to B$ be a $*$-invariant Boolean homomorphism. Then the following hold:*

(1) *There exists an element $\chi_h \in {}^*B_\sigma$, $\mu(\chi_h) = 0$, such that the equality $h(b) = 0$ holds for every $b \in {}^*B$ satisfying the condition $b \wedge \chi_h = 0$;*

(2) *The carriers of the measures μ and μ_h are disjoint.*

$\lhd$ (1): By Theorem 4.7.3, for every $n \in \mathbb{N}$, there exists an element $\chi_n \in {}^*B$ such that $\tilde{\mu}(\chi_n) \leq \frac{1}{n}$ and $h(b) = 0$ for all $b \in {}^*B$, $b \wedge \chi_n = 0$. Put $\chi_h = \bigwedge_{n=1}^{\infty} \chi_n$. It is clear that $\chi_h \in {}^*B_\sigma$ and $\tilde{\mu}(\chi_h) = 0$. Take an arbitrary $b \in {}^*B$, $b \wedge \chi_h = 0$. Then there exists an $n \in \mathbb{N}$ such that $b \wedge \chi_n = 0$. Thus, $h(b) = 0$, as required.

(2) ensues from (1). Indeed, the carriers of the measures $\tilde{\mu}$ and $\tilde{\mu}_h$ are disjoint elements χ_h and $\mathbf{1} \setminus \chi_h$ of the Boolean algebra ${}^*B_\sigma$. $\rhd$

4.8. Order Hulls of Vector Lattices

In this section, we define the order hull of a vector lattice. Some properties of order hulls are established. In particular, the question about their (r)- and (o)-completeness is studied to some extent. Some conditions are found for the order hull of a vector lattice E to be isomorphic with the initial vector lattice E and (if E is a normed lattice) for the order hull of E to be isomorphic with the nonstandard hull of E regarded as a normed vector space.

4.8.1. Let E be a vector lattice. As it was mentioned in 4.3.1, the set $\operatorname{fin}({}^*E)$ of limited elements in ${}^*E$ is a vector lattice too, while the set $\eta({}^*E)$ of (o)-infinitesimal elements in ${}^*E$ is an ideal in $\operatorname{fin}({}^*E)$. Consider the quotient vector lattice

$$(o)\text{-}E := \operatorname{fin}({}^*E)/\eta({}^*E).$$

We call (o)-E the *order hull* of E and denote by $[\varkappa]$ the coset $\varkappa + \eta({}^*E) \in (o)\text{-}E$ that contains $\varkappa \in \operatorname{fin}({}^*E)$. Define the mapping $\widehat{\eta}_E : E \to (o)\text{-}E$ by

$$\widehat{\eta}_E(x) := [x] \quad (x \in E).$$

Clearly, $\widehat{\eta}_E : E \to (o)\text{-}E$ is a Riesz homomorphism. It will be denoted by $\widehat{\eta}$ if this does not lead to ambiguity.

4.8.2. Theorem. *The set $\widehat{\eta}(E)$ is a complete vector sublattice of (o)-E.*

Before proving, we give some explanations. Let L be a vector sublattice of a vector lattice M. Recall that L is a *complete vector sublattice* of M if, for every nonempty $D \subseteq L$ and every $a \in L$, the condition $\inf_L D = a$ implies $\inf_M D = a$. It is easy to see that L is a complete vector sublattice of M if and only if, for every nonempty $D \subseteq L$, the condition $\inf_L D = 0$ implies $\inf_M D = 0$.

$\lhd$ Let $D \subseteq E$ such that $\inf_E D = 0$. Show that $\inf_{(o)\text{-}E} \widehat{\eta}(D) = 0$. Assume the contrary. Then for some $\varkappa \in \operatorname{fin}({}^*E)$ we have

$$\widehat{\eta}(D) \geq [\varkappa] > 0.$$

Since $\varkappa \notin \eta(^*E)$, there is an $a \in E$ such that

$$U(\varkappa) \geq a > 0.$$

Take an arbitrary $d \in D$. Then $\widehat{\eta}(d) \geq [\varkappa]$ and, consequently, $(\varkappa - d)_+ \in \eta(^*E)$. Thus, $\inf_E \mathscr{U} = 0$, where $\mathscr{U} := U((\varkappa - d)_+)$. Given $u \in \mathscr{U}$, note

$$d + u \geq d + (\varkappa - d)_+ \geq \varkappa.$$

Hence, $d + \mathscr{U} \subseteq U(\varkappa)$, and so $d + \mathscr{U} \geq a$. Therefore,

$$d = \inf_E (d + \mathscr{U}) \geq a.$$

Since the last inequality is valid for all $d \in D$ and $\inf_E D = 0$, we have $a = 0$. A contradiction shows that $\inf_{(o)\text{-}\overline{E}} \widehat{\eta}(D) = 0$. ▷

4.8.3. Theorem. *The order hull of a vector lattice E is Archimedean if and only if E is Archimedean.*

◁ Necessity follows from the fact that each vector lattice may be embedded as a vector sublattice into its order hull. To prove sufficiency, we consider an Archimedean vector lattice E. By Theorem 4.3.5, $\eta(^*E)$ is a relatively uniform closed ideal in $\mathrm{fin}(^*E)$. Then, according to the well-known theorem by A. I. Veksler [26] (see also [21, Theorem 60.2]), the quotient (o)-E is Archimedean. ▷

4.8.4. Theorem. *The order hull of a vector lattice is relatively uniformly complete.*

◁ Let E be a vector lattice. Since every quotient vector lattice of a relatively uniformly complete vector lattice is relatively uniform complete too (see, for example, [21, Theorem 59.4]), it is enough to establish relative uniform completeness of $\mathrm{fin}(^*E)$. Take a relatively uniformly Cauchy sequence $(\varkappa_n)_{n=1}^{\infty} \subseteq \mathrm{fin}(^*E)$. Then, there exist a sequence $(\varepsilon_n) \subseteq \mathbb{R}$, $\varepsilon_n \downarrow 0$, and an element $\delta \in \mathrm{fin}(^*E)$ such that

$$|\varkappa_m - \varkappa_k| \leq \varepsilon_n \delta \text{ for all } m, k, n \in \mathbb{N}$$

whenever $m, k \geq n$. We extend $(\varkappa_n)_{n \in \mathbb{N}}$ to an internal sequence $(\varkappa_n)_{n \in {}^*\mathbb{N}} \subseteq {}^*E$ and associate with each natural k the internal set:

$$I_k := \{m \in {}^*\mathbb{N} : |\varkappa_m - \varkappa_k| \leq \varepsilon_m \delta\}.$$

It is easy to see that the family $\{I_k\}_{k=1}^{\infty}$ has the finite intersection property. By the general saturation principle, there is a $\nu \in \bigcap_{k=1}^{\infty} I_k$. Then every $k \in \mathbb{N}$ satisfies $|\varkappa_k - \varkappa_\nu| \leq \varepsilon_k \delta$. This implies $\varkappa_\nu \in \mathrm{fin}(^*E)$ and $\varkappa_n \xrightarrow{(r)} \varkappa_\nu$. ▷

4.8.5. The matter with Dedekind completeness of the order hull of a lattice differs from that with relative uniform completeness. We show that the order hull of a Dedekind complete vector lattice containing nonatomic elements is not necessarily Dedekind complete (Theorem 4.8.7 establishes that the order hull of an atomic Dedekind complete vector lattice is Dedekind complete).

Recall that a Dedekind complete vector lattice E is called *regular* (see, for example, [29]) if the following hold:

(1) Order convergence and relative uniform convergence coincide for every sequence in E;

(2) Each ideal with a countable order basis in E is contained in some principal ideal;

(3) E is order separable.

As examples of regular vector lattices, we may take Banach lattices with order continuous norm and $L_p([0,1])$ with $0 < p < 1$.

Theorem. *The order hull of a nonatomic regular vector lattice is not Dedekind complete.*

$\lhd$ Let E be a nonatomic regular vector lattice. Then there is a nonatomic element $e \in E$, $e > 0$. Let $\nu \in {}^*\mathbb{N} \setminus \mathbb{N}$ be some illimited natural number. By Lemma 4.4.4, there exists a family $\{e_n\}_{n=1}^{\nu}$ of disjoint e-punches. Assign $D := \{[e_n]\}_{n=1}^{\infty}$. Then D is a nonempty and bounded above (for example, by the element $[e]$) subset of (o)-E. We show that this subset fails to have a least upper bound in (o)-E. By way of contradiction, assume

$$[\varkappa] = \sup_{(o)\text{-}\overline{E}} D \text{ for some } \varkappa \in \operatorname{fin}({}^*E).$$

Then for all $k \in \mathbb{N}$, we have

$$(e_k - \varkappa)_+ \in \eta({}^*E).$$

By Theorem 4.3.6, E has the property $\eta({}^*E) = \lambda({}^*E)$ (here properties (1) and (3) from the definition of a regular vector lattice are used). Hence, $(e_k - \varkappa)_+ \in \lambda({}^*E)$ holds for all $k \in \mathbb{N}$. By using (2), it is easy to see that there exists a $d \in E$ for which

$$(e_k - \varkappa)_+ \leq m^{-1}d \quad (k, m \in \mathbb{N}).$$

Applying the general saturation principle, we find $\omega, \gamma \in {}^*\mathbb{N} \setminus \mathbb{N}$ such that

$$\omega \leq \nu \text{ and } (e_\omega - \varkappa)_+ \leq \gamma^{-1}d.$$

Then

$$(e_\omega - \varkappa)_+ \in \lambda(^*E) = \eta(^*E)$$

and, consequently, $[e_\omega] \leq [\varkappa]$. At the same time, $[e_\omega] > 0$ (because $\omega \leq \nu$ while e_ω is an e-punch) and $[e_\omega] \wedge [e_k] = 0$ for all $k \in \mathbb{N}$ (since e_ω and e_k are disjoint). Hence

$$[\varkappa] > [\varkappa] - [e_\omega] > [e_k] \quad \text{for every } k \in \mathbb{N},$$

which contradicts the assumption $[\varkappa] = \sup_{(o)\text{-}\overline{E}} D$. Thus, the order hull (o)-E of E is not Dedekind complete. ▷

4.8.6. We establish one more property of order hulls concerning cardinality. We denote by card(A) the cardinality of a set A.

Lemma. *Assume that a vector lattice E is not Archimedean or not atomic. Then*

$$\text{card}(E) < \text{card}((o)\text{-}E).$$

◁ Take an arbitrary $\nu \in {}^*\mathbb{N}\setminus\mathbb{N}$. According to Lemma 4.0.4, card$(E) <$ card(ν). Therefore it is sufficient to establish that the order hull of E contains ν distinct elements.

Assume first that E is not Archimedean. Then there are elements $u, v \in E$ such that $0 < nu \leq v$ for all natural n. By the transfer principle, $0 < nu \leq v$ holds for all $n \in {}^*\mathbb{N}$. In particular, $nu \in \text{fin}(^*E)$ for all $n \in {}^*\mathbb{N}$. The inequality

$$0 < u \leq |nu - mu|$$

is valid for all $n, m \in {}^*\mathbb{N}$ such that $n \neq m$. Therefore, $[nu] \neq [mu]$ whenever $n, m \in {}^*\mathbb{N}$ and $n \neq m$. Thus, $\{[nu]\}_{n=1}^{\nu}$ is a family consisting of ν distinct elements of (o)-E.

It remains to consider the case in which the vector lattice E is Archimedean but not atomic. Then, there is a nonatomic element $e \in E$, $e > 0$. By Lemma 4.4.4, there exists a family $\{e_k\}_{k=1}^{\nu}$ of disjoint e-punches in $\text{fin}(^*E)$. It follows immediately that the elements $[e_k]$ of the order hull (o)-E are distinct for $k = 1, 2, \dots, \nu$. ▷

In the rest of the section, we study the question about conditions under which the order hull of E coincides with E or with the nonstandard hull of E (if the lattice E is assumed to be normed and regarded as a normed vector space).

4.8.7. ***Theorem.*** *For every vector lattice E, the following are equivalent:*

(1) *E is Riesz isomorphic to (o)-E;*

(2) *E is an atomic Dedekind complete vector lattice;*

(3) *$\widehat{\eta}_E$ is a Riesz isomorphism of E onto the order hull of E.*

$\lhd$ (1)$\to$(2): The vector lattice E is Archimedean and atomic by Lemma 4.8.6. So, $\mathrm{fin}(^*E) = o\text{-}\mathrm{pns}(^*E)$ by Theorem 4.4.5. Then, $(o)\text{-}E = o\text{-}\mathrm{pns}(^*E)/\eta(^*E)$, and by 4.4.1(1), the vector lattice $(o)\text{-}E$ and, consequently, E is Dedekind complete.

(2)$\to$(3): Since by Theorem 4.4.6 $\mathrm{fin}(^*E) = E + \eta(^*E)$, for every $u \in (o)\text{-}E$, there is an $x \in E$ such that $u = x + \eta(^*E)$. In particular, the image of the Riesz homomorphism $\widehat{\eta} : E \to (o)\text{-}E$ coincides with $(o)\text{-}E$. Thus, $\widehat{\eta}$ is a Riesz isomorphism.

The implication (3)$\to$(1) is obvious. $\rhd$

It is interesting to compare this theorem with Proposition 4.0.6.

4.8.8. Let (E, ρ) be a normed vector lattice. Recall that, according to 4.0.6, we may arrange the quotient vector lattice

$$\ddot{\bar{E}} := \mathrm{Fin}(^*E)/\mu(^*E)$$

with the respective quotient norm. $\ddot{\bar{E}}$ is called the *nonstandard hull* of (E, ρ). Obviously, $\ddot{\bar{E}}$ is a Banach lattice. Note that $\ddot{\bar{E}}$ depends not only on E but also on the choice of the norm ρ. Denote the coset of an element $\varkappa \in \mathrm{Fin}(^*E)$ in the quotient vector lattice $\ddot{\bar{E}}$ by $\langle\langle\varkappa\rangle\rangle$ and consider the mapping $\widehat{\mu} : E \to \ddot{\bar{E}}$ (cf. 4.0.6) such that $\widehat{\mu}(x) := \langle\langle\varkappa\rangle\rangle$ for every $x \in E$. It is easy to see that $\widehat{\mu}$ is a Riesz monomorphism.

Theorem. *Let (E, ρ) be a normed vector lattice. Then the following are equivalent:*

(1) *There exists a Riesz isomorphism π of $(o)\text{-}E$ onto $\ddot{\bar{E}}$ such that $\pi \circ \widehat{\eta} = \widehat{\mu}$;*

(2) *E is finite-dimensional.*

$\lhd$ (1)$\to$(2): Let $\pi : (o)\text{-}E \to \ddot{\bar{E}}$ be a Riesz isomorphism such that $\pi \circ \widehat{\eta} = \widehat{\mu}$. The ideal, generated by $\widehat{\eta}(E)$, coincides with $(o)\text{-}E$ and, furthermore, $\widehat{\mu}(E) = \pi(\widehat{\eta}(E))$; therefore, the ideal, generated by $\widehat{\mu}(E)$, coincides with $\ddot{\bar{E}}$. Consequently,

$$\mathrm{Fin}(^*E) = \mathrm{fin}(^*E) + \mu(^*E),$$

which implies $\mu(^*E) \subseteq \eta(^*E)$ by Theorem 4.5.2. According to Theorem 4.5.4, it remains to establish the reverse inclusion. Let $\varkappa \in \eta(^*E)$. Then $U := U(|\varkappa|)$ is directed downwards and $U \downarrow 0$. In this case $\widehat{\eta}(U) \downarrow 0$ in $(o)\text{-}E$ by Theorem 4.8.2. Hence, $\pi \circ \widehat{\eta}(U) \downarrow 0$ in the vector lattice $\ddot{\bar{E}}$. In other words,

$$\inf_{\ddot{\bar{E}}} \widehat{\mu}(U) = \inf_{\ddot{\bar{E}}} \pi \circ \widehat{\eta}(U) = 0.$$

Since $\widehat{\mu}(U) \geq \langle\langle\varkappa\rangle\rangle \geq 0$, we have $\langle\langle\varkappa\rangle\rangle = 0$ and, consequently, $\varkappa \in \mu(^*E)$. Thus, $\mu(^*E) = \eta(^*E)$, and so E is finite-dimensional by 4.5.4.

(2)$\to$(1): This is obvious. $\rhd$

4.9. Regular Hulls of Vector Lattices

Here we define and study the regular hull of a vector lattice. We will obtain a criterion for a vector lattice to be isomorphic to its regular hull. Some related questions concerning regular hulls are discussed.

4.9.1. Let E be a vector lattice. Like in the preceding section, we consider the quotient $(r)\text{-}E := \mathrm{fin}({}^*E)/\lambda({}^*E)$ and call it the *regular hull* of E. We denote the coset $\varkappa+\lambda({}^*E)$ where $\varkappa \in \mathrm{fin}({}^*E)$ by $\langle\varkappa\rangle$, and define a mapping $\widehat{\lambda}_E : E \to (r)\text{-}E$ as follows:

$$\widehat{\lambda}_E(x) := \langle x\rangle \quad (x \in E).$$

Obviously $\widehat{\lambda}_E$ is a Riesz homomorphism. We denoted it by $\widehat{\lambda}$ if this does not lead to ambiguity.

We present a criterion for a vector lattice to coincide with its regular hull. Recall that a vector lattice E is called *almost regular* if E is Dedekind complete and order separable, and if order convergence and relative uniform convergence are equivalent for every sequence in E.

Theorem. *For every vector lattice E, the following are equivalent:*

(1) *$\widehat{\lambda} : E \to (r)\text{-}E$ is a Riesz isomorphism of E onto $(r)\text{-}E$;*

(2) *E is atomic and almost regular.*

$\lhd$ (1)$\to$(2): Assume that $\widehat{\lambda}$ is a Riesz homomorphism of E onto $(r)\text{-}E$. In particular, $\widehat{\lambda}$ is injective. Thus, obviously, E is Archimedean, so $\lambda({}^*E) \subseteq \eta({}^*E)$, by Theorem 4.3.5. Since $\mathrm{fin}({}^*E) = E + \lambda({}^*E)$, we have

$$\mathrm{fin}({}^*E) = E + \eta({}^*E). \tag{1}$$

Therefore E is an atomic Dedekind complete vector lattice by Theorem 4.4.6. In order to complete the proof of the implication (1)$\to$(2), in accordance with Theorem 4.3.6, it remains to check the inclusion $\eta({}^*E) \subseteq \lambda({}^*E)$. Take an arbitrary element $\varkappa \in \eta({}^*E)$. Since $\eta({}^*E) \subseteq E + \lambda({}^*E)$ holds by (1), the element $\varkappa$ can be written as $\varkappa = e + \varkappa_1$, where $e \in E$ and $\varkappa_1 \in \lambda({}^*E)$. Then

$$e = \varkappa - \varkappa_1 \in \eta({}^*E) - \lambda({}^*E) \subseteq \eta({}^*E).$$

Consequently, $e \in E \cap \eta({}^*E) = \{0\}$, and hence $e = 0$. Finally, $\varkappa = \varkappa_1 \in \lambda({}^*E)$.

(2)$\to$(1): Let E be an almost regular vector lattice. Then, by Theorems 4.4.6 and 4.3.6,

$$\mathrm{fin}({}^*E) = E + \eta({}^*E) = E + \lambda({}^*E),$$

which immediately implies that $\widehat{\lambda}$ is onto. Moreover, since E is Archimedean, $\widehat{\lambda}$ is an injection. Thus, $\widehat{\lambda}$ is a Riesz isomorphism of E onto $(r)\text{-}E$. $\rhd$

4.9.2. According to Theorem 4.3.6, the regular hull (r)-E of an Archimedean order separable vector lattice E in which order convergence and relative uniform convergence are equivalent for every sequence coincides with the order hull (o)-E. We show that there are no other types of vector lattices with this property.

Theorem. *For an arbitrary vector lattice E, the following are equivalent:*

(1) *There exists a Riesz isomorphism π of (o)-E onto (r)-E such that $\pi \circ \widehat{\eta} = \widehat{\lambda}$;*

(2) $\eta({}^*E) = \lambda({}^*E)$;

(3) *E is an order separable Archimedean vector lattice in which order convergence and relative uniform convergence are equivalent for every sequence.*

$\lhd$ By Theorem 4.3.6 it suffices to verify (1)$\to$(2).

Let $\pi : (o)\text{-}E \to (r)\text{-}E$ be a Riesz isomorphism such that $\pi \circ \widehat{\eta} = \widehat{\lambda}$. Take elements $u, v \in E$ satisfying the condition $0 \leq nu \leq v$ for all $n \in \mathbb{N}$. It is easy to see that $\pi \circ \widehat{\eta}(u) = \widehat{\lambda}(u)$, and hence $\widehat{\eta}(u) = 0$. Since $\widehat{\eta}$ is injection, the relation $u = 0$ holds. Thus, E is Archimedean. The inclusion $\lambda({}^*E) \subseteq \eta({}^*E)$ follows now.

To complete the proof, it remains to establish the reverse inclusion: $\eta({}^*E) \subseteq \lambda({}^*E)$. Assume that there is a $\varkappa \in \eta({}^*E) \setminus \lambda({}^*E)$. We may suppose $\varkappa \geq 0$. Then $\langle\varkappa\rangle > 0$. At the same time, the condition $\varkappa \in \eta({}^*E)$ implies $\inf_E U(\varkappa) = 0$. Therefore, according to Theorem 4.8.2, $\inf_{(o)\text{-}\overline{E}} \widehat{\eta}(U(\varkappa)) = 0$. Since π is an isomorphism of (o)-E onto (r)-E, we have

$$\inf_{(r)\text{-}\overline{E}} \widehat{\lambda}(U(\varkappa)) = \inf_{(o)\text{-}\overline{E}} \pi \circ \widehat{\eta}(U(\varkappa)) = 0,$$

which contradicts the condition $\widehat{\lambda}(U(\varkappa)) \geq \langle\varkappa\rangle > 0$. Thus, we have $\eta({}^*E) \subseteq \lambda({}^*E)$. The proof of the theorem is complete. $\rhd$

4.9.3. We now discuss interrelation between the regular hull (r)-E and the nonstandard hull $\ddot{E}$ of a normed vector lattice E. Namely, we find a condition for (r)-E to coincide with $\ddot{E}$. We use the notation and terminology of 4.4.2 and 4.8.8.

Theorem. *Let (E, ρ) be a normed vector lattice. Then the following are equivalent:*

(1) *E possesses a strong order unity e such that the norm ρ is equivalent to the norm $\|\cdot\|_e$;*

(2) (r)-$E = \ddot{E}$;

(3) *There exists a Riesz isomorphism* φ *of* (r)-E *onto* $\ddot{E}$ *such that* $\varphi \circ \widehat{\lambda} = \widehat{\mu}$.

$\lhd$ (1)$\to$(2): This is immediate from Theorem 4.5.2.

(2)$\to$(3): Obvious.

(3)$\to$(1): Let $\varphi : (r)\text{-}E \to \ddot{E}$ be a Riesz isomorphism such that $\varphi \circ \widehat{\lambda} = \widehat{\mu}$. By Theorem 4.5.2, we need to establish the relation $\mathrm{Fin}({}^*E) = \mathrm{fin}({}^*E) + \mu({}^*E)$. The inclusion $\mathrm{fin}({}^*E)+\mu({}^*E) \subseteq \mathrm{Fin}({}^*E)$ is obvious. For proving the reverse inclusion, take an arbitrary $\varkappa \in \mathrm{Fin}({}^*E)$. Then $\langle\!\langle \varkappa \rangle\!\rangle = \varphi(\langle \varkappa_1 \rangle)$ for some $\varkappa_1 \in \mathrm{fin}({}^*E)$. Let $x \in E_+$ satisfy $|\varkappa_1| \leq x$. Hence

$$|\langle\!\langle \varkappa \rangle\!\rangle| = |\varphi(\langle \varkappa_1 \rangle)| = \varphi(\langle |\varkappa_1| \rangle) \leq \varphi(x) = \varphi \circ \widehat{\lambda}(x) = \widehat{\mu}(x) = \langle\!\langle x \rangle\!\rangle.$$

The inequality $|\langle\!\langle \varkappa \rangle\!\rangle| \leq \langle\!\langle x \rangle\!\rangle$ implies that the element $\varkappa$ can be written in the form $\varkappa = \xi_1 + \xi_2$, where $|\xi_1| \leq x$ and $\xi_2 \in \mu({}^*E)$. Thus, $\varkappa \in \mathrm{fin}({}^*E) + \mu({}^*E)$. $\rhd$

4.9.4. In contrast to 4.8.2, the image of the vector lattice E under $\widehat{\lambda}$ is not necessarily a complete vector sublattice of (r)-E. Indeed, consider the vector lattice l_∞ of all bounded sequences in $\mathbb{R}$, and let D be a subset of l_∞ consisting of all sequences with the property that all but finitely many coordinates are equal to 1. Then $\inf_E D = 0$, but $\widehat{\lambda}(D) \geq [e_\nu] > 0$ for all $\nu \in {}^*\mathbb{N} \setminus \mathbb{N}$, where e_ν is the internal sequence in ${}^*\mathbb{R}$ in which the only nonzero coordinate has index ν and equals 1.

4.9.5. Exactly as in the proof of Theorem 4.8.4 in which relative uniform completeness of order hulls was established, we can show that the regular hull of an arbitrary vector lattice is relatively uniformly complete. At the same time, since, by Theorem 4.3.6, the regular hull of a regular vector lattice coincides with its order hull, Theorem 4.8.5 shows that the regular hull of a nonatomic regular vector lattice is not Dedekind complete.

4.9.6. It follows from Theorems 4.3.6 and 4.3.3 that the regular hull of an order separable Archimedean vector lattice in which order convergence and relative uniform convergence are equivalent for every sequence is Archimedean too. Another case is described by the following

Theorem. *Let E be a vector lattice in which for every sequence $(x_n) \subseteq E^+$, there exists a sequence (λ_n) of strictly positive reals such that the set $\{\lambda_n x_n\}$ is order bounded. Then (r)-E is Archimedean.*

$\lhd$ We need to show that $\lambda({}^*E)$ is a relatively uniformly closed ideal in $\mathrm{fin}({}^*E)$, by Theorem 4.3.5. To this, consider $0 \leq v_n \uparrow$ and $v_n \xrightarrow{(r)} v$, where $v_n \in \lambda({}^*E)$. It is sufficient to prove that $v \in \lambda({}^*E)$.

Since $v_n \xrightarrow{(r)} v$, there exists a sequence $(\varepsilon_n) \subseteq \mathbb{R}^+$, $\varepsilon_n \to 0$, and an element $d \in E^+$ such that $|v_n - v| \leq \varepsilon_n d$ for all $n \in \mathbb{N}$. Since $v_n \in \lambda({}^*E)$, there exists $w_n \in E$ for which $0 \leq kv_n \leq w_n$ simultaneously for all $k \in \mathbb{N}$. By hypothesis, take $0 < \lambda_n \in \mathbb{R}$ and $w \in E$ such that $\lambda_n w_n \leq w$ for all $n \in \mathbb{N}$. Consequently,

$$|v| \leq |v_n - v| + |v| \leq \varepsilon_n d + \max\{\varepsilon_n, 1/n\}\lambda_n w_n \leq \max\{\varepsilon_n, 1/n\}(d + w)$$

for every $n \in \mathbb{N}$. Therefore, we have $v \in \lambda({}^*E)$, by using $\varepsilon_n \to 0$. ▷

Corollary. *The regular hull of a Banach lattice is Archimedean.*

There are non-Archimedean vector lattices whose regular hulls are non-Archimedean either. To see this, we consider an example of the vector lattice L by T. Nakayama (see [21, Example 62.2]). The ideal

$$I_0(L) := \{x \in L : (\exists y \in L)(\forall n \in \mathbb{N})\ |nx| \leq y\}$$

is not relatively uniformly closed in L. Hence, there are a sequence $(x_n) \subseteq I_0(L)$, $0 \leq x_n \uparrow$, and an element $x \in L$ such that

$$x_n \xrightarrow{(r)} x \notin I_0(L).$$

Since $I_0(L) = \lambda({}^*L) \cap L$, we have that the ideal $\lambda({}^*L)$ is not relatively uniformly closed in $\mathrm{fin}({}^*L)$. Thus, by Veksler's Theorem (see [21, Theorem 60.2]), (r)-L is non-Archimedean. The question remains open whether the regular hull of an arbitrary Archimedean vector lattice is Archimedean.

4.10. Order and Regular Hulls of Lattice Normed Spaces

In this section we define and begin studying the order and regular hulls of lattice normed spaces.

4.10.1. Let $(\mathscr{X}, \alpha, {}^*E)$ be some internal LNS normed by a standard lattice ${}^*E$. Consider the following external subspaces of the internal vector space $\mathscr{X}$:

$$\mathrm{fin}(\mathscr{X}) := \{x \in \mathscr{X} : \alpha(x) \in \mathrm{fin}({}^*E)\},$$
$$\eta(\mathscr{X}) := \{x \in \mathscr{X} : \alpha(x) \in \eta({}^*E)\},$$
$$\lambda(\mathscr{X}) := \{x \in \mathscr{X} : \alpha(x) \in \lambda({}^*E)\}.$$

The vector spaces $\eta(\mathscr{X})$ and $\lambda(\mathscr{X})$ are subspaces of $\mathrm{fin}(\mathscr{X})$. Therefore, we may arrange the following quotients:

$$(o)\text{-}\overline{\mathscr{X}} := \mathrm{fin}(\mathscr{X})/\eta(\mathscr{X}),$$
$$(r)\text{-}\overline{\mathscr{X}} := \mathrm{fin}(\mathscr{X})/\lambda(\mathscr{X}).$$

We denote by $[x]$ the coset $x+\eta(\mathscr{X})$ in (o)-$\overline{\mathscr{X}}$ and by $\langle x\rangle$ the coset $x+\lambda(\mathscr{X})$ in (r)-$\overline{\mathscr{X}}$, where $x\in \mathrm{fin}(\mathscr{X})$. Given $x\in \mathrm{fin}(\mathscr{X})$, assign

$$\overline{\alpha}([x]) := \alpha(x)+\eta({}^*E),$$
$$\overline{\alpha}_{(r)}(\langle x\rangle) := \alpha(x)+\lambda({}^*E).$$

It is easy to see that the mappings $\overline{\alpha} : (o)\text{-}\overline{\mathscr{X}} \to (o)\text{-}E$ and $\overline{\alpha}_{(r)} : (r)\text{-}\overline{\mathscr{X}} \to (r)\text{-}E$ are well defined.

DEFINITION. We call the LNS $((o)-\overline{\mathscr{X}},\overline{\alpha},(o)\text{-}E)$ $(((r)-\overline{\mathscr{X}},\overline{\alpha}_{(r)},(r)\text{-}E))$ the *order hull* (*regular hull*) of an internal LNS $(\mathscr{X},\alpha,{}^*E)$.

4.10.2. Theorem. *Let* $(\mathscr{X},\alpha,{}^*E)$ *be an internal decomposable* LNS *with a standard norm lattice* ${}^*E$. *Then its order hull and regular hull are decomposable and* (r)*-complete* LNS.

$\lhd$ Consider the external LNS $(\mathrm{fin}(\mathscr{X}),\alpha,\mathrm{fin}({}^*E))$. The proof of (r)-completeness of this LNS is almost the same as the proof of relative uniform completeness of the vector lattice $\mathrm{fin}({}^*E)$ in 4.8.4 (it suffices to replace $\mathrm{fin}({}^*E)$ by $\mathrm{fin}(\mathscr{X})$ and the modulus by the norm α). Clearly, the norm α is decomposable in $\mathrm{fin}(\mathscr{X})$. Since the order hull (regular hull) of $(\mathscr{X},\alpha,{}^*E)$ is the quotient of $(\mathrm{fin}(\mathscr{X}),\alpha,\mathrm{fin}({}^*E))$ by the ideal $\eta({}^*E)$ (respectively, the ideal $\lambda({}^*E)$) of $\mathrm{fin}({}^*E)$, we complete the proof by using Proposition 4.0.14. $\rhd$

4.10.3. Let $(E,\|\cdot\|)$ be a normed vector lattice. A Dedekind completion $\widehat{E}=\widehat{\eta}(E)$ of E is a normed vector lattice under the norm

$$\|x\| := \inf\{\|e\| : e\in E \ \&\ \widehat{\eta}(e)\geqslant |x|\}. \tag{2}$$

Now, the LNS (o)-$\overline{E}$ is a normed vector lattice under the norm $|\!|x|\!| := \|p(x)\|$. We have the direct expression for $|\!|\cdot|\!|$ as follows:

$$|\!|x|\!| := \inf\{\|e\| : e\in E \ \&\ \widehat{\eta}(e)\geqslant |x|\} \quad (x\in (o)\text{-}\overline{E})$$

which extends the norm (2) from $\widehat{E}$ to (o)-$\overline{E}$. Note that the embeddings $\widehat{\eta} : (E,\|\cdot\|)\hookrightarrow(\widehat{E},\|\cdot\|)$ and $(\widehat{E},\|\cdot\|)\subseteq((o)\text{-}\overline{E},|\!|\cdot|\!|)$ are isometric.

Recall that a normed vector lattice $(E,\|\cdot\|)$ satisfies the *weak Riesz–Fisher condition* if every sequence $(v_n)\subseteq E$ with the property $\sum_{n=1}^{\infty}\|v_n\|<\infty$ is order bounded.

Theorem. *The normed lattice* $((o)\text{-}\overline{E},|\!|\cdot|\!|)$ *is a Banach lattice if and only if* $(E,\|\cdot\|)$ *satisfies the weak Riesz–Fisher condition.*

◁ Suppose that $(E, \|\cdot\|)$ satisfies the weak Riesz–Fisher condition. Then $\widehat{E}$ is a Banach lattice under the norm (2) by [28, Theorem 101.6]. Applying [16, Theorem 4.1.2], we obtain from (r)-completeness of the LNS $((o)\text{-}\overline{E}, p, \widehat{E})$ with

$$p(x) = \inf_{\widehat{E}}\{\widehat{\eta}(e) : e \in E \;\&\; \widehat{\eta}(e) \geqslant |x|\},$$

that $((o)\text{-}\overline{E}, |\!\cdot\!|)$ is a Banach lattice.

Conversely, suppose that $((o)\text{-}\overline{E}, |\!\cdot\!|)$ is a Banach lattice and take an arbitrary sequence $(v_n) \subseteq E$ such that $\sum_{n=1}^{\infty} \|v_n\| < \infty$. Then

$$\sum_{n=1}^{\infty} |\widehat{\eta}(|v_n|)| = \sum_{n=1}^{\infty} \|v_n\| < \infty.$$

Consequently, there is an $u \in (o)\text{-}\overline{E}$,

$$u = (o)\text{-}\sum_{n=1}^{\infty} \widehat{\eta}(|v_n|) \in (o)\text{-}\overline{E}.$$

Since $\widehat{\eta}(E)$ is cofinal in $(o)\text{-}\overline{E}$, there exists an element $v \in E$ such that $\widehat{\eta}(v) \geqslant u$. Obviously, $(v_n) \subseteq [-v, v]$. Thus, $(E, \|\cdot\|)$ satisfies the weak Riesz–Fisher condition. ▷

4.10.4. In the sequel, we assume that E is Archimedean. We consider the quotient

$$\widehat{E} := o\text{-pns}({}^*E)/\eta({}^*E)$$

and recall that the vector lattice $\widehat{E}$ is a Dedekind completion of E by Theorem 4.4.1. We need some preliminary work. We start with a few lemmata:

Lemma. *Let $y \in o\text{-pns}({}^*E)$. Then*

$$[y] = \inf_{\widehat{E}} \widehat{\eta}(U(y)).$$

◁ Since $L(y) \leqslant y \leqslant U(y)$ and $\widehat{E}$ is Dedekind complete, the following holds:

$$\sup_{\widehat{E}} \widehat{\eta}(L(y)) \leqslant [y] \leqslant \inf_{\widehat{E}} \widehat{\eta}(U(y)).$$

Consequently,

$$0 \leqslant \inf_{\widehat{E}} \widehat{\eta}(U(y)) - [y] \leqslant \inf_{\widehat{E}} \widehat{\eta}(U(y)) - \sup_{\widehat{E}} \widehat{\eta}(L(y)) \leqslant \inf_{\widehat{E}} \widehat{\eta}(U(y) - L(y)).$$

Since $y \in o\text{-pns}({}^*E)$, we have $\inf_E(U(y) - L(y)) = 0$. Thus

$$\inf_{\widehat{E}} \widehat{\eta}(U(y) - L(y)) = 0,$$

because $\widehat{E}$ is a Dedekind completion of the sublattice $\widehat{\eta}(E)$. Therefore, the above-established inequality implies $[y] = \inf_{\widehat{E}} \widehat{\eta}(U(y))$. ▷

4.10.5. *Lemma.* *Each nonempty order bounded subset $\mathscr{D} \subseteq \widehat{E}$ possesses some supremum and infimum in (o)-$\overline{E}$. Moreover,*

(1) $\inf_{(o)\text{-}\overline{E}} \mathscr{D} = \inf_{\widehat{E}} \mathscr{D}$;

(2) $\sup_{(o)\text{-}\overline{E}} \mathscr{D} = \sup_{\widehat{E}} \mathscr{D}$.

$\lhd$ (1) Suppose that $\mathscr{D} \subseteq \widehat{E}$ and $\mathscr{D} \neq \varnothing$. It is sufficient to show that $\inf_{\widehat{E}} \mathscr{D} = 0$ implies $\inf_{(o)\text{-}\overline{E}} \mathscr{D} = 0$. Take a $\varkappa \in \operatorname{fin}({}^*E)$ such that $0 \leqslant \varkappa$ and $[\varkappa] \leqslant \mathscr{D}$. To complete the proof, it remains to establish that $[\varkappa] = 0$ or, in other words, to verify the condition $\inf_E U(\varkappa) = 0$. Assume that an element $a \in E$ satisfies the inequality

$$0 \leqslant a \leqslant U(\varkappa) \tag{3}$$

and take an arbitrary $d \in \mathscr{D}$. Then $d = [\delta]$ for some $\delta \in o\text{-pns}({}^*E)$. It is obvious that

$$\varkappa = \varkappa \wedge \delta + (\varkappa - \delta)_+ \leqslant U(\delta) + U((\varkappa - \delta)_+).$$

Consequently,

$$U(\delta) + U((\varkappa - \delta)_+) \subseteq U(\varkappa). \tag{4}$$

From (3) and (4), it ensues that

$$0 \leqslant a \leqslant U(\delta) + U((\varkappa - \delta)_+). \tag{5}$$

Using Lemma 4.1.2 and Dedekind completeness of $\widehat{E}$, we obtain from (5) that

$$0 \leqslant \widehat{\eta}(a) \leqslant \inf_{\widehat{E}} \widehat{\eta}(U(\delta)) + \inf_{\widehat{E}} \widehat{\eta}(U((\varkappa - \delta)_+)) = [\delta] + \inf_{\widehat{E}} \widehat{\eta}(U((\varkappa - \delta)_+)). \tag{6}$$

At the same time, $[(\varkappa - \delta)_+] = ([\varkappa] - d)_+ = 0$. Hence, $\inf_E U((\varkappa - \delta)_+) = 0$. Thus, we have $\inf_{\widehat{E}} \widehat{\eta}(U((\varkappa - \delta)_+)) = 0$. Now, (6) implies

$$0 \leqslant \widehat{\eta}(a) \leqslant [\delta] = d. \tag{7}$$

Since $d \in \mathscr{D}$ is arbitrary; therefore, in view of $\inf_{\widehat{E}} \mathscr{D} = 0$, we deduce from (7) that $a = 0$, as required.

Assertion (2) ensues immediately from (a). $\rhd$

4.10.6. Let $x \in (o)$-$\overline{E}$. Assign

$$\mathscr{U}(x) := \{e \in E : \widehat{\eta}(e) \geqslant x\},$$

$$\widehat{\mathscr{U}}(x) := \{y \in \widehat{E} : y \geqslant x\}.$$

It is clear that $\mathscr{U}(x)$ and $\widehat{\mathscr{U}}(x)$ are nonempty order bounded subsets of E and $\widehat{E}$ respectively.

Lemma. *For every $x \in (o)\text{-}\overline{E}$, the following hold:*

(1) $\inf_{\widehat{E}} \widehat{\mathscr{U}}(x) \in \widehat{\mathscr{U}}(x)$;

(2) $\inf_{\widehat{E}} \widehat{\mathscr{U}}(x) = \inf_{\widehat{E}} \widehat{\eta}(\mathscr{U}(x))$;

(3) *If an element $\varkappa \in \operatorname{fin}({}^*E)$ satisfies $x = [\varkappa]$ then*

$$\inf_{\widehat{E}} \widehat{\mathscr{U}}(x) = \inf_{\widehat{E}} \widehat{\eta}(\mathscr{U}(x)) = \inf_{\widehat{E}} \widehat{\eta}(U(\varkappa)).$$

$\lhd$ (1): By Lemma 4.10.5, $\widehat{\mathscr{U}}(x)$ has an infimum in $(o)\text{-}\overline{E}$, and $\inf_{(o)\text{-}\overline{E}} \widehat{\mathscr{U}}(x) = \inf_{\widehat{E}} \widehat{\mathscr{U}}(x)$. Hence, from $\widehat{\mathscr{U}}(x) \geqslant x$, it follows that $\inf_{(o)\text{-}\overline{E}} \widehat{\mathscr{U}}(x) \geqslant x$. Then $\inf_{\widehat{E}} \widehat{\mathscr{U}}(x) \geqslant x$, as required.

(2): Assign $x_0 := \inf_{\widehat{E}} \widehat{\mathscr{U}}(x)$. From (1) it ensues that $\mathscr{U}(x_0) \subseteq \mathscr{U}(x)$. Establish the reverse inclusion. Let $z \in \mathscr{U}(x)$. Then $\widehat{\eta}(z) \geqslant x$, and hence $\widehat{\eta}(z) \in \widehat{\mathscr{U}}(x)$. Consequently, $\widehat{\eta}(z) \geqslant x_0$ which is equivalent to $z \in \mathscr{U}(x_0)$. To complete the proof, it remains to see that the relation $x_0 = \inf_{\widehat{E}} \widehat{\eta}(\mathscr{U}(x_0))$ is valid since $\widehat{E}$ is an order completion of $\widehat{\eta}(E)$.

(3): Let $\varkappa \in \operatorname{fin}({}^*E)$ and $x = [\varkappa]$. Assign $x_0 := \inf_{\widehat{E}} \widehat{\mathscr{U}}(x)$. According to (1), $x_0 \geqslant x$. Take an element $y \in \operatorname{fin}({}^*E)$ such that $x_0 = [y]$ and $y \geqslant \varkappa$. Then we have from Lemma 4.10.4

$$x_0 = [y] = \inf_{\widehat{E}} \widehat{\eta}(U(y)) \geqslant \inf_{\widehat{E}} \widehat{\eta}(U(\varkappa)). \tag{8}$$

Next, the obvious inclusion $\widehat{\eta}(U(\varkappa)) \subseteq \widehat{\mathscr{U}}([\varkappa])$ implies

$$\inf_{\widehat{E}} \widehat{\eta}(U(\varkappa)) \geqslant \inf_{\widehat{E}} \widehat{\mathscr{U}}([\varkappa]) = x_0. \tag{9}$$

From (8), (9), and (1) we now obtain the required result. $\rhd$

4.10.7. Define the mapping $p : (o)\text{-}\overline{E} \to \widehat{E}$ as follows:

$$p(x) := \inf_{\widehat{E}} \widehat{\mathscr{U}}(|x|) \quad (x \in (o)\text{-}\overline{E}). \tag{10}$$

We list some properties of this mapping.

Theorem. *The mapping p is an $\widehat{E}$-valued norm on $(o)\text{-}\overline{E}$ such that, for all $x, y \in (o)\text{-}\overline{E}$, the following hold:*

(1) $p(x) = \inf_{\widehat{E}} \widehat{\eta}(\mathscr{U}(|x|))$;

(2) $p(x) \geqslant |x|$;

(3) $|x| \geqslant |y|$ *implies* $p(x) \geqslant p(y)$.

Moreover, an arbitrary sequence $(x_n) \subseteq (o)\text{-}\overline{E}$ *(r)-converges in the norm p to an element* $x_0 \in (o)\text{-}\overline{E}$ *(is (r)-Cauchy in the norm p) if and only if it (r)-converges to* x_0 *in the vector lattice* $(o)\text{-}\overline{E}$ *(is (r)-Cauchy in* $(o)\text{-}\overline{E}$*). The lattice normed space* $((o)\text{-}\overline{E}, p, \widehat{E})$ *is (r)-complete and decomposable.*

◁ From (10) it is straightforward that the mapping p satisfies conditions 4.0.10 (2), and 4.0.10 (3), and item (3) of the theorem. Item (2) of the theorem follows from 4.10.6 (1). Now, conditions 4.0.10 (1) ensue from (2). Hence, p is an $\widehat{E}$-valued norm on $(o)\text{-}\overline{E}$. Condition (1) is a particular instance of 4.10.6 (2).

If $x_n \xrightarrow{(r)} x_0$ in the norm p with regulator $e \in \widehat{E}$ then, in view of item (2) of the theorem, $x_n \xrightarrow{(r)} x_0$ in $(o)\text{-}\overline{E}$ with the same regulator. Conversely, if $x_n \xrightarrow{(r)} x_0$ in $(o)\text{-}\overline{E}$ with regulator $d \in (o)\text{-}\overline{E}$ then $x_n \xrightarrow{(r)} x_0$ in the norm p with regulator $p(d)$, in accordance with item (3) of the theorem. For (r)-Cauchy sequences the proof is essentially the same.

The quotient $(o)\text{-}\overline{E}$ is relatively uniformly complete in view of Theorem 4.8.4. So, as we showed, $((o)\text{-}\overline{E}, p, \widehat{E})$ is (r)-complete. Now, to verify decomposability of p, it is sufficient to establish (d)-decomposability of p by Proposition 4.0.10. Let $x \in (o)\text{-}\overline{E}$ and let $e_1, e_2 \in \widehat{E}$ be such that $p(x) = e_1 + e_2$ and $e_1 \wedge e_2 = 0$. Assign

$$x_1 := x_+ \wedge e_1 - x_- \wedge e_1; \quad x_2 := x_+ \wedge e_2 - x_- \wedge e_2.$$

It is easy to see that $p(x_1) = e_1$, $p(x_2) = e_2$, and $x = x_1 + x_2$. ▷

4.10.8. Consider the mapping $p \circ \overline{\alpha} : (o)\text{-}\overline{\mathscr{X}} \to \widehat{E}$, where $\overline{\alpha} : (o)\text{-}\overline{\mathscr{X}} \to (o)\text{-}E$ is the (o)-E-valued norm defined in 4.10.1.

Theorem. *The triple* $((o)\text{-}\overline{\mathscr{X}}, p \circ \overline{\alpha}, \widehat{E})$ *is a decomposable (r)-complete LNS.*

◁ It is easy to see that $p \circ \overline{\alpha}$ is an $\widehat{E}$-valued norm in $(o)\text{-}\overline{\mathscr{X}}$. Take an arbitrary sequence $(x_n) \subseteq (o)\text{-}\overline{\mathscr{X}}$ that is (r)-Cauchy in the norm $p \circ \overline{\alpha}$ with regulator $e \in \widehat{E}$. In view of Theorem 4.10.7 (item (2)), it is (r)-Cauchy in the norm $\overline{\alpha}$ with the same regulator. Consequently, by Theorem 4.10.2, there is an element $x_0 \in (o)\text{-}\overline{\mathscr{X}}$ such that $x_n \xrightarrow{(r)} x_0$ (e) in the norm $\overline{\alpha}$. Now, from Theorem 4.10.7 (item (3)) it ensues that $x_n \xrightarrow{(r)} x_0$ in the norm $p \circ \overline{\alpha}$ with regulator $p(e) = e$. Thus, every sequence $(x_n) \subseteq (o)\text{-}\overline{\mathscr{X}}$, that is (r)-Cauchy in the norm $p \circ \overline{\alpha}$, is (r)-convergent in $(o)\text{-}\overline{\mathscr{X}}$ with the same regulator. Thus, $(o)\text{-}\overline{\mathscr{X}}$ is (r)-complete in the norm $p \circ \overline{\alpha}$.

In view of (r)-completeness, to establish decomposability of the norm $p \circ \overline{\alpha}$, it is sufficient to verify its (d)-decomposability, by Proposition 4.0.10. Let $x \in (o)\text{-}\overline{\mathscr{X}}$ and let $e_1, e_2 \in \widehat{E}$ be such that $p \circ \overline{\alpha}(x) = e_1 + e_2$ and $e_1 \wedge e_2 = 0$. Then decomposability of the norm p implies that there are $\alpha_1, \alpha_2 \in (o)\text{-}\overline{E}$ such that $\overline{\alpha}(x) = \alpha_1 + \alpha_2$, $p(\alpha_1) = e_1$, and $p(\alpha_2) = e_2$. In view of Theorem 4.10.7 (item (2)), we have $\alpha_1 \leqslant e_1$ and $\alpha_2 \leqslant e_2$. Hence, from the conditions $\alpha_1 + \alpha_2 = \overline{\alpha}(x) \geqslant 0$ and $e_1 \wedge e_2 = 0$ it ensues that $\alpha_1 \geqslant 0$ and $\alpha_2 \geqslant 0$. It remains to use decomposability of the norm $\overline{\alpha}$ for finding elements $x_1, x_2 \in (o)\text{-}\overline{\mathscr{X}}$ such that $x_1 + x_2 = x$, $\alpha(x_1) = \alpha_1$, and $\alpha(x_2) = \alpha_2$. It is clear that $p \circ \overline{\alpha}(x_1) = e_1$ and $p \circ \overline{\alpha}(x_2) = e_2$. $\triangleright$

4.11. Associated Banach–Kantorovich Spaces

We give a nonstandard construction of an order completion of a decomposable LNS. The scheme rests on embedding the LNS into the associated Banach–Kantorovich space (BKS). We study extensions onto associated BKSs of internal dominated operators admitting standard (o)-continuous dominants. Throughout the section we suppose that $(\mathscr{X}, a, E)$ and $(\mathscr{Y}, b, F)$ are decomposable LNS in which the norm lattices E and F are Dedekind complete.

4.11.1. The lattice normed space $((o)\text{-}\overline{\mathscr{X}}, p \circ \overline{\alpha}, \widehat{E})$ defined in 4.10.8 is called *associated* with the order hull $((o)\text{-}\overline{\mathscr{X}}, \overline{\alpha}, (o)\text{-}E)$ of the LNS (X, a, E). We establish that this LNS is a Banach–Kantorovich space.

Since the vector lattice $\widehat{E}$ is a Dedekind completion of E, we have $\widehat{E} \cong E$ under our assumptions. To be more precise, the mapping $\widehat{\eta} : E \to \widehat{E}$ is a Riesz isomorphism of E onto $\widehat{E}$. Consider the mapping $\rho_E : (o)\text{-}\overline{E} \to E$ defined by the rule

$$\rho_E(x) := \inf_E \{e \in E : \widehat{\eta}(e) \geqslant |x|\} \quad (x \in (o)\text{-}\overline{E}).$$

We have the following

Lemma. *The mapping ρ_E is connected with the norm $p : (o)\text{-}\overline{E} \to \widehat{E}$ by the relation $\rho_E = \widehat{\eta}^{-1} \circ p$. Moreover, for every $x \in \operatorname{fin}({}^*E)$, we have*

$$\rho_E([\varkappa]) = \inf_E \{e \in E : e \geqslant |\varkappa|\}.$$

$\triangleleft$ The first part of the lemma ensues from the definitions of p and ρ_E, and the second from the relation

$$\rho_E([\varkappa]) = \inf_E \mathscr{U}(|[\varkappa]|)$$

$$= \widehat{\eta}^{-1} \inf_{\widehat{E}} \widehat{\eta}(\mathscr{U}(|[\varkappa]|)) = \widehat{\eta}^{-1} \inf_{\widehat{E}} \widehat{\eta}(U(|\varkappa|))$$

$$= \inf_E U(|\varkappa|) = \inf_E \{e \in E : e \geqslant |\varkappa|\},$$

in which the second and the fourth equalities are valid because $\widehat{\eta}$ is a Riesz isomorphism, and the third in view of item (3) of Lemma 4.10.6. ▷

Theorem 4.10.8 and Lemma 4.11.1 imply the following

Corollary. *The triple $((o)\text{-}\overline{\mathscr{X}}, \rho_E \circ \overline{\alpha}, E)$ is a decomposable (r)-complete lattice normed space. Moreover, for each $x \in \operatorname{fin}({}^*E)$, we have*

$$\rho_E \circ \overline{\alpha}(\langle\varkappa\rangle) = \inf_E \{e \in E : e \geqslant \alpha(\varkappa)\}. \tag{11}$$

4.11.2. Denote by $\mathscr{B}(E)$ the family of all band projections in E. Note that for each internal band projection $\tau \in {}^*\mathscr{B}(E)$, there exists a unique band projection $h(\tau)$ in $\mathscr{X}$ satisfying the condition

$$\alpha(h(\tau)\varkappa) = \tau\alpha(\varkappa) \quad (\varkappa \in \mathscr{X}).$$

This property is easily obtainable from decomposability of the internal norm $\alpha : \mathscr{X} \to {}^*E$.

Lemma. *For all $\pi \in \mathscr{B}(E)$ and $\varkappa \in \operatorname{fin}(\mathscr{X})$, we have*

$$\pi \circ \rho_E \circ \overline{\alpha}(\langle\varkappa\rangle) = \rho_E \circ \overline{\alpha}(\langle h({}^*\pi)\varkappa\rangle).$$

◁ Let $x \in \operatorname{fin}(\mathscr{X})$. Show that, for every $\pi \in \mathscr{B}(E)$, the inequality

$$\pi \circ \rho_E \circ \overline{\alpha}(\langle\varkappa\rangle) \geqslant \rho_E \circ \overline{\alpha}(\langle h({}^*\pi)\varkappa\rangle) \tag{12}$$

holds. To this end, take an $e \in E$, $e \geqslant \alpha(\varkappa)$. Then $\pi e \geqslant {}^*\pi(\alpha(\varkappa)) = \alpha(h({}^*\pi)\varkappa)$. Applying (11), obtain

$$\pi(e) \geqslant \inf_E \{f \in E : f \geqslant \alpha(h({}^*\pi)\varkappa)\} = \rho_E \circ \overline{\alpha}(\langle h({}^*\pi)\varkappa\rangle).$$

Since $e \in E$, $e \geqslant \alpha(\varkappa)$, is taken arbitrarily, we obtain from order continuity of π and (11) that

$$\pi \circ \rho_E \circ \overline{\alpha}(\langle\varkappa\rangle) = \pi \inf_E \{e \in E : e \geqslant \alpha(\varkappa)\}$$

$$= \inf_E \{\pi e : e \in E \ \& \ e \geqslant \alpha(\varkappa)\} \geqslant \rho_E \circ \overline{\alpha}(\langle h({}^*\pi)\varkappa\rangle).$$

Inequality (12) is established.

Consider an arbitrary band projection $\pi \in \mathscr{B}(E)$ and denote by $\pi^\perp$ the complementary projection to π. Then, applying (12) to π and $\pi^\perp$, we have

$$\begin{aligned}\rho_E \circ \overline{\alpha}(\langle \varkappa \rangle) &= \pi \circ \rho_E \circ \overline{\alpha}(\langle \varkappa \rangle) + \pi^\perp \circ \rho_E \circ \overline{\alpha}(\langle \varkappa \rangle)\\ &\geqslant \rho_E \circ \overline{\alpha}(\langle h({}^*\pi)\varkappa \rangle) + \rho_E \circ \overline{\alpha}(\langle h({}^*\pi^\perp)\varkappa \rangle)\\ &\geqslant \rho_E([{}^*\pi \circ \alpha(\varkappa) + {}^*\pi^\perp \circ \alpha(\varkappa)]) = \rho_E \circ \overline{\alpha}(\langle \varkappa \rangle).\end{aligned}$$

Hence,

$$\begin{aligned}&\pi \circ \rho_E \circ \overline{\alpha}(\langle \varkappa \rangle) + \pi^\perp \circ \rho_E \circ \overline{\alpha}(\langle \varkappa \rangle)\\ &= \rho_E \circ \overline{\alpha}(\langle h({}^*\pi)\varkappa \rangle) + \rho_E \circ \overline{\alpha}(\langle h({}^*\pi^\perp)\varkappa \rangle).\end{aligned}$$

Consequently, in view of (12), $\pi \circ \rho_E \circ \overline{\alpha}(\langle \varkappa \rangle) = \rho_E \circ \overline{\alpha}(\langle h({}^*\pi)\varkappa \rangle)$, as required. ▷

4.11.3. *Lemma.* *The associated lattice normed space $((o)\text{-}\overline{\mathscr{X}}, \rho_E \circ \overline{\alpha}, E)$ is disjointly complete.*

◁ Take an arbitrary partition of unity $(\pi_\xi)_{\xi \in \Xi} \subseteq \mathscr{B}(E)$ and a family $(x_\xi)_{\xi \in \Xi} \subseteq (o)\text{-}\overline{\mathscr{X}}$ bounded in the norm $\rho_E \circ \overline{\alpha}$.

Suppose that, for $e \in E$, we have

$$\rho_E \circ \overline{\alpha}(x_\xi) \leqslant e \quad (\xi \in \Xi). \tag{13}$$

Choose a $\varkappa_\xi \in \mathscr{X}$ such that $\langle \varkappa_\xi \rangle = x_\xi$ for all $\xi \in \Xi$. Using the definition of the norm $\overline{\alpha}$, the relation $\rho_E = \widehat{\eta}^{-1} \circ p$, and item (2) of Theorem 4.10.7, we rewrite inequality (13) as $[\alpha(\varkappa_\xi)] \leqslant \widehat{\eta}(e)$. Consequently, for a suitable $\eta_\xi \in \eta({}^*E)$, we have the inequality $\alpha(\varkappa_\xi) \leqslant e + \eta_\xi$ $(\xi \in \Xi)$. Hence, according to Lemma 4.0.13, there are elements $\varkappa'_\xi \in \mathrm{fin}(\mathscr{X})$ for which

$$\alpha(\varkappa_\xi - \varkappa'_\xi) \in \eta({}^*E), \quad \alpha(\varkappa'_\xi) \leqslant e \quad (\xi \in \Xi). \tag{14}$$

Fix some $\nu \in {}^*\mathbb{N} \setminus \mathbb{N}$ and denote by $\mathscr{F}$ the set of all internal mappings from ${}^*\Xi$ into $\mathscr{X}$. Let Card stand for the internal cardinality. Given $\xi \in \Xi$, define the internal subset A_ξ of $\mathscr{F}$ as follows:

$$\begin{aligned}A_\xi := \{\varphi \in \mathscr{F} : \alpha \circ \varphi({}^*\Xi) \subseteq [-e, e] \ \&\ \varphi(\xi) = \varkappa'_\xi\\ \&\ \mathrm{Card}(\{\xi \in {}^*\Xi : \varphi(\xi) \neq 0\}) \leqslant \nu\}.\end{aligned}$$

It is easy to see that the family $(A_\xi)_{\xi \in \Xi}$ has the finite intersection property. Consequently, in view of the general saturation principle, there is an element $\varphi_0 \in \bigcap\{A_\xi : \xi \in \Xi\}$. Put

$$\Theta := \{\xi \in {}^*\Xi : \varphi_0(\xi) \neq 0\}.$$

Since $\mathrm{Card}(\Theta) \leqslant \nu$, the set Θ is hyperfinite. Furthermore,

$$\Xi \subseteq \Theta \subseteq {}^*\Xi \quad \text{and} \quad \alpha(\varphi_0(\xi)) \leqslant e \quad (\xi \in \Theta).$$

For convenience, we let $\varkappa'_\xi := \varphi_0(\xi)$ whenever $\xi \in \Theta$. This does not cause difficulties, since $\varphi_0(\xi) = \varkappa'_\xi$ for all $\xi \in \Xi$ by the choice of φ_0.

Thus, the family $(\varkappa'_\xi)_{\xi \in \Xi}$ extends to a hyperfinite family $(\varkappa'_\xi)_{\xi \in \Theta} \subseteq \mathscr{X}$ such that

$$\alpha(\varkappa'_\xi) \leqslant e \quad (\xi \in \Theta). \tag{15}$$

Let $(\tau_\xi)_{\xi \in {}^*\Xi} := {}^*((\pi_\xi)_{\xi \in \Xi})$ be the nonstandard enlargement of the partition $(\pi_\xi)_{\xi \in \Xi}$ of unity. Then $(\tau_\xi)_{\xi \in {}^*\Xi}$ is an internal partition of unity in ${}^*\mathscr{B}(E)$. Furthermore, $\tau_\xi = {}^*\pi_\xi$ for all $\xi \in \Xi$. The hyperfinite sum $\varkappa := \sum_{\xi \in \Theta} h(\tau_\xi)\varkappa'_\xi$ is an element of the internal vector space $\mathscr{X}$, where $h(\tau_\xi)$'s are the band projections defined in 4.11.2. Disjointness of the set of projections τ_ξ, together with (15), implies $|\varkappa| \leqslant e$. In particular, $\varkappa \in \mathrm{fin}(\mathscr{X})$.

For every $\xi_0 \in \Xi$, consider the following chain of equalities:

$$\pi_{\xi_0} \circ \rho_E \circ \overline{\alpha}(x_{\xi_0} - \langle \varkappa \rangle) = \pi_{\xi_0} \circ \rho_E \circ \overline{\alpha}(\langle \varkappa'_{\xi_0} - \varkappa \rangle)$$

$$= \rho_E \circ \overline{\alpha}(\langle h({}^*\pi_{\xi_0})(\varkappa'_{\xi_0} - \varkappa) \rangle)$$

$$= \rho_E \circ \overline{\alpha}\Big(\Big\langle h(\tau_{\xi_0})\Big(\sum_{\xi \in \Theta \setminus \{\xi_0\}} h(\tau_\xi)\varkappa'_\xi \Big) \Big\rangle\Big) = 0.$$

The first equality holds due to the choice of the elements $\varkappa_\xi$ and (14). Validity of the second is ensured by Lemma 4.11.2. The third equality is valid by the choice of $\varkappa$ and the equality $\tau_{\xi_0} = {}^*\pi_{\xi_0}$ mentioned above. Finally, the last equality holds since τ_ξ is disjoint. Therefore, $\pi_\xi \circ \rho_E \circ \overline{\alpha}(x_\xi - \langle \varkappa \rangle) = 0$ for every $\xi \in \Xi$, and hence $\langle \varkappa \rangle = \mathrm{mix}(\pi_\xi x_\xi)_{\xi \in \Xi}$.

Thus, for every partition of unity $(\pi_\xi)_{\xi \in \Xi} \subseteq \mathscr{B}(E)$ and every bounded in the norm $\rho_E \circ \overline{\alpha}$ family $(x_\xi)_{\xi \in \Xi}$, the mixing $\mathrm{mix}(\pi_\xi x_\xi) \in (o)\text{-}\overline{\mathscr{X}}$ exists. The proof of the lemma is complete. $\rhd$

4.11.4. We are in a position to state the main result of this section.

Theorem. *The associated lattice normed space $((o)\text{-}\overline{\mathscr{X}}, \rho_E \circ \overline{\alpha}, E)$ is a Banach –Kantorovich space.*

$\lhd$ It is follows immediately from Corollary 4.11.1 and Lemma 4.11.3 by using Proposition 4.0.11. $\rhd$

Note that, whenever we take an internal normed space $\mathscr{X}$ as an internal decomposable LNS, the associated space coincides with the classical nonstandard hull

$\widetilde{\mathscr{X}}$. Furthermore, from the above-established theorem, the well-known assertion ensues that the nonstandard hull of an internal normed space is a Banach space.

If we consider an internal lattice normed space $({}^*E, |\cdot|, {}^*E)$ then the associated space is the LNS $((o)\text{-}\overline{E}, \rho_E, E)$. From the definition of the mapping $\rho_E : (o)\text{-}\overline{E} \to E$ it is clear that, for all $x, y \in (o)\text{-}\overline{E}$, the condition $|x| \leqslant |y|$ implies $\rho_E(x) \leqslant \rho_E(y)$. So, we obtain that the associated LNS $((o)\text{-}\overline{E}, \rho_E, E)$ is a Banach–Kantorovich lattice.

4.11.5. It is known that a norm completion of an normed space X can be obtained by taking the closure of the space in the nonstandard hull of X. Similarly, as we show below, an (o)-completion of a decomposable LNS can be constructed on embedding it into the associated BKS.

For simplicity, denote by $((o)\text{-}\overline{X}, \rho_E \circ \overline{a}, E)$ the BKS associated with the order hull of (X, a, E). Consider the mapping $\widehat{\eta} : X \to (o)\text{-}\overline{X}$ such that

$$\widehat{\eta}(x) := \langle x \rangle \quad (x \in X). \tag{16}$$

It is easy to see that $\widehat{\eta}$ is an isometrically isomorphic embedding of the LNS (X, a, E) into $((o)\text{-}\overline{X}, \rho_E \circ \overline{a}, E)$. Denote by $\widehat{X}$ the set of limits of all $\rho_E \circ \overline{a}$ convergent nets of elements of $\widehat{\eta}(X)$.

Lemma. *For every element $x \in (o)\text{-}\overline{X}$, the following are equivalent:*

(1) $x \in \widehat{X}$;

(2) $\inf\limits_{y \in X} \rho_E \circ \overline{a}(x - \widehat{\eta}(y)) = 0.$

$\lhd$ (1)$\to$(2): This is immediate from the definition of $\widehat{X}$.

(2)$\to$(1): Let an element $x \in (o)\text{-}\overline{X}$ satisfy condition (2). Show that $x \in \widehat{X}$. Define the relation $\prec$ on X as follows:

$$y \prec z \leftrightarrow \rho_E \circ \overline{a}(x - \widehat{\eta}(y)) \geqslant \rho_E \circ \overline{a}(x - \widehat{\eta}(z)).$$

The set X is directed upwards with respect to $\prec$. Indeed, for all $y, z \in X$, we have $y, z \leqslant h(\pi)y + h(\pi^\perp)z$, where $\pi \in \mathscr{B}(E)$ is the band projection that satisfies the condition

$$\pi \circ \rho_E \circ \overline{a}(x - \widehat{\eta}(y)) + \pi^\perp \circ \rho_E \circ \overline{a}(x - \widehat{\eta}(z))$$
$$= \rho_E \circ \overline{a}(x - \widehat{\eta}(y)) \wedge \rho_E \circ \overline{a}(x - \widehat{\eta}(z)),$$

and $h(\pi)$ and $h(\pi^\perp)$ are the corresponding band projections in (X, a, E). Consider the net $(\widehat{\eta}(y))_{y \in (X, \prec)}$. From the definition of $(X, \prec)$, in view of the condition $\inf_{g \in X} \rho_E \circ \overline{a}(x - \widehat{\eta}(y)) = 0$, it follows that the net $(\widehat{\eta}(y))_{y \in (X, \prec)}$ converges to x in $(o)\text{-}\overline{X}$. Since the net is constituted by elements of $\widehat{\eta}(X)$, we have $x \in \widehat{X}$. $\rhd$

4.11.6. *Theorem.* *The triple $(\widehat{X}, \rho_E \circ \overline{a}, E)$ is an (o)-completion of the decomposable LNS (X, a, E).*

$\lhd$ It is sufficient to verify properties 4.0.12 (1)–(3). Clearly, $(\widehat{X}, \rho_E \circ \overline{a}, E)$ is a lattice normed space. Show that this space is (o)-complete. Take an arbitrary (o)-Cauchy net (x_ξ). Then, by (o)-completeness of the associated LNS, there is an element $x \in (o)\text{-}\overline{X}$ such that $x = (o)\text{-}\lim(x_\xi)$. Show that $x \in \widehat{X}$. The conditions $x_\xi \in \widehat{X}$ and $x = (o)\text{-}\lim(x_\xi)$ imply that

$$\inf_{y \in X} \rho_E \circ \overline{a}(x_\xi - \widehat{\eta}(y)) = 0, \quad \inf_{\xi} \rho_E \circ \overline{a}(x - x_\xi) = 0. \tag{17}$$

Next, from (17) we obtain

$$\begin{aligned} 0 &\leqslant \inf_{y \in X} \rho_E \circ \overline{a}(x - \widehat{\eta}(y)) \\ &\leqslant \inf_{\xi} \inf_{y \in X} (\rho_E \circ \overline{a}(x - x_\xi) + \rho_E \circ \overline{a}(x_\xi - \widehat{\eta}(y))) \\ &\leqslant \inf_{\xi} \rho_E \circ \overline{a}(x - x_\xi) + \inf_{\xi} \inf_{y \in X} \rho_E \circ \overline{a}(x_\xi - \widehat{\eta}(y)) = 0. \end{aligned}$$

Thereby, $\inf_{y \in X} \rho_E \circ \overline{a}(x - \widehat{\eta}(y)) = 0$ and, in view of Lemma 4.11.5, we have $x \in \widehat{X}$. Thus, every (o)-Cauchy net $(x_\xi) \subseteq \widehat{X}$ is (o)-convergent. Hence, $\widehat{X}$ is (o)-complete in the norm $\rho_E \circ \overline{a}$. It is easy to verify that the norm $\rho_E \circ \overline{a}$ is (d)-decomposable on $\widehat{X}$. Consequently, taking (o)-completeness and Proposition 4.0.10 into account, we obtain decomposability of the norm $\rho_E \circ \overline{a}$ on $\widehat{X}$. Property 4.0.12 (1) is established. Property 4.0.12 (2) is obvious for the embedding $\widehat{\eta} : X \to \widehat{X}$.

To verify 4.0.12 (3), take some $x' \in \widehat{X}$ and $e \in E_+$. Assign

$$\mathscr{E} := \{\pi \in \mathscr{B}(E) : \pi \circ \rho_E \circ \overline{a}(x' - \widehat{\eta}(y)) \leqslant e \quad \text{for some} \quad x \in X\}.$$

Since $x' \in \widehat{X}$, we have

$$\inf_{x \in X} \rho_E \circ \overline{a}(x' - \widehat{\eta}(x)) = 0.$$

Hence, the set $\mathscr{E}$ is dense in the band $\mathscr{B}_e(E)$ generated by the band projection pr_e. By the exhaustion principle, there exists a partition $(\sigma_\gamma)_{\gamma \in \Omega} \subseteq \mathscr{E}$ of pr_e. In accordance with the definition of $\mathscr{E}$, there is a family $(x'_\gamma)_{\gamma \in \Omega} \subseteq X$ for which

$$\sigma_\gamma \circ \rho_E \circ \overline{a}(x' - \widehat{\eta}(x'_\gamma)) \leqslant e \quad (\gamma \in \Omega). \tag{18}$$

Take a $\gamma_0 \notin \Omega$ and assign $\Gamma := \Omega \cup \{\gamma_0\}$, $\sigma_{\gamma_0} := \mathrm{pr}_e^{\perp}$, and $x'_{\gamma_0} := 0$. For each $\gamma \in \Gamma$, due to decomposability of (X, a, E), there exists a unique band projection τ_γ in the space X which satisfies $a \circ \tau_\gamma = \delta_\gamma \circ a$. Define a family $(x_\gamma)_{\gamma \in \Gamma} \subseteq X$ so that $x_\gamma := \tau_\gamma x'_\gamma$ for every $\gamma \in \Gamma$. From (18) it follows that

$$a(x_\gamma) = a(\tau_\gamma x'_\gamma) = \sigma_\gamma \circ a(x'_\gamma) = \sigma_\gamma \circ \rho_E \circ \overline{a}\ (\widehat{\eta}(x'_\gamma))$$

$$\leqslant \sigma_\gamma \circ \rho_E \circ \overline{a}\ (x') + \sigma_\gamma \circ \rho_E \circ \overline{a}\ (x' - \widehat{\eta}(x'_\gamma)) \leqslant 2\rho_E \circ \overline{a}\ (x') + e$$

for each $\gamma \in \Gamma$. Thus, the family $(x_\gamma)_{\gamma \in \Gamma}$ is bounded with respect to the norm a. By (o)-completeness of $(\widehat{X}, \rho_E \circ \overline{a}, E)$, the mixing $\operatorname{mix}(\delta_\gamma, \widehat{\eta}(x_\gamma))_{\gamma \in \Gamma} \in \widehat{X}$ exists. Using (18) again, we see that

$$\mathrm{pr}_e \circ \rho_E \circ \overline{a}(x' - \operatorname{mix}(\sigma_\gamma, \widehat{\eta}(x_\gamma))) \leqslant e.$$

Since the choice of the elements $x' \in \widehat{X}$ and $e \in E_+$ is arbitrary, property 4.0.12 (3) is established. The proof of the theorem is complete. ▷

4.11.7. Denote the space of all regular (respectively, order continuous) operators from E into F by $L_r(E, F)$ $(L_n(E, F))$, and denote by $\mathscr{M}(\mathscr{X}, \mathscr{Y})$ the set of all internal linear operators from $\mathscr{X}$ into $\mathscr{Y}$ which admit some standard dominant ${}^*Q$, $Q \in L_r(E, F)$ (see 4.0.15). Next, let $\mathscr{M}_n(\mathscr{X}, \mathscr{Y})$ be the set of all operators in $\mathscr{M}(\mathscr{X}, \mathscr{Y})$ each of which admits a dominant of the form ${}^*S$ with $S \in L_n(E, F)$.

Lemma. *For every internal linear operator $T : \mathscr{X} \to \mathscr{Y}$, the following hold:*

(1) $T \in \mathscr{M}(\mathscr{X}, \mathscr{Y}) \to T(\operatorname{fin}(\mathscr{X})) \subseteq \operatorname{fin}(\mathscr{Y})$;

(2) $T \in \mathscr{M}_n(\mathscr{X}, \mathscr{Y}) \to T(\eta(\mathscr{X})) \subseteq \eta(\mathscr{Y})$.

◁ We verify only (1), since (2) is established similarly. According to the condition $T \in \mathscr{M}(\mathscr{X}, \mathscr{Y})$, there is an operator $Q \in L_r(E, F)$ for which

$$\beta(T\varkappa) \leqslant {}^*Q\alpha(\varkappa) \quad (\varkappa \in \mathscr{X}). \tag{19}$$

Take an arbitrary $\varkappa \in \operatorname{fin}(\mathscr{X})$. Then $\alpha(\varkappa) \leqslant e$ for some $e \in E$. By (19), we have $\beta(T\varkappa) \leqslant Q(e)$ and, consequently, $T\varkappa \in \operatorname{fin}(\mathscr{Y})$. ▷

Throughout the sequel, the vector lattice $L_n(E, F)$ is denoted by L.

4.11.8. Suppose that $T \in \mathscr{M}_n(\mathscr{X}, \mathscr{Y})$. According to Lemma 4.11.7, the mapping $\overline{T} : (o)\text{-}\overline{\mathscr{X}} \to (o)\text{-}\overline{\mathscr{Y}}$ acting as

$$\overline{T}(\langle \varkappa \rangle) := \langle T\varkappa \rangle \quad (\varkappa \in \operatorname{fin}(\mathscr{X})) \tag{20}$$

is soundly defined.

Theorem. *The mapping $\overline{T}$ is a dominated linear operator from the associated BKS $((o)\text{-}\overline{\mathscr{X}}, \rho_E \circ \overline{\alpha}, E)$ to the associated BKS $((o)\text{-}\overline{\mathscr{Y}}, \rho_F \circ \overline{\beta}, F)$. Furthermore, $⦀\overline{T}⦀ \leqslant \rho_L([⦀T⦀])$.*

Before proving, we make necessary clarifications. By $⦀\overline{T}⦀$ (by $⦀T⦀$) we denote the least (least internal) dominant of $\overline{T}$ (of T). By ρ_L we denote the L-valued norm in the LNS $((o)\text{-}\overline{L}, \rho_L, L)$.

◁ It is sufficient to establish that the operator $\rho_L([⦀T⦀]) \in L_n(E, F)$ is a dominant for $\overline{T}$. Take an arbitrary operator $S \in L$ that satisfies the condition ${}^*S \geqslant ⦀T⦀$. Then, using relation (19), for every $\varkappa \in \mathrm{fin}(\mathscr{X})$, we obtain

$$\rho_F \circ \overline{\beta}(\overline{T}\langle\varkappa\rangle) = \inf_F\{f \in F : f \geqslant \beta(T\varkappa)\}$$

$$\leqslant \inf_F\{f \in F : f \geqslant ⦀T⦀\alpha(\varkappa)\} \leqslant \inf_F\{Se : e \in E : e \geqslant \alpha(\varkappa)\}$$

$$= S\inf_F\{e \in E : e \geqslant \alpha(\varkappa)\} = S \circ \rho_E \circ \overline{\alpha}(\langle\varkappa\rangle).$$

Consequently, $S \geqslant ⦀\overline{T}⦀$. Using Lemma 4.11.1, we find

$$\rho_L([⦀T⦀]) = \inf_L\{S \in L : {}^*S \geqslant ⦀T⦀\} \geqslant ⦀\overline{T}⦀,$$

as required. ▷

4.11.9. Denote by $M_n(\mathscr{X}, \mathscr{Y})$ the set of all linear operators from $\mathscr{X}$ into $\mathscr{Y}$ which admit (o)-continuous dominants. It is clear that the conditions $T \in M_n(\mathscr{X}, \mathscr{Y})$ and ${}^*T \in M_n({}^*\mathscr{X}, {}^*\mathscr{Y})$ are equivalent. Take $T \in M_n(\mathscr{X}, \mathscr{Y})$. Then, according to (20), there exists a mapping $\overline{T} : (o)\text{-}\overline{\mathscr{X}} \to (o)\text{-}\overline{\mathscr{Y}}$ such that

$$\overline{T}(\langle\varkappa\rangle) = \langle{}^*T\varkappa\rangle \quad (\varkappa \in \mathrm{fin}({}^*\mathscr{X})).$$

Theorem. *For every $T \in M_n(\mathscr{X}, \mathscr{Y})$, the mapping $\overline{T}$ is a dominated linear operator from the BKS $((o)\text{-}\overline{\mathscr{X}}, \overline{a}, E)$ to the BKS $((o)\text{-}\overline{\mathscr{Y}}, \overline{b}, F)$. Furthermore,*

(1) $\overline{T}(\widehat{\eta}_{\mathscr{X}}(x)) = \widehat{\eta}_{\mathscr{Y}}(Tx)$ $(x \in \mathscr{X})$;

(2) $⦀\overline{T}⦀ = ⦀T⦀$,

where $\widehat{\eta}_{\mathscr{X}} : \mathscr{X} \to (o)\text{-}\overline{\mathscr{X}}$ and $\widehat{\eta}_{\mathscr{Y}} : \mathscr{Y} \to (o)\text{-}\overline{\mathscr{Y}}$ are the canonical embeddings defined by (16).

◁ The mapping $\overline{T}$ is linear by construction. Equality (1) ensues immediately from the definitions of $\widehat{\eta}_{\mathscr{X}}$, $\widehat{\eta}_{\mathscr{Y}}$, and $\overline{T}$. The fact that the operator $\overline{T}$ is dominated, as well as the inequality $⦀\overline{T}⦀ \leqslant ⦀T⦀$, is established in Theorem 5.2. It remains to verify the reverse inequality.

Let $x \in \mathscr{X}$. Taking property (1) into account, as well as the fact that $\widehat{\eta}_{\mathscr{X}}$ and $\widehat{\eta}_{\mathscr{Y}}$ are isometrically isomorphic embeddings, we obtain

$$b(T(x)) = \rho_F \circ \overline{b}(\widehat{\eta}_{\mathscr{Y}}(Tx)) = \rho_F \circ \overline{b}(\overline{T}(\widehat{\eta}_{\mathscr{X}}(x)))$$

$$\leqslant \langle\!\langle\!\langle \overline{T} \rangle\!\rangle\!\rangle (\rho_E \circ \overline{a}(\widehat{\eta}_{\mathscr{X}}(x))) = \langle\!\langle\!\langle \overline{T} \rangle\!\rangle\!\rangle a(x).$$

Since the element $x \in \mathscr{X}$ is chosen arbitrary, it follows that $\langle\!\langle\!\langle T \rangle\!\rangle\!\rangle \leqslant \langle\!\langle\!\langle \overline{T} \rangle\!\rangle\!\rangle$. ▷

Let $\widehat{\mathscr{X}}$ and $\widehat{\mathscr{Y}}$ be the (o)-completions of $\mathscr{X}$ and $\mathscr{Y}$, constructed in Theorem 4.11.6. From the previous theorem we obtain the following assertion (see [19, 14, Theorem 2.3.3]).

Corollary (A. G. Kusraev; V. Z. Strizhevskiĭ). *For every $T \in M_n(\mathscr{X}, \mathscr{Y})$, there exists a unique operator $\widehat{T} \in M_n(\widehat{\mathscr{X}}, \widehat{\mathscr{Y}})$ extending T in the sense that $\widehat{T}(\widehat{\eta}_E(x)) = \widehat{\eta}_F(Tx)$ for all $x \in \mathscr{X}$. Furthermore, $\langle\!\langle\!\langle \widehat{T} \rangle\!\rangle\!\rangle = \langle\!\langle\!\langle T \rangle\!\rangle\!\rangle$.*

◁ It is sufficient to take as $\widehat{T}$ the restriction of the operator $\overline{T}$ onto $\widehat{\mathscr{X}}$. Uniqueness of extension ensues from the requirement $\widehat{T} \in M_n(\widehat{\mathscr{X}}, \widehat{\mathscr{Y}})$ and the construction of $\widehat{\mathscr{X}}$. ▷

References

1. Aliprantis C. D. and Burkinshaw O., Positive Operators, Academic Press, New York and London (1985).
2. Albeverio S., Fenstad J. E., et al., Nonstandard Methods in Stochastic Analysis and Mathematical Physics, Academic Press, Orlando etc. (1990).
3. Bernau S. J., "Sums and extensions of vector lattice homomorphisms," Acta Appl. Math., **27**, 33–45 (1992).
4. Conshor H., "Enlargements contain various kinds of completions," in: Victoria Symposium of Nonstandard Analysis, Springer-Verlag, Berlin etc., 1974, pp. 60–79. (Lecture Notes in Math., **369**.)
5. Cozart D and Moore L. C., "The nonstandard hull of a normed Riesz space," Duke Math. J., No. 41, 263–275 (1974).
6. Emel′yanov È. Yu., "The order and regular hulls of Riesz spaces," Sibirsk. Mat. Zh., **35**, No. 6, 1243–1252 (1994).
7. Emel′yanov È. Yu., "Banach–Kantorovich spaces associated with the order hulls of decomposable lattice normed spaces," Sibirsk. Mat. Zh., **36**, No. 1, 72–85 (1995).
8. Emel′yanov È. Yu., "The infinitesimal approach to the representation of vector lattices by spaces of continuous functions on a compactum," Dokl. Akad. Nauk, **344**, No. 1, 9–11 (1995).

9. Emel′yanov È. Yu., "Infinitesimal analysis and vector lattices," Siberian Adv. Math., **6**, No. 1, 19–70 (1996).

10. Emel′yanov È. Yu., "Invariant homomorphisms of nonstandard enlargements of Boolean algebras and vector lattices," Sibirsk. Mat. Zh., **38**, No. 2, 286–296 (1997).

11. Henson C. W. and Moore L. C., "Nonstandard analysis and the theory of Banach spaces," in: Nonstandard Analysis: Recent Developments, Springer-Verlag, Berlin etc., 1983, pp. 27–112. (Lecture Notes in Math., **983**.)

12. Hurd A. E. and Loeb P. A., An Introduction to Nonstandard Real Analysis, Academic Press, Orlando etc. (1985).

13. Kusraev A. G., "Dominated operators. I," Siberian Adv. Math., **4**, No. 3, 51–82 (1994).

14. Kusraev A. G., "Dominated operators. II," Siberian Adv. Math., **4**, No. 4, 24–59 (1994).

15. Kusraev A. G., "Dominated operators. III," Siberian Adv. Math., **5**, No. 1, 49–76 (1995).

16. Kusraev A. G., "Dominated operators. IV," Siberian Adv. Math., **5**, No. 2, 99–121 (1995).

17. Kusraev A. G. and Kutateladze S. S., Nonstandard Methods of Analysis, Kluwer Academic Publishers, Dordrecht (1994).

18. Kusraev A. G. and Kutateladze S. S., "Nonstandard methods for Kantorovich spaces," Siberian Adv. Math., **2**, No. 2, 114–152 (1992).

19. Kusraev A. G. and Strizhevskiĭ V. Z., "Lattice normed spaces and dominated operators," in: Studies on Geometry and Functional Analysis. Vol. 7 [in Russian], Trudy Inst. Mat. (Novosibirsk), Novosibirsk, 1987, pp. 132–158.

20. Luxemburg W. A. J., "A general theory of monads," in: Applications of Model Theory to Algebra, Analysis, and Probability, Holt, Rinehart, and Winston, New York, 1969, pp. 18–86.

21. Luxemburg W. A. J. and A. C. Zaanen A. C., Riesz Spaces. Vol. 1, North-Holland, Amsterdam and London (1971).

22. Meyer-Nieberg P., Banach Lattices, Springer-Verlag, Berlin etc. (1991).

23. Robinson A., "Non-standard analysis," Proc. Roy. Acad. Amsterdam Ser. A, No. 64, 432–440 (1961).

24. Schaefer H. H., Banach Lattices and Positive Operators, Springer-Verlag, Berlin etc. (1974).

25. Sikorski R., Boolean Algebras, Springer-Verlag, Berlin etc. (1964).

26. Veksler A. I., "The Archimedes principle in the homomorphs of l-groups and vector lattices," Izv. Vyssh. Uchebn. Zaved. Mat., No. 5 (53), 33–38 (1966).

27. Wolff M. P. H., "An introduction to nonstandard functional analysis. Nonstandard analysis (Edinburgh, 1996)," NATO Adv. Sci. Inst. Ser. C. Math. Phys. Sci., **493**; Kluwer Academic Publishing, Dordrecht, 1997, pp. 121–151.
28. Zaanen A. C., Riesz Spaces. Vol. 2, North-Holland, Amsterdam etc. (1983).

Chapter 5

Vector Measures and Dominated Mappings

BY

A. G. Kusraev and S. A. Malyugin

S.S. Kutateladze (ed.), Nonstandard Analysis and Vector Lattices, 231–299.

The modern vector measure theory contains two weakly interacting directions of research.

The first consists in studying measures that take values in normed or locally convex spaces and stems from the classical contributions of the second half of the 1930s by S. Bochner, I. M. Gelfand, N. Dunford, and B. Pettis. It is now a beautiful theory reach in applications and exposed fairly in the monographs; for instance, see the books by N. Dinculeanu [1] and J. Diestel and J. J. Uhl [2] on vector measures as well as the relevant volume of the treatise by N. Bourbaki.

The second direction deals with measures ranging in ordered vector spaces. Some convergence emerging from the order takes the place of the topology of the target space, while order completeness takes the place of its topological counterpart. These measures seem to become an object of independent research because of the problem of finding an analytical representation for linear operators acting into ordered vector spaces; cf. the book [3] by L. V. Kantorovich, B. Z. Vulikh, and A. G. Pinsker. There is no denying that the measures ranging in vector lattices had appeared implicitly long before, disguised for instance as homomorphisms of abstract Boolean algebras and spectral characteristics of selfadjoint operators in Hilbert space. Moreover, studying selfadjoint operators from a standpoint of order analysis leads to the concepts of measure and integral with values in an ordered vector space; see the books [7] by A. I. Plesner and [8] by B. Z. Vulikh.

The questions innate to the classical measure theory were abstracted to Boolean measures starting from the 1950s under the influence of the theory of ordered vector spaces (V. I. Sobolev, B. Z. Vulikh, D. A. Vladimirov, et al.). J. D. M. Wright intensively studied lattice-valued measures along the same lines starting from the late 1960s. His publications arouse a keen interest in measures ranging in ordered vector spaces. Many valuable results have been put in stock since then; however, no comprehensive survey is available yet.

This chapter suggests a unified approach to the two directions in measure theory, resting on the fundamental concept of lattice normed space (LNS). The crux of the matter is in the fact that normed spaces and locally convex spaces are particular instances of lattice normed spaces. Therefore, all vector measures in the literature are listed among LNS-valued measures.

It is worth observing that this new theory of vector measures is not a mechanical matching of the miscellany available. On the contrary, rather different ideas and methods, serving to the topological and order-analytical approaches to vector measures, intertwine tightly, leading to new analytical tools and results. The time is not ripe for declaring a new theory, but we may speak of significant progress in a few problems.

Section 5.1 gives the definition of measure of bounded vector variation and sets forth a Lebesgue-type integral with respect to such a measure.

Section 5.2 studies quasi-Radon measures which are the closest analogs of scalar Radon measures in the case when the norm lattice fails to obey the weak σ-distributive law. We prove that a measure is quasi-Radon if and only if so is its vector variation.

Section 5.3 contains a criterion for a dominated operator to be represented as an integral with respect to a quasi-Radon measure. We also consider the problem of extending a quasi-Radon measure from a dense subalgebra (a vector analog of the Prokhorov Theorem).

Section 5.4 gives a version of the Fubini Theorem for the product of vector measures.

In Section 5.5 we solve an analog of the Hausdorff moment problem for a dominated sequence of vectors in a lattice normed space. Sections 5.6 and 5.7 treat the Hamburger moment problem in a vector setting.

In Section 5.8 we introduce the concept of dominated mapping on a locally compact Abelian group which is a vector analog of the concept of positive-definite mapping. In Section 5.9 we prove a "vector" Bochner Theorem on representation of a dominated mapping by a quasi-Radon measure on the Borel σ-algebra of the dual group. In Section 5.10 we define convolution for quasi-Radon measures given a bilinear mapping. We also prove here a representation theorem for a lattice-valued homomorphism of a locally compact group and a vector analog of the Bochner Theorem for positive-definite mapping with values in a monotonically complete ordered vector space. Section 5.11 contains a Boolean valued interpretation of the Wiener Lemma.

5.1. Vector Measures

Let $\mathfrak{X}$ be a completely regular topological space. Assume further that $\mathscr{T}$ and $\mathscr{F}$ stand for the collections of open and closed subsets of $\mathfrak{X}$; $\mathscr{T}_0$ and $\mathscr{F}_0$ are the collections of functionally open and functionally closed subsets of $\mathfrak{X}$; $\mathscr{B} := \mathscr{B}(\mathfrak{X}) := \mathscr{B}_{\mathfrak{X}}$ and $\mathscr{K}$ are the collections of Borel and compact subsets of $\mathfrak{X}$; $\mathscr{M}(\mathfrak{X})$ is the space of Borel functions on $\mathfrak{X}$; $\mathscr{M}_b(\mathfrak{X})$ ($C_b(\mathfrak{X})$) is the space of bounded Borel (continuous) functions on $\mathfrak{X}$. We let the symbol $C_{00}(\mathfrak{X})$ ($C_0(\mathfrak{X})$) stand for the space of compactly supported continuous functions $f : \mathfrak{X} \to \mathbb{R}$ (functions such that $\inf\{\sup\{|f(x)| : x \in \mathfrak{X} \setminus K\} : K \in \mathscr{K}\} = 0$). If $\mathfrak{C}$ is a family of subsets of $\mathfrak{X}$, then we denote by $\mathscr{A}(\mathfrak{C})$ ($\Sigma(\mathfrak{C})$) the least algebra (σ-algebra) generated by $\mathfrak{C}$. The articles [8–10] contain all conventional concepts we need in regard to Kantorovich spaces, lattice normed spaces, and dominated operators.

We say that a K_σ-space F (a σ-complete Boolean algebra $\mathbf{B}$) obeys the weak σ-distributive law if, given a bounded double sequence $\{x_{ij} : i, j \in \mathbb{N}\}$ of members of F (of $\mathbf{B}$) for which the sequence x_{ij} decreases to zero as $j \to \infty$ for all $i \in \mathbb{N}$, we have

$$\bigwedge\left\{\bigvee_{i=1}^{\infty} x_{i\varphi(i)} : \varphi \in \mathbb{N}^{\mathbb{N}}\right\} = 0.$$

We say that a Kantorovich space F (a complete Boolean algebra $\mathbf{B}$) obeys the weak (σ, ∞)-distributive law if, given a bounded sequence of vanishing nets $\{x_{i,\xi} : \xi \in \Xi_i\}$ $(i \in \mathbb{N})$ of members of F (of $\mathbf{B}$), we have

$$\bigwedge\left\{\bigvee_{i=1}^{\infty} x_{i,\varphi(i)} : \varphi \in \prod_{i=1}^{\infty} \Xi_i\right\} = 0.$$

We always let Y stand for an o-complete lattice normed space whose norm lattice is a Kantorovich space F. The F-norm of an element $y \in Y$ is denoted by $|y|$. Also, $\mathfrak{G}(e)$ is the Boolean algebra of fragments of a positive element $e \in F$.

Assume given an algebra $\mathscr{A}_0$ of subsets of $\mathfrak{X}$. A *measure* $\mu : \mathscr{A}_0 \to Y$ is an additive mapping from $\mathscr{A}_0$ to Y. We say that a *measure* μ *has bounded variation* if there is a positive measure $\nu : \mathscr{A}_0 \to F$ such that $|\mu(A)| \leqslant \nu(A)$ $(A \in \mathscr{A}_0)$. In the Kantorovich space $\mathrm{ba}(\mathscr{A}_0, F)$ of bounded measures from $\mathscr{A}_0$ to F there is a unique least element ν satisfying the above inequality. We call ν the *vector variation* of μ and denote it by $|\mu|$. We denote by $F - \mathrm{ba}(\mathscr{A}_0, Y)$ the space of measures from $\mathscr{A}_0$ to Y of bounded vector variation. The vector variation $|\mu|$ may be calculated by the formula:

$$|\mu|(A) = \bigvee\Bigg\{\sum_{i=1}^{n} |\mu(A_i)| : (A_i)_{i=1}^{n} \subseteq \mathscr{A}_0,$$

$$A_i \cap A_j = \varnothing\ (i \neq j), \bigcup_{i=1}^{n} A_i = A\Bigg\} \quad (A \in \mathscr{A}_0).$$

We denote by $F - \mathrm{bca}(\mathscr{A}_0, Y)$ the space of σ-additive measures from $\mathscr{A}_0$ to Y.

Let $S(\mathscr{A}_0)$ stand for the space of $\mathscr{A}_0$-simple functions on $\mathfrak{X}$; i.e. $g \in S(\mathscr{A}_0)$ means that $g = \sum_{i=1}^{n} c_i \chi_{A_i}$ for some $(c_i)_{i=1}^{n} \subseteq \mathbb{R}$ and disjoint $(A_i)_{i=1}^{n} \subseteq \mathscr{A}_0$. Consider a measure $\mu \in F - \mathrm{ba}(\mathscr{A}_0, Y)$. Given $g \in S(\mathscr{A}_0)$, define

$$\int g\,d\mu = \sum_{i=1}^{n} c_i \mu(A_i).$$

This integral extends to the uniform closure $\overline{S(\mathscr{A}_0)}$ of the space $S(\mathscr{A}_0)$ by uniform continuity. In case $\mathscr{A}_0 = \mathfrak{B}$ and $\mu \in F - \mathrm{bca}(\mathscr{A}_0, Y)$ we may extend this integral to a much wider class of functions. We say that $f \in \mathscr{M}(\mathfrak{X})$ is a μ-*integrable* function if the set $\{\int g\,d|\mu| : g \in S(\mathscr{A}_0),\ 0 \leqslant g \leqslant |f|\}$ is bounded in F. We denote the space of μ-integrable functions by $L_1(\mu)$. Take $f \in L_1(\mu), f \geqslant 0$, and assume that some sequence of functions $(g_n)_{n=1}^{\infty} \subseteq S(\mathscr{A}_0)$ increases and converges pointwise to f. Then the sequence of integrals of g_n is o-Cauchy, and we put

$$\int f d\mu = bo\text{-}\lim_n \int g_n d\mu.$$

This integral extends to the whole space $L_1(\mu)$ by additivity. The resultant integral possesses all main properties of the Lebesgue integral. In particular, the Dominated Convergence Theorem is valid:

Lebesgue Theorem. *Assume that $f_n, g \in L_1(\mu)$, $|f_n| \leqslant g$ $(n \in \mathbb{N})$ and f_n converges to f pointwise. Then $f \in L_1(\mu)$ and*

$$\int f d\mu = bo\text{-}\lim_n \int f_n d\mu.$$

We note that this integral was considered for a positive measure with values in a Stone space in [14].

5.2. Quasi-Radon and Quasiregular Measures

Let $E \in \mathscr{A}_0$. Consider the directed sets $\mathscr{K}_E = \{K : K \in \mathscr{K} \cap \mathscr{A}_0, K \subseteq E\}$ and $\mathscr{F}_E = \{F : F \in \mathscr{F} \cap \mathscr{A}_0, F \subseteq E\}$ whose elements are assumed inclusion-ordered.

5.2.1. DEFINITION. We call a measure $\mu : \mathscr{A}_0 \to Y$ a *Radon (quasi-Radon) measure* provided that for all $E \in \mathscr{A}_0$ (for all $E \in \mathscr{T} \cap \mathscr{A}_0$) the equality holds: $\mu(E) = bo\text{-}\lim\{\mu(K) : K \in \mathscr{K}_E\}$.

5.2.2. DEFINITION. A measure $\mu : \mathscr{A}_0 \to Y$ is *regular (quasiregular)* provided that for all $E \in \mathscr{A}_0$ (for all $E \in \mathscr{T} \cap \mathscr{A}_0$) the equality holds $\mu(E) = bo\text{-}\lim\{\mu(F) : F \in \mathscr{F}_E\}$.

If $\mathfrak{X}$ is a compact space then Definitions 5.2.1 and 5.2.2 are equivalent. Moreover, we can routinely prove the following

5.2.3. Theorem. *A measure $\mu \in F-\mathrm{ba}(\mathscr{A}_0, Y)$ is Radon (regular) if and only if its vector variation is Radon (regular).*

The purpose of this section is to prove a similar theorem for quasi-Radon measures.

5.2.4. Theorem. *Let $\mu \in F - \mathrm{ba}(\mathscr{A}_0, Y)$ be a measure satisfying either of the following conditions:*

(1) $\mathscr{A}_0 = \mathscr{A}(\mathscr{F} \cap \mathscr{A}_0)$;

(2) $\mathscr{A}_0 = \sum(\mathscr{F} \cap \mathscr{A}_0)$ and $\mu \in F - \mathrm{bca}(\mathscr{A}_0, Y)$.

Then μ is a quasi-Radon (quasiregular) measure if and only if so is the vector variation $|\mu|$ of μ.

$\lhd$ We will consider only the quasi-Radon property on assuming (2). Case (1) is settled by analogy.

Suppose that μ is a quasi-Radon measure whereas the vector variation $|\mu|$ of μ is not quasi-Radon. Then there is a set $U \in \mathscr{T} \cap \mathscr{A}_0$ such that $|\mu|(U) = \bigvee\{|\mu|(K) : K \in \mathscr{K} \cap \mathscr{A}_0,\ K \subseteq U\} > 0$. Let $K \in \mathscr{K} \cap \mathscr{A}_0, K \subseteq U$ and $e = |\mu|(U)$. There are $\varepsilon_0 > 0$, $0 < e_0 \in \mathfrak{E}(e)$ and a finite family $(E_i)_{i=1}^n \subseteq \mathscr{A}_0$ satisfying

$$\bigcup_{i=1}^n E_i = U \backslash K,\ E_i \cap E_j = \varnothing\ (i \neq j),\quad \sum_{i=1}^n |\mu(E_i)| \geqslant \varepsilon_0 e_0.$$

Denote by ω_1 the first uncountable ordinal. For some countable ordinal $\alpha_0 < \omega_1$ all E_i $(i = 1, \dots, n)$ belong to the Baire class $\mathfrak{B}_{\alpha_0}(\mathscr{A}(\mathscr{F} \cap \mathscr{A}_0))$ constructed from the algebra $\mathscr{A}(\mathscr{F} \cap \mathscr{A}_0)$ (see [6]). We may assume α_0 to be a nonlimit ordinal. Each member of the Baire class $\mathfrak{B}_\alpha = \mathfrak{B}_\alpha(\mathscr{A}(\mathscr{F} \cap \mathscr{A}_0))$ is a countable union or countable intersection of some members of the preceding Baire classes. Therefore, there are $\alpha_1 < \alpha_0$ and sequences $(E_{i,k})_{k=1}^\infty \subseteq \mathfrak{B}_{\alpha_1}$ $(i = 1, \dots, n)$ such that for all i the sequence $(E_{i,k})_{k=1}^\infty$ converges monotonously to E_i. Moreover, we may assume that for all i, k we have the inclusion $E_{i,k} \subseteq U \backslash K$. For every $\delta > 0$ there are $e_1 \in \mathfrak{E}(e)$, $0 < e_1 \leqslant e_0$, and an index k_1 such that

$$\sum_{i=1}^n \left|\mu\left(E_{i,k_1} \backslash \bigcup_{j<i} E_{j,k_1}\right)\right| \geqslant \left(\varepsilon_0 - \frac{\delta}{4}\right) e_1.$$

This follows from the σ-additivity of μ.

Put $E_i^1 = E_{i,k_1}$ $(i = 1, \dots, n)$. Repeating the above procedure sufficiently many times, we arrive at a decreasing sequence of ordinals $\alpha_m < \alpha_{m-1} < \dots < \alpha_0$, a sequence of elements $(e_k)_{k=1}^m \subseteq \mathfrak{E}(e)$ and sequences $(E_i^n)_{i=1}^n \subseteq \mathfrak{B}_{\alpha_k}$ $(k = 0, 1, \dots, m)$ such that $0 < e_m \leqslant e_{m-1} \leqslant \dots \leqslant e_1$ and

$$\sum_{i=0}^n \left|\mu\left(E_i^k \backslash \bigcup_{j<i} E_j^k\right)\right| \geqslant \left(\varepsilon_0 - \frac{\delta}{4} - \dots - \frac{\delta}{2^{k+1}}\right) e_k \quad (k = 0, 1, \dots, m).$$

Since the ordinal are well-ordered, we see that this process terminates. Consequently, we may assume that $\alpha_m = 0$. Putting

$$F_i = E_i^m \backslash \bigcup_{j<i} E_j^m \ \ (i = 1, \dots, n), \quad F_0 = (U \backslash K) \backslash \bigcup_{i=1}^n F_i, \quad g = e_m,$$

we have $0 < g \leqslant e$, $g \in \mathfrak{G}(e)$ and $\sum_{i=0}^{n}|\mu(F_i)| \geqslant (\varepsilon_0 - \frac{\delta}{2})g$; moreover, $(F_i)_{i=0}^{n} \subseteq \mathscr{A}(\mathscr{F} \cap \mathscr{A}_0)$, $F_i \cap F_j = \varnothing$ $(i \neq j)$ and $\bigcup_{i=0}^{n} F_i = U \backslash K$. Without loss of generality, we may assume that each member F_i has the form $F_i = U_i \backslash V_i$ with $U_i, V_i \in \mathscr{T} \cap \mathscr{A}_0, U_i \cup V_i \subseteq U \backslash K$ $(i = 0, 1, \dots, n)$, since each element of the algebra $\mathscr{A}(\mathscr{F} \cap \mathscr{A}_0)$ may be written as a disjoin union of finitely many elements of the form $U_i \backslash V_i$, with $U_i, V_i \in \mathscr{T} \cap \mathscr{A}_0$.

Enumerate all U_i and V_i into a sole sequence $(W_i)_{i=1}^{m}$. Put $M = \{1, \dots, m\}$. Given $J \subseteq M$, assign $H_J = \bigcap\{W_i : \ i \in M \backslash J\}$. Obviously, $H_J \cap H_{J'} = H_{J \cap J'}$. Since μ is a quasi-Radon measure $\varepsilon' > 0$, there are a compact set $K_\varnothing \in \mathscr{K} \cap \mathscr{A}_0$, $K_\varnothing \subseteq H_\varnothing$ and an element $g_0 \in \mathfrak{G}(e)$, $0 < g_0 \leqslant g$, such that for all $K' \in \mathscr{K} \cap \mathscr{A}_0$, $K_\varnothing \subseteq K' \subseteq H_\varnothing$, we have

$$g_0|\mu(K' \backslash K_\varnothing)| \leqslant \varepsilon' e, \quad g_0|\mu(H_\varnothing \backslash K_\varnothing)| \leqslant \varepsilon' e$$

(on assuming multiplication in the ideal $F(e)$ in which e is a ring-unity; cf. [8]). For each $i \in M$ there are $K_{\{i\}} \in \mathscr{K} \cap \mathscr{A}_0$, $g_i \in \mathfrak{G}(e)$ such that $K_{\{i\}} \subseteq H_{\{i\}} \backslash K_\varnothing$, $0 < g_i \leqslant g_0$ and, for all $K' \in \mathscr{K} \cap \mathscr{A}_0$, satisfying $K_{\{i\}} \subseteq K' \subseteq H_{\{i\}} \backslash K_\varnothing$, we have

$$g_i|\mu(K' \backslash K_{\{i\}})| \leqslant \varepsilon' e, \quad g_i|\mu((H_{\{i\}} \backslash K_\varnothing) \backslash K_{\{i\}})| \leqslant \varepsilon' e.$$

We may assume all g_i ordered, for instance, as follows: $g_m \leqslant g_{m-1} \leqslant \dots \leqslant g_1$. For $i \neq j$ we have $K_{\{i\}} \cap K_{\{j\}} = H_\varnothing$, and so $g_0|\mu(K_{\{i\}} \cap K_{\{j\}})| \leqslant \varepsilon' e$. We then make the induction step: Assume that, for some $k \leqslant m$, our construction is implemented for all $J \subset M$ with $\mathrm{card} J < k$. In particular, for all $J \subset M$, $\mathrm{card} J < k$, we have $K_J \in \mathscr{K} \cap \mathscr{A}_0$. Let $J \subset M$ and $\mathrm{card} J = k$. Take $K_J \in \mathscr{K} \cap \mathscr{A}_0$, $g_J \in \mathfrak{G}(e)$ so that $K_J \subseteq H_J \backslash \cup \{K_{J'} : J' \subset J\}$, $0 < g_J \leqslant \bigwedge\{g_{J'} : \ \mathrm{card} J' < k\}$, and for all $K' \in \mathscr{K} \cap \mathscr{A}_0$ satisfying $K_J \subseteq K' \subseteq H_J \backslash \cup \{K_{J'} : J' \subset J\}$ the following hold: $g_J|\mu(K' < K_J)| \leqslant \varepsilon' e$, $g_J|\mu((H_J \backslash \bigcup\{K_{J'} : J' \subseteq J\}) \backslash K_J)| \leqslant \varepsilon' e$. We may assume that all g_J with $\mathrm{card} J \leqslant k$ are totally ordered. Moreover, given the subsets $J \subset M$, $J' \subset M$, $J \ni J'$, such that $\mathrm{card} J = k$, $\mathrm{card} J' \leqslant k$, or $\mathrm{card} J \leqslant k$, $\mathrm{card} J' = k$, we obtain $K_J \cap K_{J'} \subseteq H_J \cap H_{J'} = H_{J \cap J'}$. If $J'' \subseteq J \cap J'$, then $K_J \cap K_{J''} = \varnothing$ or $K_{J'} \cap K_{J''} = \varnothing$ by construction. This means that, for all $K' \in \mathscr{K} \cap \mathscr{A}_0$, $K' \subseteq K_J \cap K_{J'}$, we have $\overline{g}|\mu(K')| \leqslant \varepsilon' e$, with $\overline{g} = \bigwedge\{g_{J'} : \ \mathrm{card} J' \leqslant k\}$. This induction terminates when $k = m - 1$. We now draw some conclusions.

Given $\sigma : M \to \{0, 1\}$, put $W^\sigma = \bigcup_{i=1}^{m} W_i^{\sigma(i)}$, with $W_i^0 = (U \backslash K) \backslash W_i$, $W_i^1 = W_i$ $(i \in M)$. We have proven that for all $\varepsilon' > 0$ there are $\overline{g} \in \mathfrak{G}(e)$, $0 < \overline{g} \leqslant q$ and $K^\sigma \in \mathscr{K} \cap \mathscr{A}_0$ $(\sigma \in \{0, 1\}^M)$ satisfying $K^\sigma \subseteq U \backslash K$ and

$$\overline{g}|\mu(W^\sigma) - \mu(K^\sigma)| \leqslant \varepsilon' \overline{g}, \ \overline{g}|\mu(K')| \leqslant \varepsilon' \overline{g} \tag{2.1}$$

$$(K' \in \mathscr{K} \cap \mathscr{A}_0, \ K' \subseteq K^\sigma \cap K^{\sigma'}, \sigma, \sigma' \in \{0, 1\}^M, \ \sigma \neq \sigma').$$

This means that for $K_1 = \cup\{K^\sigma : \sigma \in \{0,1\}^M\}$ we have

$$\overline{g}|\mu|(K_1) \geqslant \overline{g}\sum\Big\{\Big|\mu\Big(K^\sigma\setminus\bigcap_{\sigma'\neq\sigma}K^{\sigma'}\Big)\Big| : \sigma\in\{0,1\}^M\Big\}$$

$$+\overline{g}|\mu(\cup\{K^\sigma\cap K^{\sigma'} : \sigma,\sigma'\in\{0,1\}^M,\ \sigma\neq\sigma'\})| \geqslant \overline{g}\sum_\sigma|\mu(K^\sigma)|$$

$$-\overline{g}\sum_\sigma\Big|\mu\Big(\bigcup_{\sigma'\neq\sigma}K^\sigma\cap K^{\sigma'}\Big)\Big|.$$

Given $\sigma \in \{0,1\}^M$, estimate $\overline{g}|\mu(\bigcup_{\sigma'\neq\sigma}K^\sigma\bigcap K^{\sigma'})|$. To this end, put $\mathscr{M}_\sigma = \{0,1\}^M\setminus\{\sigma\}$, $L_{\sigma'} = K^\sigma\cap K^{\sigma'}$ and consider the identity

$$\mu\Big(\bigcup_{\sigma'}L_{\sigma'}\Big) = \sum_{k=1}^{l}(-1)^{k+1}\Big(\sum_{\sigma_1<\dots<\sigma_k}\mu(L_{\sigma_1}\cap\dots\cap L_{\sigma_k})\Big).$$

Here $\sigma',\sigma_1,\dots,\sigma_k \in \mathscr{M}_\sigma$, $l = 2^m - 1$ and we also assume that the members of $\mathscr{M}_\sigma$ are totally ordered somehow. The number of terms on the right-hand side of the sum equals $2^l - 1$; moreover, we may estimate every summand by (2.1). Then

$$\overline{g}|\mu|(K_1) \geqslant \overline{g}\sum_\sigma|\mu(W^\sigma)| - 2^{l+m}\varepsilon'\overline{g} \geqslant \overline{g}\sum_{i=0}^{n}|\mu(F_i)| - 2^{l+m}\varepsilon'\overline{g} \geqslant (\varepsilon_0-\delta)\overline{g}.$$

The last inequality becomes valid if we take ε' sufficiently small.

We have thus proven the following: For all $e_1 \in \mathfrak{C}(e)$, $K \in \mathscr{K}\cap\mathscr{A}_0, \delta > 0$ satisfying $0 < e_1 \leqslant e_0$, $K \subseteq U$, there are $e_2 \in \mathfrak{C}(e)$, $K_1 \in \mathscr{K}\cap\mathscr{A}_0$ such that $0 < e_2 \leqslant e_1$, $K_1 \subseteq U\setminus K$ and $|\mu|(K_1) \geqslant (\varepsilon_0-\delta)e_2$. We first put $K = \varnothing$. We now find K_1 with the above properties. We then set $K = K_1$ and find $K_2 \in \mathscr{K}\cap\mathscr{A}_0$, $e_3 \in \mathfrak{C}(e)$ satisfying $K_2 \subseteq U\setminus K_1, 0 < e_3 \leqslant e_2$ and $|\mu|(K_2) \geqslant (\varepsilon_0 - \delta - \delta/2)e_3$. Proceeding likewise, we find some sequences $(K_n)_{n=1}^\infty \subseteq \mathscr{K}\cap\mathscr{A}_0$ and $(e_n)_{n=1}^\infty \subseteq \mathfrak{C}(e)$, enjoying the properties

$$K_n \subseteq \bigcup_{i=1}^{n-1}K_1, \quad 0 < e_{n+1} \leqslant e_n,$$

$$|\mu|(K_n) \geqslant (\varepsilon_0 - \delta - \dots - \delta/2^{n-1})e_{n+1} \quad (n\in\mathbb{N}).$$

This means that for all $n \in \mathbb{N}$ the following holds:

$$|\mu|(U) \geqslant n(\varepsilon_0 - 2\delta)e_{n+1}.$$

We so arrive at a contradiction with the inequality $e_{n+1} \leqslant e = |\mu|(U)$ valid for all $n \in \mathbb{N}$.

Conversely, assuming that the vector variation $|\mu|$ of μ is quasi-Radon, we immediately infer that μ is quasi-Radon using the inequality

$$|\mu(E) - \mu(F)| \leqslant |\mu|(E \backslash F) \quad (E, F \in \mathscr{A}_0, F \subseteq E). \ \triangleright$$

5.2.5. Corollary. *Suppose that $\mu \in F - \mathrm{ba}(\mathscr{A}_0, Y)$ satisfies either of the conditions (1) and (2) of Theorem 5.2.4. The following definitions of the quasi-Radon property for μ are equivalent:*

(1) *the equality $\mu(E) = bo\text{-}\lim\{\mu(K) : K \in \mathscr{K}_E\}$ holds for all $E \in \mathscr{T} \cap \mathscr{A}_0$;*

(2) *the equality of (1) holds for all $E \in \mathscr{A}(\mathscr{F} \cap \mathscr{A}_0)$;*

(3) *the equality $|\mu|(E) = \bigvee\{|\mu|(K) : K \in \mathscr{K}_E\}$ holds for all $E \in \mathscr{T} \cap \mathscr{A}_0$;*

(4) *the equality of (3) holds for all $E \in \mathscr{A}(\mathscr{F} \cap \mathscr{A}_0)$.*

5.2.6. Corollary. *With $\mathscr{F}_E$ substituted for $\mathscr{K}_E$, conditions (1)–(4) are equivalent definitions of the quasi-Radon property for μ.*

5.2.7. Corollary. *If $\mu \in F - \mathrm{ba}(\mathscr{A}_0, Y)$ is a quasi-Radon measure satisfying either of the conditions (1), (2) of Theorem 5.2.4, then μ is quasiregular.*

In this connection the following obvious fact is worth mentioning: if $\mu : \mathscr{A}_0 \to F$ is a positive quasi-Radon measure then μ is quasiregular. Moreover, we have

5.2.8. Lemma. *Let $\mu \in F - \mathrm{ba}(\mathscr{A}_0, Y)$ be a measure such that the vector variation of $|\mu|$ is quasi-Radon and the equality holds*

$$|\mu|(\mathfrak{X}) = \bigvee \{|\mu|(K) : K \in \mathscr{K} \cap \mathscr{A}_0\}. \tag{2.2}$$

Then μ and $|\mu|$ are quasi-Radon.

5.2.9. Theorem. *If $\mu \in F - \mathrm{ba}(\mathscr{A}_0, Y)$ is a quasi-Radon measure then the restriction of μ to the algebra $\mathscr{A}(\mathscr{F} \cap \mathscr{A}_0)$ is a σ-additive measure.*

$\triangleleft$ Preserve the denotation μ for the restriction of μ to $\mathscr{A}(\mathscr{F} \cap \mathscr{A}_0)$. The vector variation of μ with respect to this new algebra may change and we denote it by $|\mu|_0$. Prove that $|\mu|_0$ is σ-additive (then the σ-additivity of the restriction of μ to $\mathscr{A}(\mathscr{F} \cap \mathscr{A}_0)$ will follow from the σ-additivity of $|\mu|_0$). Assume that some sequence $(A_k)_{k=1}^{\infty} \subseteq \mathscr{A}(\mathscr{F} \cap \mathscr{A}_0)$ decreases to the empty set. Put $e = |\mu|_0(\mathfrak{X})$. Suppose that $\bigwedge_{n=1}^{\infty} |\mu|_0(A_n) > 0$. Then $|\mu|_0(A_n) \geqslant \varepsilon_0 e_0$ $(n \in \mathbb{N})$ for some $\varepsilon_0 > 0$ and $0 < e_0 \in$

$\mathfrak{E}(e)$. By Theorem 5.2.4 $|\mu|_0$ is a quasi-Radon measure. Consequently, there are $K_1 \in \mathscr{K} \cap \mathscr{A}_0$, $e_1 \in \mathfrak{E}(e_0)$ such that $K_1 \subseteq A_1$, $0 < e_1 \leqslant e_0$ and $e_1|\mu|_0(A_1 \backslash K_1) \leqslant (\varepsilon_0/4)e_0$. Proceeding by induction, come to some sequences $(K_n)_{n=1}^{\infty} \subseteq \mathscr{K} \cap \mathscr{A}_0$ and $(e_n)_{n=1}^{\infty} \subseteq \mathfrak{E}(e)$ such that $0 \leqslant e_{n+1} \leqslant e_n$, $K_n \subseteq A_n$ and $e_n|\mu|_0(A_n \backslash K'_n) \leqslant (\varepsilon_0/2^{n+1})e_0$ $(n \in \mathbb{N})$. Putting $K'_n = \bigcap_{i=1}^{n} K_i$, find that $e_n|\mu|_0(A_n \backslash K'_n) \leqslant (\varepsilon_0/2)e_0$ $(n \in \mathbb{N})$. This implies the equality $|\mu|_0(K'_n) = |\mu|_0(A_n) - |\mu|_0(A_n \backslash K'_n) \geqslant (\varepsilon_0/2)e_n$. Since $K'_n \searrow \varnothing$, we have $K'_{n_0} = \varnothing$ for some $n_0 \in \mathbb{N}$. This contradiction completes the proof of the theorem. $\triangleright$

An analogous fact for quasi-Radon measures is valid without any constraints.

5.2.10. *Theorem.* *If $\mu \in F - \mathrm{ba}(\mathscr{A}_0, Y)$ is a quasi-Radon measure then μ is σ-additive.*

The next example shows that the requirement of σ-additivity of μ in (2) of Theorem 5.2.4 is essential even in the case of real measures.

5.2.11. EXAMPLE. Let $\mathbb{Q}$ stand for the rationals. Furnish $\mathbb{Q}$ with the topology $\mathscr{T}_{\mathbb{Q}}$ induced by the natural topology $\mathscr{T}$ on $\mathbb{R}$. Take the Borel σ-algebra $\mathfrak{B}(\mathbb{R})$ of the real axis $\mathbb{R}$ and consider a probability measure λ whose zeros are meager Borel sets (see [12]). This measure induces some measure ρ on the algebra $\mathscr{A}(\mathscr{T}_{\mathbb{Q}})$ as follows: Let $A \in \mathscr{A}(\mathscr{T}_{\mathbb{Q}})$. Then there is some B in $\mathscr{A}(\mathscr{T})$ such that $A \in B \cap \mathbb{Q}$. Put $\rho(A) = \lambda(B)$. It is easy to see that this definition is sound. If a compact set K belongs to $\mathbb{Q}$ then it is meager in $\mathbb{R}$. Therefore $\rho(K) = 0$, and ρ is not a quasi-Radon measure. Split $\mathbb{Q}$ into two sets Q_1 and Q_2 each dense in $\mathbb{R}$. By the Łoš–Marczewski Theorem it is possible to extend the measure ρ on the boolean $2^{\mathbb{Q}}$ of $\mathbb{Q}$ (see [13]). Consider two extensions ρ_1, ρ_2 satisfying $\rho_1(Q_1) = 1 = \rho_2(Q_2)$. The reason behind this possibility is as follows: If $A \in \mathscr{A}(\mathscr{T}_{\mathbb{Q}})$ and $A \supseteq Q_1$ $(A \supseteq Q_2)$, then the closure of A coincides with $\mathbb{R}$, implying $\rho(A) = 1$. Consider the measure $\mu = \rho_1 - \rho_2$ on the σ-algebra $\sum(\mathscr{T}_{\mathbb{Q}}) = 2^{\mathbb{Q}}$. Since the restriction of μ to the algebra $\mathscr{A}(\mathscr{T}_{\mathbb{Q}})$ is zero; therefore, μ is a quasi-Radon measure. However, the variation of μ, equal to $\rho_1 + \rho_2$, is not a quasi-Radon measure.

5.3. Integral Representations and Extension of Measures

Given a completely regular topological space $\mathfrak{X}$, consider some function vector lattice $\mathfrak{L} \subseteq C_b(\mathfrak{X})$. We denote by $\mathscr{T}(\mathfrak{L})$ the weakest topology making every member of $\mathfrak{L}$ continuous. If $\mathscr{T}(\mathfrak{L})$ coincides with the original topology $\mathscr{T}$ on $\mathfrak{X}$ we say that $\mathfrak{L}$ *generates* $\mathscr{T}$.

A linear operator $T : \mathfrak{L} \to Y$ is *dominated* if there is a positive operator $S : \mathfrak{L} \to F$ satisfying $|Tf| \leqslant S|f|$ $(f \in \mathfrak{L})$. We call the least of these operators S the *dominant* of T and denoted it by $|T|$.

5.3.1. Theorem. *Assume that a vector lattice $\mathfrak{L} \subseteq C_b(\mathfrak{X})$ contains the identically one function $1_{\mathfrak{X}}$ and generates the topology $\mathscr{T}$. To each dominated operator $T : \mathfrak{L} \to Y$ there is a unique quasi-Radon measure $\mu \in F - \text{bca}(\mathscr{B}, Y)$ satisfying*

$$Tf = \int f d\mu \quad (f \in \mathfrak{L}), \tag{3.1}$$

if and only if

$$|T|(1) = \bigvee \Big\{ \bigwedge \{|T|g : \ g \in \mathfrak{L}, g \geqslant \chi_K\} : \ K \in \mathscr{K} \Big\}. \tag{3.2}$$

$\lhd$ Assume that (3.2) holds and some net $(f_\alpha)_{\alpha \in \mathrm{A}}$ of functions in $\mathfrak{L}$ decreases pointwise to zero. Fix $\alpha_0 \in \mathrm{A}$ and $\varepsilon > 0$. Then, for some $M > 0$ and all $\alpha \geqslant \alpha_0$ we have $0 \leqslant x_\alpha \leqslant M 1_{\mathfrak{X}}$. Let $K \in \mathscr{K}$. There is $\alpha_1 \geqslant \alpha_0$ such that, for all $\alpha \geqslant \alpha_1$, we have

$$0 \leqslant |T| f_\alpha \leqslant \varepsilon |T|(1_{\mathfrak{X}}) + M a_K. \tag{3.3}$$

Here, we put $a_K = |T|(1_{\mathfrak{X}}) - \bigwedge \{|T| f : \ f \in \mathfrak{L}, f \geqslant \chi_K\}$. The net $\{a_K : \ K \in \mathscr{K}\}$ decreases to zero by (3.2). By the Realization Theorem, the band $(|T|(1_{\mathfrak{X}}))^{\perp\perp}$ of the Kantorovich space F is isomorphic with an order-dense ideal of some space $C_\infty(Q)$, with Q an extremally disconnected compact space; see 1.7.10. We may assume that this isomorphism sends $|T|(1_{\mathfrak{X}})$ to the identically one function on Q. Since a_K decreases to zero in order, there is a comeager set $Q_0 \subseteq Q$ such that the numeric net $a_K(q)$ vanishes for all $q \in Q_0$. From (3.3) it follows that if a net $(f_\alpha)_{\alpha \in \mathrm{A}} \subseteq \mathfrak{L}$ decreases to zero then the net $\omega_q(f_\alpha) = (|T| f_\alpha)(q)$ vanishes for all $q \in Q_0$. By the well-known Daniell–Stone Theorem the positive functional ω_q extends to a sequentially o-continuous positive functional $\bar{\omega}_q : \mathscr{M}_b(\mathfrak{X}) \to \mathbb{R}$ (some Kantorovich space versions of this theorem were studied in [3–5]). Consider the mapping $V : \mathscr{M}_b(\mathfrak{X}) \to \mathbb{R}^Q$, defined by the formulas

$$(Vf)(q) = \begin{cases} \bar{\omega}_q(f), \ q \in Q_0, \\ 0, \ q \notin Q_0 \quad (f \in \mathscr{M}_b(\mathfrak{X})). \end{cases}$$

Clearly, $Vf \in \mathscr{M}_b(Q)$ for all $f \in \mathfrak{L}$. Furthermore, the sequential o-continuity of $\bar{\omega}_q$ implies that if $Vf_n \in \mathscr{M}_b(Q)$ for some sequence $f_n \in \mathscr{M}_b(\mathfrak{X})$ convergent in order to f then $Vf \in \mathscr{M}_b(Q)$. This enables us to conclude that $Vf \in \mathscr{M}_b(Q)$ for all $f \in \mathscr{M}_b(\mathfrak{X})$. Assign $W = \jmath \circ V$, where $\jmath : \mathscr{M}_b(Q) \to F$ is a Birkhoff–Ulam homomorphism. Then $W : \mathscr{M}_b(\mathfrak{X}) \to F$ is a sequentially o-continuous extension of W. We extend the operator T as follows: Take $f \in \mathscr{M}_b(\mathfrak{X})$ and suppose that there is a bounded net $(g_\alpha)_{\alpha \in \mathrm{A}} \subseteq \mathfrak{L}$, increasing pointwise to f. The estimate $|Tg_\alpha - Tg_\beta| \leqslant W(|g_\alpha - g_\beta|)$ $(\alpha, \beta \in \mathrm{A})$ implies that Tg_α is a bo-Cauchy net. Put

$T_0 f = bo\text{-}\lim T g_\alpha$. Letting $C_b^\uparrow$ stand for the cone of bounded lower semicontinuous functions on $\mathfrak{X}$, we then extend T to a dominated operator $T_0 : \mathscr{M}_0 \to Y$, where $\mathscr{M}_0 = C_b^\uparrow - C_b^\uparrow$.

Further extension of T_0 is carried out by transfinite induction up to the first uncountable ordinal ω_1. Furthermore, the estimate for the norm remains the same:

$$|T_0 f| \leqslant W(|f|) \quad (f \in \mathscr{M}_0).$$

Assume that for all ordinals $\beta < \alpha < \omega_1$ we have already defined some linear sublattices $\mathscr{M}_\beta \subset \mathscr{M}_b(\mathfrak{X})$ and linear operators $T_\beta : \mathscr{M}_\beta \to Y$ satisfying

$$|T_\beta f| \leqslant W(|f|) \quad (f \in \mathscr{M}_\beta)$$

and such that $\mathscr{M}_\beta \subset \mathscr{M}_\gamma$, $T_\gamma|_{\mathscr{M}_\beta} = T_\beta$ for $\beta < \gamma < \alpha$. If α is a limit ordinal we then put $\mathscr{M}_\alpha := \cup\{\mathscr{M}_\beta : \beta < \alpha\}$ and define a linear operator $T_\alpha : \mathscr{M}_\alpha \to Y$ so that $T_\alpha|_{\mathscr{M}_\beta} = T_\beta$ $(\beta < \alpha)$. If α is a nonlimit ordinal, we consider the set $\mathscr{M}_{\alpha-1}^\sigma$ of all $x \in \mathscr{M}_b(\mathfrak{X})$, presenting suprema of bounded countable subsets of $\mathscr{M}_{\alpha-1}$. If a countable sequence $(f_n)_{n\in\mathbb{N}}$ lies in $\mathscr{M}_{\alpha-1}$ and $\sup_n f_n = f \in \mathscr{M}_{\alpha-1}^\sigma$, then by analogy we conclude that $(T_{\alpha-1} f_n)_{n\in\mathbb{N}}$ is an o-Cauchy sequence. We thus may put $T_{\alpha-1}^\sigma f := bo\text{-}\lim T_{\alpha-1} f_n$. Since W is sequentially o-continuous, this soundly defines the operator $T_{\alpha-1}^\sigma : \mathscr{M}_{\alpha-1}^\sigma \to Y$ that satisfies the inequalities

$$|T_{\alpha-1}^\sigma f| \leqslant W f \quad (0 \leqslant f \in \mathscr{M}_{\alpha-1}^\sigma).$$

Similar reasoning implies that $T_{\alpha-1}^\sigma$ extends by additivity to the linear operator $T_\alpha : \mathscr{M}_\alpha \to Y$ where $\mathscr{M}_\alpha := \mathscr{M}_{\alpha-1}^\sigma - \mathscr{M}_{\alpha-1}^\sigma$ is a linear sublattice of $\mathscr{M}_b(\mathfrak{X})$. It is easy that $\mathscr{M}_b(\mathfrak{X}) = \mathscr{M}_{\omega_1}$ and the operator $T_1 := T_{\omega_1}$ is sequentially o-continuous extension of T to $\mathscr{M}_b(\mathfrak{X})$. This operator is dominated in view of the estimate

$$|T_1 f| \leqslant W(|f|) \quad (f \in \mathscr{M}_b(\mathfrak{X})).$$

We now define μ by the equality $\mu(E) = T_1(\chi_E)$ $(E \in \mathscr{B})$. Prove that μ is a quasiregular measure. Let $U \in \mathscr{T}$. Then χ_U is a lower semicontinuous function and, by construction, $|\mu|(U) = |T|(\chi_U) = \bigvee\{|T|f : f \in \mathfrak{L}^+, f \leqslant \chi_U\}$. Given $\varepsilon > 0$ and $f \leqslant \chi_U$, $f \in \mathfrak{L}$, put $F = \{x \in \mathfrak{X} : f(x) \geqslant \varepsilon\}$. Then $f \leqslant \chi_F + \varepsilon 1_{\mathfrak{X}}$ and $|T|f \leqslant |\mu|(F) + \varepsilon|\mu|(U)$. This means that $|\mu|(U) = \bigvee\{|\mu|(F) : F \in \mathscr{F}, F \subseteq U\}$. Using (3.1), prove by analogy that $|\mu|(\mathfrak{X}) = \bigvee\{|\mu|(K) : K \in \mathscr{K}\}$. By Lemma 5.2.8 $|\mu|$ and μ are quasi-Radon measures.

Conversely, assume that T admits the integral representation (3.1) with $\mu \in F - \mathrm{bca}(\mathscr{B}, Y)$ a quasi-Radon measure. If $K \in \mathscr{K}$ then $|\mu|(\mathfrak{X}\backslash K) = \bigvee\{|\mu|(K') : K' \in \mathscr{K}, K' \subseteq U\backslash K\}$. Since for all $K' \subseteq U\backslash K$ there is some $f \in \mathfrak{L}$ satisfying $0 \leqslant f \leqslant 1_{\mathfrak{X}}$ and $f[K'] = \{0\}, f[K] = \{1\}$; therefore, $|\mu|(K) = \bigvee\{|T|f : f \in \mathfrak{L}, f \geqslant \chi_K\}$. Consequently, we arrive at (3.2). $\triangleright$

To illustrate application of this theorem, we first consider the problem of extending a quasi-Radon measure which is a version of the well-known Prokhorov Theorem.

5.3.2. DEFINITION. An algebra $\mathscr{A}_0 \subseteq 2^{\mathfrak{X}}$ is called *dense* provided that the following are satisfied:

(1) For all $V \in \mathscr{A}_0 \cap \mathscr{T}_0$ there is a function $\varphi \in \overline{S(\mathscr{A}_0)} \cap C_b(\mathfrak{X})$ such that $V = \{x \in \mathfrak{X} : \varphi(x) > 0\}$;

(2) The vector lattice $\mathfrak{L} = \overline{S(\mathscr{A}_0)} \cap C_b(\mathfrak{X})$ generates $\mathscr{T}$.

5.3.3. *Theorem.* *Let $\mu_0 \in F - \mathrm{ba}(\mathscr{A}_0, Y)$ be a quasi-Radon measure on a dense algebra $\mathscr{A}_0 = \mathscr{A}(\mathscr{F}_0 \cap \mathscr{A}_0)$. Then there is a unique quasi-Radon measure $\mu \in F - \mathrm{bca}(\mathscr{B}, Y)$ extending μ_0.*

$\triangleleft$ Consider a dominated operator $T : \mathfrak{L} \to Y$ defined on the vector lattice $\mathfrak{L} = \overline{S(\mathscr{A}_0)} \cap C_b(\mathfrak{X})$ by the equalities $Tf = \int f d\mu_0$ $(f \in \mathfrak{L})$. Since $\mathfrak{L}$ separates compact subsets of $\mathfrak{X}$, the dominant $|T|$ enjoys the condition (3.2). Therefore, there is a unique quasi-Radon measure $\mu \in F - \mathrm{bca}(\mathscr{B}, Y)$ such that the integral representation (3.1) holds. Show that μ extends μ_0. To this end, it suffices to check that if $U \in \mathscr{T}_0 \backslash \mathscr{A}_0$ then $\mu_0(U) = \mu(U)$. Since $\mathscr{A}_0$ is a dense algebra, there is a function $\varphi \in \mathfrak{L}$ satisfying $U = \{x \in \mathfrak{X} :\ \varphi(x) > 0\}$. Put $\varphi_n = (n\varphi) \wedge 1_{\mathfrak{X}}$ $(n \in \mathbb{N})$. Then $\varphi_n \nearrow \chi_U$ and the σ-additivity of μ and μ_0 implies $\mu_0(U) = bo\text{-}\lim \int \varphi_n d\mu_0 = bo\text{-}\lim \int \varphi_n d\mu = \mu(U)$. $\triangleright$

The extension theorems for σ-additive measures ranging in an ordered vector space, available in the literature, presume that the target obeys the weak (σ, ∞)-distributive law (see [14–16]). Theorem 5.3.1 implies that the same result is valid for some classes of noncompact spaces.

5.3.4. *Corollary.* *Let $\mathfrak{X}$ be a locally compact σ-compact topological space. Assume given $\mathscr{A}_0 = \mathscr{A}(\mathscr{F}_0 \cap \mathscr{A}_0)$. Then each measure $\mu_0 \in F - \mathrm{bca}(\mathscr{A}_0, Y)$ extends uniquely to a quasi-Radon measure $\mu \in F - \mathrm{bca}(\mathscr{B}, Y)$.*

Observe that it is essential to require in Corollary 5.3.4 that $\mathfrak{X}$ is locally compact. Moreover, we have

5.3.5. *Lemma.* *Assume that a σ-complete Boolean algebra $\mathbf{B}$ does not obey the weak σ-distributive law. There is an F_σ-subset $\mathfrak{X}$ in the Cantor discontinuum $\{0,1\}^\omega$ together with a σ-homomorphism $\mu_0 : \mathscr{U}(\mathfrak{X}) \to \mathbf{B}$ on the algebra $\mathscr{U}(\mathfrak{X})$ of clopen subsets of $\mathfrak{X}$, which does not extend to a σ-homomorphism from the Borel σ-algebra of $\mathfrak{X}$ to $\mathbf{B}$.*

$\triangleleft$ Since $\mathbf{B}$ fails to obey the weak σ-distributive law, there are nonzero elements $e \in \mathbf{B}$, $e_{i,j} \in \mathbf{B}$ $(i, j \in \mathbb{N})$ such that $e_{i,j} \searrow 0$ as $j \to \infty$ $(i \in \mathbb{N})$ and for all $\varphi : \mathbb{N} \to \mathbb{N}$ we have $\bigvee_{i=1}^{\infty} e_{i,\varphi(i)} = e$. Using the Stone Theorem, realize $\mathbf{B}$ as the algebra $\mathrm{Clop}(Q)$ of clopen subsets of a quasiextremal compact space Q. Let h be the Stone transform of $\mathbf{B}$ to Clop(Q). Let

$$E = \left(\bigcup_{i=1}^{\infty} \bigcap_{j=1}^{\infty} h(e_{i,j}) \right) \cup (Q \backslash h(e)).$$

Consider a countable algebra $\mathscr{E}$ in E generated by the sets $Q\backslash h(e)$, $h(e_{i,j}) \cap E$ $(i, j \in \mathbb{N})$. Take $\mathscr{E}$ as a base for a new topology on E. The resultant topological space $(E, \mathscr{T})$ is regular. Assume that $\widehat{E}$ is the factor space of E by the partition into the closures of singletons. The space $\widehat{E}$ with the factor topology $\widehat{\mathscr{T}}$ is a separated, regular, σ-compact totally disconnected space with a countable base. By the Alexandroff Theorem it is homeomorphic with an F_σ-subset $\mathfrak{X}$ in the Cantor discontinuum $\{0, 1\}^\omega$. Let α be a homeomorphism from $\widehat{E}$ onto $\mathfrak{X}$, and let p stand for the canonical projection of E onto $\widehat{E}$. Consider an arbitrary clopen subset U of $\mathfrak{X}$. Then $(\alpha \circ p)^{-1}(U)$ is a clopen subset of E. There is an open set $V \subseteq Q$ such that $V \cap E = (\alpha \circ p)^{-1}(U)$. If $x \in U$ then $(\alpha \circ p)^{-1}(x)$ is the intersection of a countable sequence of the elements $h(e)$, $h(e_{i,j})$ and their complements. Consequently, there is a clopen V_x such that $(\alpha \circ p)^{-1}(x) \subseteq V_x \subseteq V$. Since each V_x is a finite intersection of members of the countable family $Q\backslash h(e)$, $h(e_{i,j})$, $h(e)\backslash h(e_{i,j})$, there is an open F_σ-set $V_1 \subseteq V$ satisfying $V \cap E = V_1 \cap E$. By quasiextremality, the closure $W = \mathrm{cl}(V_1)$ is clopen in Q, meeting the equality $W \cap E = (\alpha \circ p)^{-1}(U)$. We now put $\mu_0(U) = h^{-1}(W)$. The resultant σ-homomorphism $\mu_0 : \ \mathscr{U}(\mathfrak{X}) \to \mathbf{B}$ of the algebra $\mathscr{U}(\mathfrak{X})$ of clopen subsets of $\mathfrak{X}$ to $\mathbf{B}$ does not extend to a σ-homomorphism defined on the Borel σ-algebra $\mathscr{B}$ of the space $\mathfrak{X}$. Suppose to a contradiction that such an extension $\mu : \ \mathscr{B} \to \mathbf{B}$ exists. Then, for $E_{i,j} = h(e_{i,j}) \cap E$ we would have

$$e = \mu\left(\bigcup_{i=1}^{\infty} \bigcap_{j=1}^{\infty} \alpha \circ p(E_{i,j}) \right) = \bigvee_{i=1}^{\infty} \bigwedge_{j=1}^{\infty} e_{i,j} = 0.$$

This contradiction proves the lemma. ▷

By analogy, we demonstrate

5.3.6. Lemma. *Let a complete Boolean algebra* $\mathbf{B}$ *fail to obey the weak* (σ, ∞)*-distributive law. Then there is an* F_σ*-subset of* $\mathfrak{X}$ *in the generalized Cantor discontinuum together with a* σ*-homomorphism* $\mu_0 : \ \mathscr{U}(\mathfrak{X}) \to \mathbf{B}$ *failing to extend to a quasi-Radon* σ*-homomorphism of the Borel* σ*-algebra of the space* $\mathfrak{X}$ *to* $\mathbf{B}$.

We cannot waive the requirement that $\mathfrak{X}$ is σ-compact in Corollary 5.3.4 either. This is easy to see taking as $\mathfrak{X}$ an uncountable discrete space.

We now turn to the representation theorem for a dominated operator on a vector lattice without unity.

We call a linear operator $\Phi : C_0(\mathfrak{X}) \to Y$ *dominated* provided that there is a positive operator $\Psi : C_0(\mathfrak{X}) \to F$ satisfying

$$|\Phi(f)| \leqslant \Psi(|f|) \quad (f \in C_0(\mathfrak{X})). \tag{3.4}$$

In this case there is a least positive operator Ψ meeting (3.4) which is denoted by $|\Phi|$.

5.3.7. Theorem. *Let a dominated operator $\Phi : C_0(\mathfrak{X}) \to Y$ satisfy the inequality*

$$|\Phi|(f) \leqslant b\|f\|_\infty \quad (f \in C_0(\mathfrak{X})_+)$$

for some $b \in F_+$. Then there is a unique measure μ in $\mathrm{qca}(\mathfrak{X}, Y)$ *such that*

$$\Phi(f) = \int_{\mathfrak{X}} f(\chi)\mu(d\chi) \quad (f \in C_0(\mathfrak{X})).$$

$\lhd$ Let Ψ stand for the restriction of $|\Phi|$ to $C_{00}(\mathfrak{X})$. By the Wright Theorem (cf. [17, Theorem 1]) there is a unique quasiregular measure $\nu : \mathscr{B}(\mathfrak{X}) \to F_+ \cup \{\infty\}$ order bounded on the compact subsets of $\mathfrak{X}$ and satisfying

$$\Psi(f) = \int_{\mathfrak{X}} f(\chi)\nu(d\chi) \quad (f \in C_{00}(\mathfrak{X})).$$

This representing measure is order bounded in the case we study since the inequality

$$\Psi(f) \leqslant b\|f\|_\infty \quad (f \in C_{00}(\mathfrak{X})_+)$$

implies the estimate

$$\nu(\mathfrak{X}) = \sup\{\Psi(f) : f \in C_{00}(\mathfrak{X}),\ 0 \leqslant f \leqslant 1_{\mathfrak{X}}\} \leqslant b.$$

Therefore $\nu \in \mathrm{qca}(\mathfrak{X}, F)_+$. Assume that $(f_\alpha)_{\alpha\in\mathrm{A}}$ is a net in $C_{00}(\mathfrak{X})$ increasing pointwise to $1_{\mathfrak{X}}$. From the estimate

$$|\Phi(f_\alpha) - \Phi(f_\beta)| \leqslant \Psi(|f_\alpha - f_\beta|) = \int_{\mathfrak{X}} |f_\alpha(\chi) - f_\beta(\chi)|\nu(d\chi)$$

it follows that $\{\Phi(f_\alpha) : \alpha \in \mathrm{A}\}$ is a *bo*-Cauchy net. If $f = f_0 + a1_{\mathfrak{X}}$ where $f \in C_{00}(\mathfrak{X})$, $a \in \mathbb{R}$, then we put

$$\bar{\Phi}(f) = bo\text{-}\lim_{\alpha\in\mathrm{A}} \Phi(f_0 + af_\alpha).$$

This definition is clearly sound. We exclude from consideration the case in which $\mathfrak{X}$ is itself a compact space. Therefore, we have defined an extension of Φ to the linear operator $\ddot{\Phi}$ on the function vector lattice $\mathscr{L} = C_{00}(\mathfrak{X}) \oplus \mathbb{R} \cdot 1_{\mathfrak{X}}$ with order-unity. Check that $\ddot{\Phi}$ is a dominated operator. Let $f \in \mathscr{L}_+$ and $f = f_0 + a1_{\mathfrak{X}}$ $(f_0 \in C_{00}(\mathfrak{X}),$ $a \in \mathbb{R})$. We then have $a \geqslant 0$ and $f_0^- := (-f_0) \bigvee 0 \leqslant a1_{\mathfrak{X}}$. For $a > 0$ the net $g_\alpha = f_\alpha \vee \left(f_0^-/a\right)$ increases to $1_{\mathfrak{X}}$ too. Using the dominant inequality (3.4), we obtain

$$|\ddot{\Phi}(f)| = o\text{-}\lim_{\alpha \in \mathrm{A}} |\Phi(f_0) - a\Phi(g_\alpha)| \leqslant \Psi\left(f_0^+\right) + o\text{-}\lim_{\alpha \in \mathrm{A}} \Psi\left(ag_\alpha - f_0^-\right) = \ddot{\Psi}(f).$$

Consequently, $\ddot{\Phi}$ is a dominated operator on a vector lattice meeting all hypotheses of Theorem 5.3.1. By this theorem, there is a unique measure $\mu \in \operatorname{qca}(\mathfrak{X}, Y)$ such that

$$\ddot{\Phi}(f) = \int_{\mathfrak{X}} f(\chi)\mu(d\chi) \quad (f \in \mathscr{L}).$$

Since the operator $\ddot{\Phi}$ extends Φ; therefore,

$$\Phi(f) = \int_{\mathfrak{X}} f(\chi)\mu(d\chi) \quad (f \in C_{00}(\mathfrak{X})).$$

Since μ is quasi-Radon, the same representation holds for all f in $C_0(\mathfrak{X})$. $\rhd$

5.4. The Fubini Theorem

Consider *bo*-complete lattice normed spaces Y, Y', and Z whose norm lattices are some Kantorovich spaces F, F', and G. Assume given a bilinear mapping $\times : Y \times Y' \to Z$ that is dominated by a positive o-continuous bilinear mapping $\circ : F \times F' \to G$; i.e.,

$$|y \times y'| \leqslant |y| \circ |y'| \quad (y \in Y,\, y' \in Y').$$

Furthermore, let $\mathfrak{X}$ and $\mathfrak{X}'$ be two Čech-complete topological spaces. Consider two Borel measures $\mu : \mathscr{B}_{\mathfrak{X}} \to Y$ and $\mu' : \mathscr{B}_{\mathfrak{X}'} \to Y'$. Put $\mathfrak{C} := \mathscr{B}_{\mathfrak{X} \times \mathfrak{X}'}$. Denote by $\mathscr{A} \square \mathscr{A}'$ the algebra of subsets which is generated by the "rectangles" $A \times A'$, with $A \in \mathscr{A}$ and $A' \in \mathscr{A}'$. The problem is to construct a Borel measure $\mu \otimes \mu' : \mathfrak{C} \to Z$ such that

$$(\mu \otimes \mu')(B \times B') = \mu(B) \times \mu'(B') \quad (B \in \mathscr{B}_{\mathfrak{X}},\, B' \in \mathscr{B}_{\mathfrak{X}'}).$$

We call $\mu \otimes \mu'$ the *product-measure* or simply *product* of μ and μ'.

A few examples follow.

EXAMPLES. **(1)** Let $Y := F$ and $Y' := F'$ where F and F' are order-dense ideals of the same universally complete Kantorovich space $mF = mF'$, with a fixed order-unity $\mathbf{1}$. Then mF admits a unique multiplication making it into an ordered ring with unity $\mathbf{1}$. Take another order-dense ideal $G \subset mF$ satisfying $F \cdot F' \subset G$. In this case to each pair of elements $x \in F$ and $x' \in F'$ there corresponds their product $x \cdot x' \in G$. The o-continuity of this multiplication is well-known together with the equality $|x \cdot x'| = |x| \cdot |x'|$ $(x \in F, x' \in F')$, which enables us to speak of the product of Borel measures $\mu : \mathscr{B}_{\mathfrak{X}} \to F$ and $\mu' : \mathscr{B}_{\mathfrak{X}'} \to F'$. If $F = F' = G$ then F is an ordered ring, and each of the measures μ, μ', $\mu \otimes \mu'$ ranges in the same Kantorovich space F.

(2) Assume that $Y = F = \mathrm{Orth}(F')$ and the vector norm of Y' is decomposable. It is well known that in this case Y' admits the structure of an F-module; i.e., there is a bilinear mapping from $F \times Y'$ to Y' such that $|a \cdot y| = |a| \cdot |y|$ $(a \in F$, $y \in Y')$ (see [9]). Given two Borel measures $\mu : \mathscr{B}_{\mathfrak{X}} \to F$ and $\mu' : \mathscr{B}_{\mathfrak{X}'} \to Y'$, we may speak of their product $\mu \otimes \mu' : \mathfrak{C} \to Y'$. In case $F = \mathbb{R}$ we arrive at the conventional multiplication of a vector measure by a scalar measure.

(3) Assume that F and F' are order-dense ideals of the same universally complete Kantorovich space $mF = mF'$ in which an order-unity is fixed, determining multiplication in mF. Assume further that $G := \mathrm{Orth}(F) \cap \mathrm{Orth}(F')$ and the product $a \cdot a' \in mF$ is defined for all $a \in F$ and $a \in F'$. We also suppose that the vector norms of the spaces Y and Y' are decomposable. By [9] we then may define the structure of a G-module on Y and Y'. Let $Y \otimes_G Y'$ stand for the algebraic tensor product of the G-modules Y and Y'; cf. [18]. Consider the lattice seminorm on $Y \otimes_G Y'$ ranging in mF and defined by the formula

$$|z| := \inf \left\{ \sum_{k=1}^{n} |y_k| \cdot |y'| \right\},$$

where inf in the Kantorovich space mF is taken over all representations of z of the form $\sum_{k=1}^{n} y_k \otimes y'_k$, $y_k \in Y$, $y'_k \in Y'$ $(k := 1, \dots, n)$. Since G is a commutative ring, the spaces Y and Y' carry some bimodule structure. Consequently, their tensor product $Y \otimes_G Y'$ is a G-module too (see [18, Section 10.2.2]). Therefore, we have the equality

$$|g \cdot z| = |g||z| \quad (g \in G,\ z \in Y \otimes_G Y').$$

In particular, this vector norm is decomposable.

Distinguish the subspace $Z := \{z \in Y \otimes_G Y' : |z| = 0\}$ of $Y \otimes_G Y'$. In line with [9], we may construct a bo-completion of the factor space $(Y \otimes_G Y')/Z$ with respect to the vector norm.

Such a completion is naturally called the *projective tensor product* of Y and Y'. We denote it by $Y \widehat{\otimes}_G Y'$. We also let $y \otimes y'$ stand for the tensor product of two elements $y \in Y$ and $y' \in Y'$. Obviously, $|y \otimes y'| = |y| \cdot |y'|$ $(y \in Y,\ y' \in Y')$ (we preserve the same symbol $|\cdot|$ for the factor norm on $Y \widehat{\otimes}_G Y'$). Also, it is clear that the bilinear mapping $\otimes : Y \times Y' \to Y \widehat{\otimes} Y'$ is *bo*-continuous. Therefore, we may introduce into consideration the tensor product $\mu \otimes \mu' : \mathfrak{C} \to Y \widehat{\otimes}_G Y'$ of two Borel measures $\mu : \mathscr{B}_{\mathfrak{X}} \to Y$ and $\mu' : \mathscr{B}_{\mathfrak{X}'} \to Y'$. By analogy we may define the concept of inductive tensor product. There is another way for constructing the tensor product of lattice normed spaces using the technique of Boolean valued analysis; see [9].

We say that the multiplications $\times : Y \times Y' \to Z$ and $\circ : F \times F' \to G$ are tied up with *cross-equality* whenever

$$|y \times y'| = |y| \circ |y'| \quad (y \in Y,\ y' \in Y').$$

Examples **(1)**, **(2)**, and **(3)** below present such multiplications.

5.4.1. Lemma. *If two Borel measures $\mu : \mathscr{B}_{\mathfrak{X}} \to Y$ and $\mu' : \mathscr{B}_{\mathfrak{X}'} \to Y'$ give rise to a product $\mu \otimes \mu' : \mathfrak{C} \to Z$ that is a quasi-Radon measure, then this product is unique.*

◁ Take an open set $U \subset \mathfrak{X} \times \mathfrak{X}'$ and a compact set $K \subset U$. Then there are finite collections of open sets $U_k \subset \mathfrak{X}, U'_k \subset \mathfrak{X}'$ $(k := 1, \dots, n)$ satisfying $K \subset \bigcup_{k=1}^{n} (U_k \times U'_k) \subset U$. Since $\mu \otimes \mu'$ is a quasi-Radon measure, we have

$$|\mu \otimes \mu'|(U) = \sup\{|\mu \otimes \mu'|(K) : K \subset U, K \in \mathscr{K}_{\mathfrak{X} \times \mathfrak{X}'}\}$$

$$= \sup\{|\mu \otimes \mu'|(V) : V \subset U, V \in \mathscr{B}_{\mathfrak{X}} \square \mathscr{B}_{\mathfrak{X}'}\}.$$

Therefore,

$$\mu \otimes \mu'(U) = bo\text{-}\lim\{\mu \otimes \mu'(V) : V \subset U, V \in \mathscr{B}_{\mathfrak{X}} \square \mathscr{B}_{\mathfrak{X}'}\}.$$

However, the measure $\mu \otimes \mu'$ on the algebra $\mathscr{B}_{\mathfrak{X}} \square \mathscr{B}_{\mathfrak{X}'}$ is uniquely determined from the values $\mu \otimes \mu'(B \times B')$ $(B \in \mathscr{B}_{\mathfrak{X}},\ B' \in \mathscr{B}_{\mathfrak{X}'})$. Consequently, the same uniqueness holds for $\mu \times \mu'(U)$. By σ-additivity we immediately deduce uniqueness for all values $\mu \otimes \mu'(C)$ $(C \in \mathfrak{C})$. ▷

5.4.2. Theorem. *Let $\mathfrak{X}$ and $\mathfrak{X}'$ be Čech-complete topological spaces, with $\mu : \mathscr{B}_{\mathfrak{X}} \to Y$ and $\mu' : \mathscr{B}_{\mathfrak{X}'} \to Y'$ quasi-Radon measures. Then there is a product $\mu \otimes \mu' : \mathfrak{C} \to Z$ that is a quasi-Radon measure. The corresponding vector variations maintain the equality $|\mu \otimes \mu'| \leqslant |\mu| \otimes |\mu'|$. If the multiplications $\times$ and $\circ$ are tied up with cross-equality then $|\mu \otimes \mu'| = |\mu| \otimes |\mu'|$.*

◁ Denote by $\beta\mathfrak{X}$ and $\beta\mathfrak{X}'$ the Čech–Stone compactifications of $\mathfrak{X}$ and $\mathfrak{X}'$. Consider the measures $\overline{\mu} : \mathscr{B}_{\beta\mathfrak{X}} \to Y$ and $\overline{\mu}' : \mathscr{B}_{\beta\mathfrak{X}'} \to Y'$ determined by the equalities $\overline{\mu}(B) = \mu(B \cap \mathfrak{X})$, $\overline{\mu}'(B') = \mu'(B' \cap \mathfrak{X}')$ $(B \in \mathscr{B}_{\beta\mathfrak{X}},\ B' \in \mathscr{B}_{\beta\mathfrak{X}'})$. There is a unique measure $\lambda : \mathscr{A}_0 \Box \mathscr{A}_0' \to Z$ satisfying $\lambda(A \times A') = \overline{\mu}(A) \times \overline{\mu}'(A')$ $(A \in \mathscr{A}_0,\ A' \in \mathscr{A}_0')$, where $\mathscr{A}_0$ and $\mathscr{A}_0'$ are algebras of subsets generated by the functionally open subsets of $\mathfrak{X}$ and $\mathfrak{X}'$ respectively. It is easy that the measure λ has bounded vector variation, and $|\lambda| \leqslant |\mu| \otimes |\mu'|$. Show that λ meets the denseness conditions of Definition 5.3.2. Condition (1) is obvious. We may check condition (2) for an arbitrary set $A \times A'$, where $A \in \mathscr{A}_0'$ and $A' \in \mathscr{A}_0$. Let $(A_k)_{k\in\mathbb{N}} \subset \mathscr{A}_0'$ and $(A_k')_{k\in\mathbb{N}} \in \mathscr{A}_0$ be sequences such that $A = \bigcup_{k=1}^{\infty} A_k$, $A' = \bigcup_{k=1}^{\infty} A_k'$, and $\mathrm{cl}(A_k) \subset A, \mathrm{cl}(A_k') \subset A'$ $(k \in \mathbb{N})$. Then $\mathrm{cl}(A_k \times A_k') \subset A \times A'$ and

$$\inf_k \{|\lambda|(A \times A' \backslash A_k \times A_k')\} \leqslant \inf\{|\overline{\mu}|(A \backslash A_k) \circ |\overline{\mu}'|(A' + |\overline{\mu}|(A) \circ |\overline{\mu}'|(A' \backslash A_k'))\} = 0.$$

Theorem 5.3.3 implies existence of some quasiregular measure $\overline{\mu} \otimes \overline{\mu}' : \mathscr{B}_{\beta\mathfrak{X} \times \beta\mathfrak{X}'} \to Z$ extending λ. By definition $\overline{\mu}(A) \times \overline{\mu}'(A') = \overline{\mu} \otimes \overline{\mu}'(A \times A')$ for all $A \in \mathscr{A}_0$ and $A' \in \mathscr{A}_0'$. Since the measures $\overline{\mu}$, $\overline{\mu}'$, and $\overline{\mu} \otimes \overline{\mu}'$ are quasiregular and the multiplication $\times$ is o-continuous, the same equality holds for all $A \in \mathscr{T}$ and $A' \in \mathscr{T}'$. From σ-additivity and the Monotone Class Lemma it follows that this equality holds also for all $A \in \mathscr{B}_{\beta\mathfrak{X}}, A' \in \mathscr{B}_{\beta\mathfrak{X}'}$. This implies that the measure $\overline{\mu} \otimes \overline{\mu}'$ is in fact the product of $\overline{\mu}$ and $\overline{\mu}'$. Since the values of $\overline{\mu} \otimes \overline{\mu}'$ at all Borel subsets of $(\beta\mathfrak{X} \times (\beta\mathfrak{X}' \backslash \mathfrak{X}')) \cup ((\beta\mathfrak{X} \backslash \mathfrak{X}) \times \beta\mathfrak{X}')$ equal zero; considering the restriction of $\overline{\mu} \otimes \overline{\mu}'$ to the Borel subsets of the space $\mathfrak{X} \times \mathfrak{X}' \subset \beta\mathfrak{X} \times \beta\mathfrak{X}'$, arrive at the sought product $\mu \times \mu'$ of μ and μ', which is a quasiregular Borel measure. The vector variations $|\mu|$ and $|\mu'|$ satisfy the hypotheses of the same theorem, and so their product exist: $|\mu| \otimes |\mu'| : \mathfrak{C} \to G$. Moreover, it is easy that $|\mu \otimes \mu'| \leqslant |\mu| \otimes |\mu'|$. The quasi-Radon property of the measure $\mu \otimes \mu'$ ensues from this inequality, the quasiregularity of $\mu \otimes \mu'$, and the relations

$$\inf\{|\mu \otimes \mu'|(\mathfrak{X} \times \mathfrak{X}' \backslash K \times K') : K \in \mathscr{K}_{\mathfrak{X}}, K' \in \mathscr{K}_{\mathfrak{X}'}\}$$

$$\leqslant \inf\{|\mu|(\mathfrak{X}) \circ |\mu'|(\mathfrak{X}' \backslash K') + |\mu|(\mathfrak{X} \backslash K) \circ |\mu'|(\mathfrak{X}')\} = 0.$$

Suppose that the multiplications $\times$ and $\circ$ are tied up with cross-equality. Let $(C_k)_{k=1}^n \subset \mathscr{B}_{\mathfrak{X}}$ and $(C_l')_{l=1}^m \subset \mathscr{B}_{\mathfrak{X}'}$ be arbitrary partitions of the sets $C \in \mathscr{B}_{\mathfrak{X}}$ and $C' \in \mathscr{B}_{\mathfrak{X}'}$. Then $(C_k \times C_l')_{k=1,l=1}^{n\ \ m}$ is a partition of $C \times C'$ and so

$$\sum_{k=1}^{n} |\mu(C_k)| \circ \sum_{l=1}^{m} |\mu'(C_l')| = \sum_{k=1}^{n} \sum_{l=1}^{m} |\mu \otimes \mu'(C_k \times C_l')| \leqslant |\mu \otimes \mu'|(C \times C').$$

This implies the inequality

$$|\mu| \otimes |\mu'|(C \times C') \leqslant |\mu \otimes \mu'|(C \times C'),$$

which may be extended by additivity to arbitrary finite unions of the sets like $C \times C'$, ($C \in \mathscr{B}_{\mathfrak{X}}$, and $C' \in \mathscr{B}_{\mathfrak{X}'}$); i.e., this inequality is valid for the sets in $\mathscr{B}_{\mathfrak{X}} \square \mathscr{B}_{\mathfrak{X}'}$. Since $|\mu \otimes \mu'|$ is quasi-Radon and σ-additive, we may extend this inequality to all sets in $\mathfrak{C}$. Therefore, $|\mu| \otimes |\mu'| \leqslant |\mu \otimes \mu'|$. As was already mentioned, the reverse inequality is always true. Consequently, $|\mu \otimes \mu'| = |\mu| \otimes |\mu'|$. ▷

5.4.3. REMARK. In fact, Theorem 5.4.2 holds if we assume that $\mathfrak{X}$ and $\mathfrak{X}'$ are arbitrary completely regular spaces presenting Borel subsets of some of their compactifications.

5.4.4. REMARK. The product of Borel measures on locally compact spaces ranging in a monotonically complete ordered vector space was constructed in [19]. Theorem 5.4.2 contains this result for Kantorovich spaces. On the other hand, if we put $F = F' = \mathbb{R}$ in Example 3 then Theorem 5.4.2 provides tensor product for Banach-space-valued measures (cf. [20–22]).

We now pass to the problems relevant to the Fubini Theorem. To this end, it is necessary to define a new integral of a vector function with respect to a vector measure.

Let $\mathfrak{X}$ and $\mathfrak{X}'$ be Čech-complete topological spaces. Denote by $\mathscr{M}(\mathfrak{X}', Y)$ the space of functions $f : \mathfrak{X}' \to Y$ representable as $f = y_1 g_1 + \dots + y_n g_n$ where $y_k \in Y$ and $g_k : \mathfrak{X}' \to \mathbb{R}$ are bounded Borel-measurable functions. Assume also that $\mu : \mathscr{B}_{\mathfrak{X}} \to Y$ and $\mu' : \mathscr{B}_{\mathfrak{X}'} \to Y'$ are quasi-Radon measures. Given $f \in \mathscr{M}(\mathfrak{X}', Y)$ admitting the above presentation, we define

$$\int f d\mu = \sum_{k=1}^{n} y_i \times \int g_k d\mu'.$$

The routine arguments show that this definition is sound. Furthermore, we easily deduce the estimate

$$\left| \int f d\mu \right| \leqslant |f|_\infty \circ |\mu'|(\mathfrak{X}'),$$

with $|f|_\infty := \sup\{|f(t)| : t \in \mathfrak{X}'\}$. Let $\overline{\mathscr{M}}(\mathfrak{X}', Y)$ stand for the r-closure of the space $\mathscr{M}(\mathfrak{X}', Y)$ with respect to the norm $|\cdot|_\infty$; i.e., $f \in \overline{\mathscr{M}}(\mathfrak{X}', Y)$ if and only if there are a sequence of functions $(f_n)_{n \in \mathbb{N}} \subset \mathscr{M}(\mathfrak{X}', Y)$ and a regulator $b \in F^+$ satisfying $|f - f_n|_\infty \leqslant n^{-1} b$. By continuity with respect to this norm, we may define the integral of each function $f \in \overline{\mathscr{M}}(\mathfrak{X}', Y)$ with respect to μ', while preserving the above normative inequality.

Denote by $\overline{\mathscr{M}}(\mathfrak{X} \times \mathfrak{X}')$ the space of real functions on $\mathfrak{X} \times \mathfrak{X}'$ which are the uniform limits of the functions h on $\mathfrak{X} \times \mathfrak{X}'$ representable as $h = \sum_{i=1}^n g_i g_i'$, with $g_i : \mathfrak{X} \to \mathbb{R}$ and $g_i' : \mathfrak{X}' \to \mathbb{R}$ bounded Borel-measurable functions. For all $t' \in \mathfrak{X}'$ the function $h(\cdot, t')$ is Borel-measurable, and so we have the integral

$$\int h(\cdot, t') d\mu = f(t') = \sum_{i=1}^n y_i g_i'(t'),$$

where $y_i = \int g_i d\mu$ $(i = 1, \dots, n)$. Consequently, $\int h d\mu \in \mathscr{M}(\mathfrak{X}', Y)$ and

$$\int \left(\int h d\mu \right) \times d\mu' = \sum_{i=1}^n \left(\int g_i d\mu \right) \times \left(\int g_i' d\mu' \right) = \int h d(\mu \otimes \mu').$$

Passing to the relevant r-limits in these equalities, we arrive at the following Fubini Theorem.

5.4.5. *Theorem.* *Let $h \in \overline{\mathscr{M}}(\mathfrak{X} \times \mathfrak{X}')$. Then*

$$\int h d\mu \in \overline{\mathscr{M}}(\mathfrak{X}, Y), \quad \int h d\mu' \in \overline{\mathscr{M}}(\mathfrak{X}, Y')$$

and

$$\int \left(\int h d\mu \right) \times d\mu' = \int h d(\mu \otimes \mu') = \int d\mu \times \left(\int h d\mu' \right).$$

The function class $\overline{\mathscr{M}}(\mathfrak{X} \times \mathfrak{X}')$ is not rather wide; however, in case $\mathfrak{X}$ and $\mathfrak{X}'$ are compact spaces, the inclusion holds $C(\mathfrak{X} \times \mathfrak{X}') \subset \mathscr{M}(\mathfrak{X} \times \mathfrak{X}')$. It turns out in fact that the Fubini Theorem fails in general for arbitrary bounded measurable functions on $\mathfrak{X} \times \mathfrak{X}'$.

EXAMPLE. Assume that $\mathfrak{X} = \mathfrak{X}' = [0,1]$, and let $Y = F = Y' = F' = M[0,1]$ stand for the space of cosets of bounded Borel functions on $[0,1]$ with respect to Lebesgue measure. Given a Borel set $A \subset [0,1]$, denote the coset of the characteristic function χ_A by $\mu(A) = [\chi_A]$. As multiplication in $M[0,1]$ we take the usual multiplication of functions. Obviously,

$$\mu \otimes \mu(A \times B) = [\chi_{A \cap B}] \quad (A, B \in \mathscr{B}_{[0,1]}).$$

Let Δ be the diagonal of the square $[0,1] \times [0,1]$; i.e., $\Delta := \{(t,t) : t \in [0,1]\}$. Put $P(t,u) = t$. We may now write down the measure $\mu \otimes \mu$ explicitly. Namely,

$$(\mu \otimes \mu)(C) = \mu(P(\Delta \cap C)) = [\chi_{P(\Delta \cap C)}]$$

for all Borel sets $C \subset [0,1] \times [0,1]$. Then

$$(\mu \otimes \mu)(\Delta) = \mathbf{1} = \int \chi_\Delta d(\mu \otimes \mu).$$

However, we have

$$\int \chi_\Delta(\cdot, t')d\mu = \mu(\{t'\}) = 0$$

for each fixed $t' \in [0,1]$. Therefore, with whatever sound definition of the integral of a function $f : [0,1] \to M[0,1]$ we would have

$$\int \Big(\int \chi_\Delta d\mu \Big) d\mu = 0 \neq \int \chi_\Delta d(\mu \otimes \mu).$$

This example demonstrated the Fubini Theorem may fail in principle for the function χ_Δ.

The integral of a vector function $\varphi : G \to Y$ with respect to a Radon measure $\lambda : \mathscr{B}(G) \to \mathbb{R}$ is constructed by analogy with the Bochner integral. Given simple functions of the form

$$s = \sum_{j=1}^{n} y_j 1_{B_j} \quad (y_1, \dots, y_n \in Y,\ B_1, \dots, B_n \in \mathscr{B}(G)),$$

we as usual let

$$\int_G s(g)\lambda(dg) = \sum_{j=1}^{n} y_j \lambda(B_j).$$

We further define the space of integrable functions $L^\infty(G, Y)$ by letting $\varphi \in L^\infty(G, Y)$ if and only if there is a sequence $(s_\alpha)_{\alpha \in \mathrm{A}}$ of simple functions such that

$$\sup\{|s_\alpha(g)| : \alpha \in \mathrm{A},\ g \in G\} \leqslant b \in F^+,$$

$$\inf_{\alpha \in \mathrm{A}} \sup\{|\varphi(g) - s_\beta(g)| : \beta \geqslant \alpha,\ g \in K\} = 0 \quad (K \in \mathscr{K}(G)).$$

We then agree that

$$\int_G \varphi(g)\lambda(dg) = bo\text{-}\lim_{\alpha \in \mathrm{A}} \int_G s_\alpha(g)\lambda(dg).$$

A vector function $\varphi : G \to Y$ is *uniformly bo-continuous on a set* K provided that

$$\inf_{U \in \mathscr{U}_0} \sup\{|\varphi(g_1) - \varphi(g_2)| : g_1, g_2 \in K,\ g_1 - g_2 \in U\} = 0.$$

Clearly, $L^\infty(G, Y)$ contains order bounded functions uniformly *bo*-continuous on all compact sets $K \subseteq \mathfrak{X}$.

Suppose that $\lambda_1 : \mathscr{B}(G_1) \to \mathbb{C}$ and $\lambda_2 : \mathscr{B}(G_2) \to \mathbb{C}$ are two Radon measures defined on the Borel σ-algebras of locally compact groups G_1 and G_2. The product $\lambda = \lambda_1 \times \lambda_2 : \mathscr{B}(G_1 \times G_2) \to \mathbb{C}$ is a Radon measure too.

5.4.6. Theorem. *Let $\varphi : G_1 \times G_2 \to Y$ be an order bounded mapping uniformly bo-continuous on compact sets, and*

$$\varphi_1(g_1) = \int\limits_{G_2} \varphi(g_1, g_2)\, \lambda_2(dg_2), \quad \varphi_2(g_2) = \int\limits_{G_1} \varphi(g_1, g_2)\, \lambda_1(dg_1).$$

Then the mappings $\varphi_1 : G_1 \to Y$ and $\varphi_2 : G_2 \to Y$ are order bounded and uniformly bo-continuous on compact sets; moreover,

$$\int\limits_{G_1 \times G_2} \varphi(g)\lambda(dg) = \int\limits_{G_1} \varphi_1(g_1)\lambda_1(dg_1) = \int\limits_{G_2} \varphi_2(g_2)\lambda_2(dg_2).$$

$\triangleleft$ It is clear that φ_1 and φ_2 are order bounded. Take $K_1 \in \mathscr{K}(G_1)$ and prove that φ_1 is uniformly *bo*-continuous on K_1. Put

$$b = \sup\{|\varphi(g)| : g \in G_1 \times G_2\}.$$

Let $U = U_1 \times U_2$, where U_1 and U_2 are arbitrary zero neighborhoods in the groups G_1 and G_2. Given $K_2 \in \mathscr{K}(G_2)$, assign

$$b_U = \sup\{|\varphi(g) - \varphi(h)| : g, h \in K_1 \times K_2,\ g - h \in U\}.$$

The net b_U decreases to zero by hypothesis. We have the estimate

$$|\varphi_1(g_1) - \varphi_1(h_1)| \leqslant |\lambda_2|(G_2)b_U + |\lambda_2|(G_2 \setminus K_2)b,$$

implying that

$$\inf_{U_1} \sup\{|\varphi_1(g_1) - \varphi_1(h_1)| : g_1, h_1 \in K_1,\ g_1 - h_1 \in U_1\} \leqslant |\lambda_2|(G_2 \setminus K_2)b.$$

Since $K_2 \in \mathscr{K}(G_2)$ is arbitrary and λ_2 is a Radon measure, the above estimate ensures uniform *bo*-continuity for φ_1 on every compact set $K_1 \in \mathscr{K}(G_1)$. The same holds for φ_2.

Take arbitrary zero neighborhoods U_1 and U_2 in the groups G_1 and G_2 together with $K_1 \in \mathscr{K}(G_1)$ and $K_2 \in \mathscr{K}(G_2)$. Consider two partitions of unity

$$\{u_j\}_{j=1}^n \subset C_{00}(G_1)_+, \quad \{v_k\}_{k=1}^m \subset C_{00}(G_2)_+$$

(for K_1 and K_2, respectively) such that from $g_1, h_1 \in \operatorname{supp} u_j$ and $g_2, h_2 \in \operatorname{supp} v_k$ it follows

$$g_1 - h_1 \in U_1, \quad g_2 - h_2 \in U_2 \quad (j = 1, \dots, n,\ k = 1, \dots, m).$$

Choose some elements $g_{1,j} \in \operatorname{supp} u_j$ and $g_{2,k} \in \operatorname{supp} v_k$ ($j = 1, \ldots, n$, $k = 1, \ldots, m$). We have the estimates

$$\left| \int\limits_{K_1 \times K_2} \varphi(g)\lambda(dg) - \sum_{j,k} \int\limits_{K_1 \times K_2} \varphi(g_{1,j}, g_{2,k}) u_j(g_1) v_k(g_2) \lambda(dg) \right|$$

$$\leqslant |\lambda|(G_1 \times G_2) b_U, \quad \left| \int\limits_{K_1} \varphi_1(g_1)\lambda_1(dg_1) \right.$$

$$\left. - \sum_j \int\limits_{K_1 \times K_2} \varphi(g_{1,j}, g_{2,k}) u_j(g_1) v_k(g_2) \lambda_1(dg_1)\lambda_2(dg_2) \right|$$

$$\leqslant \sum_{j,k} \int\limits_{K_1} \Big(\int\limits_{K_2} |\varphi(g_1, g_2) - \varphi(g_{1,j}, g_{2,k})| v_k(g_2) |\lambda_2|(dg_2) \Big) u_j(g_1) |\lambda_1|(dg_1)$$

$$+ \int\limits_{K_1} \Big(\int\limits_{G_2 \setminus K_2} |\varphi(g_1, g_2)| |\lambda_2|(dg_2) \Big) |\lambda_1|(dg_1)$$

$$\leqslant |\lambda_1|(G_1)|\lambda_2|(G_2) b_U + |\lambda_1|(G_1)|\lambda_2|(G_2 \setminus K_2) b.$$

Here $g = (g_1, g_2)$ and $U = U_1 \times U_2$, while b and b_U are the same as above. Moreover,

$$\left| \int\limits_{G_1 \times G_2} \varphi(g)\lambda(dg) - \int\limits_{K_1 \times K_2} \varphi(g)\lambda(dg) \right| \leqslant |\lambda|(G_1 \times G_2 \setminus K_1 \times K_2) b,$$

$$\left| \int\limits_{G_1} \varphi_1(g_1)\lambda_1(dg_1) - \int\limits_{K_1} \varphi_1(g_1)\lambda_1(dg_1) \right| \leqslant |\lambda_1|(G_1 \setminus K_1)|\lambda_2|(G_2) b.$$

Consequently,

$$\left| \int\limits_{G_1 \times G_2} \varphi(g)\lambda(dg) - \int\limits_{G_1} \varphi_1(g_1)\lambda_1(dg_1) \right| \leqslant 2|\lambda|(G_1 \times G_2) b_U$$

$$+ 2(|\lambda_1|(G_1 \setminus K_1)|\lambda_2|(G_2) + |\lambda_1|(G_1)|\lambda_2|(G_2 \setminus K_2)) b.$$

Passing to the bo-limit along the decreasing net b_U and the increasing nets $K_1 \in \mathscr{K}(G_1)$ and $K_2 \in \mathscr{K}(G_2)$, arrive at the sought equality

$$\int\limits_{G_1 \times G_2} \varphi(g)\lambda(dg) = \int\limits_{G_1} \varphi_1(g_1)\lambda_1(dg_1).$$

The second equality results by analogy:

$$\int\limits_{G_1\times G_2} \varphi(g)\lambda(dg) = \int\limits_{G_2} \varphi_2(g_2)\lambda_2(dg_2). \ \triangleright$$

5.5. The Hausdorff Moment Problem

The classical problem still attracts attention of determining a Borel measure given a moment sequence. It is called the *moment problem*; for instance, see [3, 23, 24]. This is witnessed by the recent articles [25–27]. One of the interesting generalizations of this problem is connected with considering a vector valued or operator valued moment sequence [28–32]. This implies the following vector formulation: Given a sequence $(a_k)_{k=0}^{\infty}$ in a lattice normed space Y, find a Y-valued Borel measure on the interval $[0,1]$ whose kth moment coincides with a_k $(k \in \omega = \{0,1,2,\dots\})$.

Two particular cases of this moment problem are worthy of emphasizing in which Y is a Kantorovich space.

Let T be a (possibly unbounded) selfadjoint operator in Hilbert space. The problem consists in finding some spectral measure μ satisfying

$$T^k = \int\limits_{\mathbb{R}} \lambda^k d\mu(\lambda) \quad (k \in \omega).$$

Clearly, this paraphrases the spectral decomposition problem. Close statements were treated in [28–31].

We now assume that $(a_k)_{k=0}^{\infty}$ is a given sequence of random variables on a probability space (Q, Σ, P). The problem consists in finding a random measure $(\mu_t)_{t\in Q}$ on the Borel σ-algebra $\mathscr{B}(\mathbb{R})$ such that the following equalities hold

$$a_k(t) = \int\limits_{\mathbb{R}} \lambda^k d\mu_t(\lambda) \quad (k \in \omega)$$

for almost all $t \in Q$. By a *random measure* we mean a family of countably-additive measures $(\mu_t)_{t\in Q}$ such that the mapping

$$\mu_{(\cdot)}(B) : t \mapsto \mu_t(B) \quad (t \in Q)$$

is measurable for all $B \in \mathscr{B}(\mathbb{R})$. To a random measure $(\mu_t)_{t\in Q}$ there corresponds the unique measure μ determined from the condition that $\mu(E)$ is the coset of the measurable function $\mu_{(\cdot)}(E)$.

5.5.1. DEFINITION. A sequence of vectors $(a_k)_{k=0}^{\infty} \subseteq F$ is called (Hausdorff) *positive-definite* provided that

$$\sum_{k=0}^{n}(-1)^k C_n^k a_{k+l} \geqslant 0 \quad (n, l \in \omega).$$

5.5.2. DEFINITION. A sequence of vectors $(y_k)_{k=1}^{\infty} \subseteq Y$ is called (Hausdorff) *dominated* provided that there is a sequence $(a_k)_{k=0}^{\infty} \subseteq F$ satisfying

$$\left|\sum_{k=0}^{n}(-1)^k C_n^k y_{k+l}\right| \leqslant \sum_{k=0}^{n}(-1)^k C_n^k a_{k+l} \quad (n, l \in \omega).$$

5.5.3. *Theorem.* *To each sequence $(y_k)_{k=0}^{\infty} \subseteq Y$ there is a unique Borel measure $\mu : \mathscr{B}([0,1]) \to Y$ having bounded vector variation and satisfying the equalities*

$$y_k = \int \lambda^k d\mu \quad (k \in \omega)$$

if and only if $(y_k)_{k=0}^{\infty}$ is a (Hausdorff) dominated sequence.

$\triangleleft$ Define a linear operator $\mathscr{U}$ from the space of polynomials $\mathscr{P}([0,1])$ to F as follows:

$$\mathscr{U}(p) = \sum_{k=0}^{n} p_k a_k, \text{ with } p(\lambda) = \sum_{k=0}^{n} p_k \lambda^k.$$

Since $(a_k)_{k=0}^{\infty}$ is a positive-definite sequence; for the polynomials $p_{n,l}(\lambda) = \lambda^l(1-\lambda)^n$ we have the inequalities $\mathscr{U}(p_{n,l}) \geqslant 0$ $(n, l \in \omega)$. If $p \in \mathscr{P}([0,1])$ is a positive polynomial on $[0,1]$ then by the Weierstrass Theorem p may be uniformly approximated by polynomials of degree at most n which take the form

$$\sum_{k=0}^{m} p\left(\frac{k}{m}\right) C_m^k p_{m-k,k}(\lambda).$$

This implies that $\mathscr{U}(p) \geqslant 0$. Using the r-continuity of $\mathscr{U}$, we may then extend $\mathscr{U}$ to the positive operator $\overline{\mathscr{U}} : C([0,1]) \to F$.

We now pass to constructing the measure μ. By analogy, define the operator $T : \mathscr{P}([0,1]) \to Y$ by the formulas

$$T(p) = \sum_{k=0}^{n} p_k y_k, \text{ with } p(\lambda) = \sum_{k=0}^{n} p_k \lambda^k.$$

By condition $|T(p_{n,l})| \leqslant \mathscr{U}(p_{n,l})$ $(n, l \in \omega)$. Analogous arguments show that $|T(p)| \leqslant \mathscr{U}(p)$ for all $p \in \mathscr{P}([0,1]), p(\lambda) \geqslant 0$ $(\lambda \in [0,1])$.

We now let $p \in \mathscr{P}([0,1])$ stand for an arbitrary polynomial. For each $\varepsilon > 0$ there is a polynomial q_ε satisfying

$$|p(\lambda)| \leqslant q_\varepsilon(\lambda) \leqslant |p(\lambda)| + \varepsilon \quad (\lambda \in [0,1]).$$

So,

$$\begin{aligned}|T(p)| = |T(p + q_\varepsilon - q_\varepsilon)| &\leqslant \mathscr{U}(p + q_\varepsilon) + \mathscr{U}(q_\varepsilon) \\ &\leqslant \overline{\mathscr{U}}(3|p| + 2\varepsilon) = 3\overline{\mathscr{U}}(|p|) + 2\varepsilon a_0.\end{aligned}$$

Since ε is arbitrary, we arrive at the inequality

$$|T(p)| \leqslant 3\overline{\mathscr{U}}(|p|) \quad (p \in \mathscr{P}([0,1])).$$

By r-continuity, T extends to the dominated operator $T_1 : C([0,1]) \to Y$. By Theorem 5.3.1 there is a Borel measure $\mu : \mathscr{B}([0,1]) \to Y$ of bounded vector variation such that

$$T(p) = \int p(\lambda)\,d\mu(\lambda) \quad (p \in \mathscr{P}([0,1])).$$

Recalling the definition of T, we see that the sought equalities are valid. $\triangleright$

It is worth noting that in case $Y = F = \mathbb{R}$ the dominance condition on the sequence $(y_k)_{k=0}^{\infty}$ amounts to the conventional *Hausdorff condition* (see [3, 23]):

$$\sum_{k=0}^{n} C_n^k \left| \sum_{i=0}^{n-k} C_{n-k}^i y_{i+k} \right| \leqslant \text{const} \quad (n \in \omega).$$

This is not so in the vector situation even in the case when Y is a Banach space. Strictly speaking, if some sequence in a Banach space satisfies the Hausdorff condition then the corresponding moment problem may be solved by a measure of unbounded vector variation. Only the semivariation of such a measure is always bounded.

We have mentioned that the spectral decomposition of a selfadjoint operator in Hilbert space may be obtained as solution to the moment problem. To demonstrate, we need the following

5.5.4. Theorem. *Let F be a monotonically complete ordered vector space. For a given sequence $(a_k)_{k=0}^{\infty} \subseteq F$ there is a unique positive Borel measure $\mu : \mathscr{B}([0,1]) \to F$ satisfying the equalities*

$$a_k = \int \lambda^k d\mu(\lambda) \quad (k \in \omega) \tag{5.1}$$

if and only if $(a_k)_{k=0}^{\infty}$ is a (Hausdorff) positive-definite sequence.

◁ Denote by $\widehat{F}$ a Dedekind completion of the ideal $F(a_0)$. This $\widehat{F}$ is a Kantorovich space. Applying Theorem 5.5.3 with $Y = \widehat{F}$ and $y_k = a_k$ $(k \in \omega)$, obtain a unique positive Borel measure $\mu : \mathscr{B}([0,1]) \to \widehat{F}$ satisfying (5.1). We are left with verifying that the values of μ belong to the original space. This is done routinely on using the Monotone Class Lemma (cf. [33]). ▷

Let $\mathscr{A}(\mathscr{H})$ be the vector space of bounded selfadjoint operators in some Hilbert space $\mathscr{H}$. We order $\mathscr{A}(\mathscr{H})$ in the routine manner: Given $S, T \in \mathscr{A}(\mathscr{H})$, we agree that the inequality $S \leqslant T$ amounts to the condition $(Sx, x) \leqslant (Tx, x)$ $(x \in \mathscr{H})$. It is well known that $(\mathscr{A}(\mathscr{H}), \leqslant)$ becomes a monotonically complete ordered vector space (for instance, see [28]). Let $T \in \mathscr{A}(\mathscr{H})$. Without loss of generality, we may assume that $0 \leqslant T \leqslant I$, with I the identity operator. We now prove that the sequence $(T^k)_{k=0}^{\infty}$ is Hausdorff positive-definite. To this end it suffices to check that

$$\sum_{k=0}^{n} (-1)^k C_n^k T^{k+l} = T^l (I - T)^n \geqslant 0 \quad (n, l \in \omega).$$

Distinguishing the factor of even degree, reduce the required inequality to the three cases:

$$T \geqslant 0; \quad I - T \geqslant 0; \quad T(I - T) \geqslant 0.$$

The first two inequalities hold by definition. The third ensues from the following chain of inequalities:

$$(T^2 x, x)^2 \leqslant (Tx, Tx)_T (x, x)_T \leqslant (T^2 x, x)(Tx, x),$$

with $(x, y)_T := (Tx, y)$. It is worth observing that we are done without the lemma claiming existence of the square root of a positive selfadjoint operator. Using Theorem 5.5.3, we now derive

5.5.5. Corollary. *To each $T \in \mathscr{A}(\mathscr{H})$ there is a unique projection-valued measure μ satisfying*

$$T^k = \int\limits_{[0,1]} \lambda^k d\mu(\lambda) \quad (k \in \omega).$$

5.6. The Hamburger Moment Problem

In this section we consider the Hamburger moment problem for a sequence of vectors $\{s_k\}_{k=0}^{\infty}$ in a K_σ-space F (as regards the definitions stemming from ordered vector spaces, see [3, 8]). This problem was studied in [30] in the case when F is a Kantorovich space by using the Kantorovich Theorem on extension of a positive operator which involves the order completeness of the range of the operator (see [8, Theorem X.3.1]). There are two reasons behind our intention to eliminate order completeness. First, the analogous moment problem in the Hausdorff setting (the case of a bounded interval) may obviously be solved in an arbitrary K_σ-space F (cf. Theorem 5.5.3). Second, it stands to reason to acquire a solution to the Hamburger moment problem formulated as follows:

Given a positive-definite sequence of measurable functions $s_n : \Omega \to \mathbb{R}$ $(n \geqslant 0)$ on some measure space $(\Omega, \mathscr{B}, \nu)$. Find a mapping $\mu : \mathscr{B}(\mathbb{R}) \times \Omega \to \mathbb{R}^+$ enjoying the next properties:

(1) For each $\omega \in \Omega$ the function $\mu(\,\cdot, \omega)$ is a Borel measure and

$$s_n(\omega) = \int\limits_{\mathbb{R}} u^n \mu(du, \omega) \quad (n = 0, 1, 2, \dots);$$

(2) for each Borel set $A \subseteq \mathbb{R}$ the function $\mu(A, \cdot\,)$ is $\mathscr{B}$-measurable.

A mapping μ, satisfying (1) for ν-almost all $\omega \in \Omega$, is usually referred to as a *random measure*. Since the scalar Hamburger problem may have several solutions, the question we are interested in reads in fact as follows: How should we choose solutions of the scalar problem for each $\omega \in \Omega$ so that these chosen solutions produce a measurable solution in the variable ω to the vector problem? If we require that (1) and (2) hold to within a ν-negligible set then this problem is partially solved in [34]. In the other cases we need some extra tricks since the space of $\mathscr{B}$-measurable functions is only a K_σ-space, and the space of polynomials $\mathscr{P}(\mathbb{R})$ may fail to be dense in $L_1(\mu)$. We also mention that various vector statements of the moment problem were under study, for instance, in [30–32, 36, 38].

In the sequel we assume that

F is an arbitrary K_σ-space;

$\mathscr{P}(\mathbb{R})$ is the space of polynomials on $\mathbb{R}$;

$\mathscr{P}_{\mathbb{Q}}(\mathbb{R})$ is the set of polynomial on $\mathbb{R}$ with rational coefficients; $C_b(\mathbb{R})$ is the space of bounded continuous functions on $\mathbb{R}$;

$C_{\mathscr{P}}(\mathbb{R})$ is the space of polynomially bounded continuous functions on $\mathbb{R}$; i.e., $\varphi \in C_{\mathscr{P}}(\mathbb{R})$ means that $\varphi \in C(\mathbb{R})$ and $|\varphi| \leqslant p$ for some polynomial $p \in \mathscr{P}(\mathbb{R})$.

A σ-additive mapping $\mu : \mathscr{B}(\mathbb{R}) \to F_+$ is a *Borel measure*. A *spectral measure* is a σ-order-continuous homomorphism from $\mathscr{B}(\mathbb{R})$ to F. We will also use the vector analog of Lebesgue measure, now with respect to an F-valued measure (cf. Section 5.1).

5.6.1. DEFINITION. A sequence $\{s_k\}_{k=0}^{\infty}$ in F is *positive-definite* provided that

$$\sum_{k,l=0}^{n} s_{k+l}\sigma_k\sigma_l \geqslant 0 \quad (\{\sigma_k\}_{k=0}^{n} \subset \mathbb{R},\ n = 0, 1, 2, \dots).$$

5.6.2. Theorem. *To a given sequence $\{s_k\}_{k=0}^{\infty} \subset F$ there is a positive Borel measure $\mu : \mathscr{B}(\mathbb{R}) \to F_+$ satisfying*

$$s_k = \int_{\mathbb{R}} u^k \mu(du) \quad (k = 0, 1, 2, \dots), \tag{6.1}$$

if and only if $\{s_k\}_{k=0}^{\infty}$ is a positive-definite sequence.

To describe our situation in more detail, define some positive linear operator $U : \mathscr{P}(\mathbb{R}) \to F$ by the formulas

$$U(p) = \sum_{k=0}^{n} a_k s_k, \quad \text{with} \quad p(u) = \sum_{k=0}^{n} a_k u^k. \tag{6.2}$$

Given $\varphi : \mathbb{R} \to \mathbb{R}$, put

$$U^{\vee}(\varphi) = \sup\{U(p) : p \in \mathscr{P}(\mathbb{R}),\ p \leqslant \varphi\};$$

$$U^{\wedge}(\varphi) = \inf\{U(p) : p \in \mathscr{P}(\mathbb{R}),\ p \geqslant \varphi\}. \tag{6.3}$$

Fixing an arbitrary complex number $\lambda \in \mathbb{C} \setminus \mathbb{R}$, consider the function

$$R_\lambda(u) = \operatorname{Re}\frac{1}{u-\lambda}, \quad I_\lambda(u) = \operatorname{Im}\frac{1}{u-\lambda}. \tag{6.4}$$

In F we define the following vectors:

$$a^+ = U^{\wedge}(R_\lambda),\ a^- = U^{\vee}(R_\lambda), \quad b^+ = U^{\wedge}(I_\lambda),\ b^- = U^{\vee}(I_\lambda). \tag{6.5}$$

Consider the vector

$$C = \frac{a^+ + a^-}{2} + i\,\frac{b^+ + b^-}{2} \tag{6.6}$$

in the complexification $F_{\mathbb{C}}$ of the space F. Put

$$R = \frac{a^+ - a^-}{2} = \frac{b^+ - b^-}{2}. \tag{6.7}$$

It is possible to furnish $F_{\mathbb{C}}$ with the vector norm $|w| = \{(\operatorname{Re} w)^2 + (\operatorname{Im} w)^2\}^{1/2}$ and then to define the vector Weyl–Hamburger circle (cf. [37])

$$K_\infty(\lambda) = \{w \in F_{\mathbb{C}} : |w - C| \leqslant R\}, \tag{6.8}$$

where C is the center of this circle and R is its radius.

Denote by s_k^H the projection of the sequence s_k to the band $\{R\}^{\perp\perp}$; and by s_k^O the projection of s_k to the complementary band $\{R\}^{\perp}$. Observe that all members of a positive-definite sequence $\{s_k\}_{k=0}^{\infty}$ belong to the band $\{s_0\}^{\perp\perp}$ (see [35, Lemma 2]). Furnish $\{s_0\}^{\perp\perp}$ with the unique partial multiplication by specifying s_0 as an order-unity. Consider the minors

$$\mathscr{D}_{-1} = 0, \quad \mathscr{D}_n = \begin{vmatrix} s_0 & s_1 & \dots & s_n \\ s_1 & s_2 & \dots & s_{n+1} \\ \vdots & \vdots & \ddots & \vdots \\ s_n & s_{n+1} & \dots & s_{2n} \end{vmatrix} \quad (n = 0, 1, 2, \dots).$$

Clearly, $\mathscr{D}_n \in F$ and $\{\mathscr{D}_{n-1}\}^{\perp\perp} \supseteq \{\mathscr{D}_n\}^{\perp\perp}$ $(n = 1, 2, \dots)$. Denote by π_n the band projection to $\{\mathscr{D}_{n-1}\}^{\perp\perp} \cap \{\mathscr{D}_n\}^{\perp}$. We may define the band $\{\mathscr{D}_n\}^{\perp\perp}$ without appealing to multiplication by putting

$$E(\sigma_0, \sigma_1, \dots, \sigma_n) = \left\{ \sum_{k,l=0}^{n} s_{k+l}\sigma_k\sigma_l \right\}^{\perp\perp} \quad (\{\sigma_k\}_{k=0}^{n} \subset \mathbb{R})$$

to obtain

$$\{\mathscr{D}_n\}^{\perp\perp} = \bigcap \{E(\sigma_0, \sigma_1, \dots, \sigma_n) : \sigma_0^2 + \sigma_1^2 + \dots + \sigma_n^2 > 0\}.$$

5.6.3. Theorem. *Given a positive-definite sequence $\{s_k\}_{k=0}^{\infty}$ and a complex number $\lambda \in \mathbb{C} \setminus \mathbb{R}$, define the operator U, vectors C, R, and circle $K_\infty(\lambda)$ by the formulas (6.2)–(6.8). The following are valid:*

(1) *For each $w \in K_\infty(\lambda)$ there is a positive measure $\mu : \mathscr{B}(\mathbb{R}) \to F_+$ solving the moment problem for $\{s_k\}_{k=0}^{\infty}$ (satisfying (6.1)) and such that*

$$w = \int_{\mathbb{R}} \frac{1}{u - \lambda} \, \mu(du). \tag{6.9}$$

Conversely, to each solution $\mu : \mathscr{B}(\mathbb{R}) \to F_+$ to this moment problem, the vector w, defined by (6.9), belongs to $K_\infty(\lambda)$.

(2) *In each nonzero principal band $E \subseteq \{R\}^{\perp\perp}$ there is a nonunique solution of the moment problem for the projections s_k^H to E.*

(3) *The moment problem for the projections $\{s_k^O\}_{k=0}^{\infty}$ in $\{R\}^{\perp}$ has a unique solution $\mu_0 : \mathscr{B}(\mathbb{R}) \to F_+$.*

(4) *The equality holds $\pi_0 \circ \mu = 0$, and $\{\mathscr{D}_{n-1}\}^{\perp\perp} \cap \{\mathscr{D}_n\}^{\perp} \subseteq \{R\}^{\perp}$ for all $n \geqslant 1$. Moreover, the measure $\pi_n \circ \mu_0$ is the sum of n disjoint spectral measures. If there are two distinct representations*

$$\pi_n \circ \mu_0 = \sum_{i=1}^{n} \mu_i^{(n)} = \sum_{j=1}^{m} \nu_j^{(m)}$$

in the form of the sum of disjoint spectral measures $\{\mu_i^{(n)}\}_{i=1}^n$ and $\{\nu_j^{(m)}\}_{j=1}^m$, then $m = n$ and there is a matrix with entries band projections $\{\pi_{ij}\}_{i,j=1}^n$ such that

$$\pi_{ij} \circ \pi_{ik} = 0, \quad \pi_{ji} \circ \pi_{ki} = 0 \quad (j \neq k),$$

$$\sum_{j=1}^{n} \pi_{ij} = \sum_{j=1}^{n} \pi_{ji} = \pi_n, \quad \nu_i^{(n)} = \sum_{j=1}^{n} \pi_{ij} \circ \mu_j^{(n)} \quad (i = 1, 2, \dots, n).$$

The following Interpolation Lemma plays a key role in what follows.

5.6.4. Lemma. *Assume that $\varphi \in C_b(\mathbb{R})$ and a polynomial $p \in \mathscr{P}(\mathbb{R})$ satisfies the inequality $p(u) \geqslant \varphi(u) + \varepsilon$ $(u \in \mathbb{R})$ for some $\varepsilon > 0$. Then there is another polynomial $q \in \mathscr{P}_Q(\mathbb{R})$ having the same degree and maintaining the relations $p(u) > q(u) > \varphi(u)$ $(u \in \mathbb{R})$.*

$\triangleleft$ The proof is elementary and thus omitted. $\triangleright$

5.6.5. Lemma. *Let $\{s_k\}_{k=0}^{\infty}$ be a positive-definite real sequence. Given a complex number $\lambda \in \mathbb{C}\backslash\mathbb{R}$, construct the corresponding Weyl–Hamburger circle $K_\infty(\lambda)$ with center C and radius R. Then, for all $\alpha \in \mathbb{C}$, $|\alpha| = 1$ the equality holds*

$$\mathrm{Re}(\overline{\alpha}C) + R = U^\wedge(\varphi_\alpha), \tag{6.10}$$

where U and $U^\wedge$ are determined from (6.2) and (6.3) and

$$\varphi_\alpha(u) = \mathrm{Re}\, \frac{\overline{\alpha}}{u - \lambda} \quad (u \in \mathbb{R}).$$

$\triangleleft$ There is a measure $\mu : \mathscr{B}(\mathbb{R}) \to \mathbb{R}_+$ solving the moment problem for the sequence $\{s_k\}_{k=0}^{\infty}$ and satisfying

$$C + \alpha R = \int_{\mathbb{R}} \frac{1}{u - \lambda}\, \mu(du)$$

(cf. [23, Theorem 2.2.4]). Extend U to a positive functional $V : C_{\mathscr{P}}(\mathbb{R}) \to \mathbb{R}$ so that to have

$$V(\varphi_\alpha) = U^\wedge(\varphi_\alpha) \tag{6.11}$$

(cf. [8, Theorem X.3.1]). There is a measure $\nu : \mathscr{B}(\mathbb{R}) \to \mathbb{R}_+$ solving the moment problem for the sequence $\{s_k\}_{k=0}^\infty$ and satisfying

$$V(\varphi) = \int_{\mathbb{R}} \varphi(u)\,\nu(du) \quad (\varphi \in C_{\mathscr{P}}(\mathbb{R})).$$

This follows, for instance, from Theorem 1 of [35] in the case $Y = \mathbb{R}$. By Theorem 2.2.4 of [23],

$$w = \int_{\mathbb{R}} \frac{1}{u-\lambda}\,\nu(du) \in K_\infty(\lambda).$$

Therefore,

$$\operatorname{Re}(\overline{\alpha} w) = V(\varphi_\alpha) \leqslant \operatorname{Re}(\overline{\alpha} C) + R = \int_{\mathbb{R}} \varphi_\alpha(u)\,\nu(du) \leqslant U^\wedge(\varphi_\alpha).$$

Comparing this with (6.11), arrive at (6.10). The proof of the lemma is complete. ▷

Lemma 5.6.5, in particular, justifies the formulas (6.3)–(6.8) in the case $F = \mathbb{R}$, while Lemma 5.6.4 demonstrated soundness of Definitions (6.3)–(6.8) in the vector case.

5.6.6. *Lemma.* *For each positive-definite sequence of vectors $\{s_k\}_{k=0}^\infty \subset F$, the vector a^+, defined by (6.5), exists and may be calculated by the formula*

$$a^+ = \inf\{U(q) : q \in \mathscr{P}_Q(\mathbb{R}),\ q \geqslant R_\lambda\}.$$

Similar assertions hold for the vectors a^-, b^+, and b^-.

◁ Let $p \in \mathscr{P}(\mathbb{R})$ and $p \geqslant R_\lambda$. By Lemma 5.6.4 to each $\varepsilon > 0$ there is a polynomial $q \in \mathscr{P}_Q(\mathbb{R})$ satisfying $p(u) + \varepsilon > q(u) > R_\lambda(u)$ $(u \in \mathbb{R})$. Then $U(p) \geqslant U(q) - \varepsilon U(\mathbf{1})$. Put $a^+ = \inf\{U(q) : q \in \mathscr{P}_Q(\mathbb{R}),\ q \geqslant R_\lambda\}$. Obviously, $U(p) \geqslant a^+ - \varepsilon U(\mathbf{1})$. Since $\varepsilon > 0$ is arbitrary; therefore, $U(p) \geqslant a^+$ for all $p \in \mathscr{P}(\mathbb{R})$, $p \geqslant R_\lambda$. Consequently, $U^\wedge(R_\lambda)$ exists and equals a^+. The same is checked by analogy for the vectors a^-, b^+, and b^-. ▷

5.6.7. Lemma. *If $\{s_k\}_{k=0}^{\infty}$ is a positive-definite sequence then its every element belongs to the band $\{s_0\}^{\perp\perp}$.*

$\triangleleft$ By Theorem 2 of [35] there is a solution $\mu : \mathscr{B}(\mathbb{R}) \to F_+$ of this moment problem with range in the Dedekind completion $\widehat{F}$ of the K_σ-space F. The claim of the lemma is now immediate from the equality

$$s_0 = \int_{\mathbb{R}} \mu(du) = \mu(\mathbb{R}). \ \triangleright$$

5.6.8. *Corollary.* *For the scalar Hamburger problem to be definite it is necessary and sufficient that either of the equalities*

$$U^{\wedge}(R_\lambda) = U^{\vee}(R_\lambda), \quad U^{\wedge}(I_\lambda) = U^{\vee}(I_\lambda)$$

be valid for some $\lambda \in \mathbb{C} \setminus \mathbb{R}$.

Theorem 5.6.2 is a brief form of a more expanded Theorem 5.6.3; therefore, we will prove only Theorem 5.6.3. It is worth observing that Theorem 5.6.2 is of interest in its own right, since it claims only existence of a solution which is easy to construct in this case as compared with Theorem 5.6.3.

Proof of Theorem 5.6.3. Show the converse part of (1). Consider a solution $\mu : \mathscr{B}(\mathbb{R}) \to F_+$ of the moment problem for a positive-definite sequence of vectors $\{s_k\}_{k=0}^{\infty} \subset F$. Realize the band $\{s_0\}^{\perp\perp}$ as an order-dense ideal in the space $C_\infty(Q)$, with Q a quasiextremal compact space so that s_0 becomes the identically one function. By Lemma 5.6.7 all entries s_k $(k = 0, 1, 2, \dots)$ are also realized by continuous functions on Q. By Lemmas 5.6.5 and 5.6.6 there is a meager Borel set $E_0 \subset Q$ such that, for all $q \in Q \setminus E_0$, the numbers $C(q)$ and $R(q)$ serve as the center and radius of the scalar Weyl–Hamburger circle for the positive-definite numeric sequence $\{s_n(q)\}_{n=0}^{\infty}$. Fixing $\lambda \in \mathbb{C} \setminus \mathbb{R}$, we may calculate all integrals

$$w = \int_{\mathbb{R}} \frac{1}{u - \lambda}\, \mu(du), \quad s_n = \int_{\mathbb{R}} u^n \mu(du) \quad (n \geqslant 0)$$

as the r-limits of the Stieltjes sums with respect to a fixed sequence of countable partitions of $\mathbb{R}$. Therefore, we may assume that the set E_0 satisfies the following condition: For all $q \in Q \setminus E_0$ we have

$$w(q) = \int_{\mathbb{R}} \frac{1}{u - \lambda}\, d\sigma_q(u), \quad s_n(q) = \int_{\mathbb{R}} u^n\, d\sigma_q(u) \quad (n \geqslant 0),$$

where $\sigma_q(u) = \mu((-\infty, u))(q)$ $(u \in \mathbb{R})$ are the distribution function with respect to the measure μ. By Theorem 2.2.4 of [23], $|w(q) - C(q)| \leqslant R(q)$ for these q. Continuity of the functions w, C, and R implies that these inequalities hold for all $q \in Q$. Consequently, we have the vector inequality $|w - C| \leqslant R$.

Given a positive-definite sequence $\{s_k\}_{k=0}^{\infty} \subset F$ and some $\lambda \in \mathbb{C} \setminus \mathbb{R}$, define the operator U and vectors C and R by (6.2)–(6.7). Consider any vector $w \in F_{\mathbb{C}}$, satisfying the inequality $|w - C| \leqslant R$. By Lemma 5.6.7 we may confine exposition to the band $\{s_0\}^{\perp\perp}$ and furnish it with some partial multiplication, taking s_0 as an order-unity. Then there are two vectors w_1, $w_2 \in \{s_0\}^{\perp\perp}$ such that $|w_1 - C| = |w_2 - C| = R$ and $w = c_1 w_1 + c_2 w_2$ for some c_1, $c_2 \in F$, $c_1 \geqslant 0$, $c_2 \geqslant 0$, $c_1 + c_2 = s_0$. It suffices to solve the problem for w_1 as well as for w_2. There is some α in $F_{\mathbb{C}}$, $|\alpha| = s_0$ such that $w_1 = C + \alpha R$ and $w_2 = C - \alpha R$. Let $\alpha = \alpha_1 + i\alpha_2$, where α_1, $\alpha_2 \in \{s_0\}^{\perp\perp}$. Consider the spectral functions $\{e_u^{(1)} : u \in \mathbb{R}\}$ and $\{e_u^{(2)} : u \in \mathbb{R}\}$ of the elements α_1 and α_2 with respect to the order-unity s_0. Let $\mathscr{A}$ stand for the algebra of unit elements (with respect to s_0) generated by the countable system $\{e_u^{(1)}, e_u^{(2)} : u \in \mathbb{Q}\}$. Denote by $\mathscr{P}_{-1}(\mathbb{R}, \mathscr{A})$ the space of functions from $\mathbb{R}$ to $\{s_0\}^{\perp\perp}$ of the form

$$\psi(u) = \sum_{j=1}^{n} e_j p_j(u) \quad (\{e_j\}_{j=1}^{n} \subset \mathscr{A},\ \{p_j\}_{j=1}^{n} \subset \mathscr{P}(\mathbb{R})). \tag{6.12}$$

Consider the space $S(\mathscr{A})$ of the simple elements like

$$a = \sum_{j=1}^{n} \beta_j e_j \quad (\{\beta_j\}_{j=1}^{n} \subset \mathbb{R},\ \{e_j\}_{j=1}^{n} \subset \mathscr{A}).$$

Look at the operator $U_{-1} : \mathscr{P}_{-1}(\mathbb{R}, \mathscr{A}) \to F$ whose value at a function of the shape (6.12) is as follows:

$$U_{-1}(\psi) = \sum_{j=1}^{n} e_j U(p_j).$$

If we order $\mathscr{P}_{-1}(\mathbb{R}, \mathscr{A})$ pointwise then U_{-1} becomes a positive operator enjoying *$S(\mathscr{A})$-linearity*, i.e.,

$$U_{-1}(a\psi) = aU_{-1}(\psi) \quad (a \in S(\mathscr{A}),\ \psi \in \mathscr{P}_{-1}(\mathbb{R}, \mathscr{A})).$$

Choose a uniformly dense set $\{\varphi_j\}_{j=1}^{\infty}$ in $C_0(\mathbb{R})$. The set $F(\mathbb{R})$ of functions from $\mathbb{R}$ to $\{s_0\}^{\perp\perp}$, furnished with the pointwise operators, is a K_σ-space. Make it into an $S(\mathscr{A})$-module by defining multiplication by $a \in S(\mathscr{A})$ pointwise. Now, the scene is ready for extending the operator U_{-1}, which we will do by induction. Assume that for some n we have constructed a positive operator $U_n : \mathscr{P}_n(\mathbb{R}, \mathscr{A}) \to F$ that extends U_{-1} and is $S(\mathscr{A})$-linear

$$U_n(a\psi_n) = aU_n(\psi_n) \quad (a \in S(\mathscr{A}),\ \psi_n \in \mathscr{P}_n(\mathbb{R}, \mathscr{A}))$$

on the space $\mathscr{P}_n(\mathbb{R}, \mathscr{A})$ of functions from $\mathbb{R}$ to $\{s_0\}^{\perp\perp}$ of the form

$$\psi_n = \psi + \sum_{j=0}^{n} a_j\varphi_j, \quad \psi \in \mathscr{P}(\mathbb{R}),\ \{a_j\}_{j=1}^{n} \subset S(\mathscr{A}),$$

$$\varphi_0(u) = \operatorname{Re} \frac{\overline{\alpha}}{u - \lambda} \quad (u \in \mathbb{R}).$$

If $\varphi_{n+1}s_0 \notin \mathscr{P}_n(\mathbb{R}, \mathscr{A})$ then we put

$$U_{n+1}(\varphi_{n+1}s_0) = \inf\{U_n(\psi_n) : \psi_n \in \mathscr{P}_n(\mathbb{R}, \mathscr{A}),\ \psi_n \geqslant \varphi_{n+1}s_0\}. \tag{6.13}$$

We will show that the infimum exists in the formula (6.13). To this end, we prove an analog of the Interpolation Lemma 5.6.4. Take an arbitrary rational $\varepsilon > 0$. If $\psi_n \in \mathscr{P}_n(\mathbb{R}, \mathscr{A})$ and $\psi_n \geqslant \varphi_{n+1}s_0$ then there is a partition of unity s_0 into disjoint unit elements $\{e_k\}_{k=1}^{m} \subset \mathscr{A}$ such that

$$e_k\psi_n = e_k\left(p_k + \sum_{j=0}^{n} \beta_{kj}\varphi_j\right) \quad (k = 1, 2, \dots, m)$$

for some $p_k \in \mathscr{P}(\mathbb{R})$, $\beta_{kj} \in \mathbb{R}$ $(k = 1, \dots, m;\ j = 0, 1, \dots, n)$. Refining the partition $\{e_k\}_{k=1}^{m}$ if need be, we may assume that, for $\beta_{k0} \neq 0$, we have the estimates

$$\left|\alpha_1 e_k - \lambda_k^{(1)} e_k\right| \leqslant \frac{\varepsilon|\operatorname{Im}\lambda|}{2|\beta_{k0}|}s_0, \quad \left|\alpha_2 e_k - \lambda_k^{(2)} e_k\right| \leqslant \frac{\varepsilon|\operatorname{Im}\lambda|}{2|\beta_{k0}|}s_0 \tag{6.14}$$

for some real $\left|\lambda_k^{(1)}\right| \leqslant 1$, $\left|\lambda_k^{(2)}\right| \leqslant 1$ $(k = 1, \dots, m)$. In result, we come to simultaneous numeric inequalities:

$$p_k(u) + \beta_{k0}\left(\lambda_k^{(1)} R_\lambda(u) + \lambda_k^{(2)} I_\lambda(u)\right) + \sum_{j=1}^{n} \beta_{kj}\varphi_j(u) + \varepsilon \geqslant \varphi_{n+1}(u)$$
$$(k = 1, \dots, m;\ u \in \mathbb{R}).$$

In case $n = 0$, these inequalities take the form

$$p_k(u) + \varepsilon \geqslant \lambda_k^{(1)} R_\lambda(u) + \lambda_k^{(2)} I_\lambda(u) \quad (k = 1, \dots, m;\ u \in \mathbb{R}),$$

where the functions R_λ and I_λ are determined from (6.4). By the Interpolation Lemma there are polynomials $q_k \in \mathscr{P}_Q(\mathbb{R})$, satisfying

$$p_k(u) > q_k(u) > \varphi_{n+1}(u) - \beta_{k0}\lambda_k^{(1)} R_\lambda(u) - \beta_{k0}\lambda_k^{(2)} I_\lambda(u) - \sum_{j=1}^{n} \beta_{kj}\varphi_j(u) - 2\varepsilon \tag{6.15}$$

$$(k = 1, \dots, m;\ u \in \mathbb{R}).$$

There are rational numbers $\gamma_{kj} \in \mathbb{Q}$ such that the inequalities still hold in (6.1) on substituting γ_{kj} for β_{kj} throughout. Moreover, we may assume that

$$\frac{2|\beta_{k0} - \gamma_{k0}|}{|\operatorname{Im}\lambda|} + \left|\sum_{j=1}^{n} (\beta_{kj} - \gamma_{kj})\varphi_j(u)\right| < \varepsilon \quad (u \in \mathbb{R}).$$

We thus obtain

$$p_k + \beta_{k0}(\lambda_k R_\lambda + \mu_k I_\lambda) + \sum_{j=1}^{n} \beta_{kj}\varphi_j + 3\varepsilon$$

$$\geqslant q_k + \gamma_{k0}(\lambda_k R_\lambda + \mu_k I_\lambda) + \sum_{j=1}^{n} \gamma_{kj}\varphi_j + 2\varepsilon \geqslant \varphi_{n+1} \quad (k = 1, \dots, m).$$

Using (6.14) once again, we come to the estimate

$$\psi_n + 6\varepsilon s_0 \geqslant \sum_{k=1}^{m} \left(e_k q_k + e_k \sum_{j=0}^{n} \gamma_{kj}\varphi_j \right) + 3\varepsilon s_0 \geqslant \varphi_{n+1} s_0.$$

Define the set $\mathscr{P}_n(\mathbb{R}, \mathscr{A}, \mathbb{Q})$ of functions from $\mathbb{R}$ to $\{s_0\}^{\perp\perp}$ of the form

$$\chi_n = \sum_{k=1}^{m} e_k \left(q_k + \sum_{j=0}^{n} \gamma_{kj}\varphi_j \right),$$

with $e_k \in \mathscr{A}$, $q_k \in \mathscr{P}_Q(\mathbb{R})$, $\gamma_{kj} \in \mathbb{Q}$ $(j = 0, 1, \dots, n;\ k = 1, \dots, m)$. Obviously, $\mathscr{P}_n(\mathbb{R}, \mathscr{A}, \mathbb{Q})$ is a countable set. We have proven that to all $\psi_n \in \mathscr{P}_n(\mathbb{R}, \mathscr{A})$, $\psi_n \geqslant \varphi_{n+1} s_0$, and $\varepsilon > 0$ there is $\chi_n \in \mathscr{P}_n(\mathbb{R}, \mathscr{A}, \mathbb{Q})$ satisfying

$$\psi_n + \varepsilon s_0 \geqslant \chi_n \geqslant \varphi_{n+1} s_0. \tag{6.16}$$

Assign

$$f_{n+1} = \inf\{U_n(\chi_n) : \chi_n \in \mathscr{P}_n(\mathbb{R}, \mathscr{A}, \mathbb{Q}),\ \chi_n \geqslant \varphi_{n+1} s_0\}. \tag{6.17}$$

From (6.16) we then deduce $U_n(\psi_n) \geqslant f_n - \varepsilon s_0$. Since $\varepsilon > 0$ is arbitrary; therefore, the inequality $U_n(\psi_n) \geqslant f_{n+1}$ holds for all $\psi_n \in \mathscr{P}_n(\mathbb{R}, \mathscr{A})$ such that $\psi_n \geqslant \varphi_{n+1} s_0$. We have thus proved that the infimum in (6.13) exists and equals f_{n+1}. Now, given $\psi_n \in \mathscr{P}_n(\mathbb{R}, \mathscr{A})$ and $a \in S(\mathscr{A})$, put

$$U_{n+1}(\psi_n + a\varphi_{n+1}) = U_n(\psi_n) + a f_n. \tag{6.18}$$

The operator $U_{n+1} : \mathscr{P}_{n+1}(\mathbb{R}, \mathscr{A}) \to F$, defined by (6.18), extends U_n and is $S(\mathscr{A})$-linear. Show its positivity. Assume that $\psi_n + a\varphi_{n+1} \geqslant 0$. Denote the supports of the elements $a^+, a^- \in S(\mathscr{A})$ by $e_+, e_- \in \mathscr{A}$. We have

$$\psi_n(s_0 - e_+ - e_-) \geqslant 0, \quad -(a^+)^{-1} \cdot \psi_n \leqslant \varphi_{n+1}, \quad (a^-)^{-1} \cdot \psi_n \geqslant \varphi_{n+1}.$$

Using the definition of U_{n+1} and the positivity of U_n, find

$$-(a^+)^{-1} U_n(\psi_n) \leqslant U_{n+1}(\varphi_{n+1}), \quad (a^-)^{-1} U_n(\psi_n) \geqslant U_{n+1}(\varphi_{n+1}),$$

$$(s_0 - e_+ - e_-) U_n(\psi_n) \geqslant 0.$$

Multiplying both sides of the first inequality by a^+; of the second, by $-a^-$; and summing up all three inequalities, we have

$$U_{n+1}(\psi_n + a\varphi_{n+1}) \geqslant 0.$$

Assign $\mathscr{P}_\infty(\mathbb{R}, \mathscr{A}) = \bigcup_{n=0}^\infty \mathscr{P}_n(\mathbb{R}, \mathscr{A})$. We have constructed some positive $S(\mathscr{A})$-linear operator $U_\infty : \mathscr{P}_\infty(\mathbb{R}, \mathscr{A}) \to F$ that extends U_{-1} and enjoys the equalities of (6.13). Using uniform r-continuity, we may extend the operator U_∞ to the positive $\overline{S(\mathscr{A})}$-linear operator $\overline{U}_\infty$ on the uniform closure $\overline{\mathscr{P}}_\infty$ of the space $\mathscr{P}_\infty(\mathbb{R}, \mathscr{A})$, where $\overline{S(\mathscr{A})}$ is the uniform closure of the space $S(\mathscr{A})$ with respect to the regulator s_0. In particular, $\alpha_1, \alpha_2 \in \overline{S(\mathscr{A})}$. Given $\varphi \in C_0(\mathbb{R})$, put $V(\varphi) = \overline{U}_\infty(\varphi s_0)$ to obtain some positive operator $V : C_0(\mathbb{R}) \to F$. The vector lattice $C_0(\mathbb{R})$ generates the conventional topology on $\mathbb{R}$ (that is, the coarsest topology, keeping all functions in $C_0(\mathbb{R})$ continuous, coincides with the natural topology on $\mathbb{R}$). Show that the *quasi-Radon property* (3.2) also holds in Theorem 5.3.1. To this end, we should check that

$$\sup_n \inf\{V(\varphi) : \varphi \in C_0(\mathbb{R}),\ \varphi \geqslant \chi_{[-n,n]}\} = s_0. \tag{6.19}$$

We will not expatiate on proving existence for the infima in (6.19). This may be done in exactly the same manner as in Theorem 8 of [34] on substituting $C_0(\mathbb{R})$ for the space $C_b(\mathbb{R})$. By Theorem 5.3.1 there is a measure $\mu_1 : \mathscr{B}(\mathbb{R}) \to F_+$ satisfying

$$V(\varphi) = \int_{\mathbb{R}} \varphi(u)\, \mu_1(du) \quad (\varphi \in C_0(\mathbb{R})).$$

Slightly modifying the proof of Theorem 8 in [34], we easily show that μ_1 is a solution to moment problem for the sequence $\{s_k\}_{k=0}^{\infty}$. Demonstrate the equality

$$w_1 = \int_{\mathbb{R}} \frac{1}{u-\lambda}\, \mu_1(du).$$

From (6.13) and (6.17) with $n = -1$ it follows that

$$\begin{aligned} &\alpha_1 V(R_\lambda) + \alpha_2 V(I_\lambda) \\ &= \inf\{U_{-1}(\psi) : \psi \in \mathscr{P}(\mathbb{R}, \mathscr{A}, \mathbb{Q}), \psi \geqslant \alpha_1 R_\lambda + \alpha_2 I_\lambda\}. \end{aligned} \tag{6.20}$$

Realize $\{s_0\}^{\perp\perp}$ as the band of continuous functions on a quasiextremal compact space Q in which s_0 becomes the identically one function on Q. There is a meager set $E_0 \subset Q$ such that the infimum in (6.20) is calculated pointwise on $Q \setminus E_0$. We may presume also that all integrals with respect to μ_1 of the functions $R_\lambda(u)$, $I_\lambda(u)$, u^k $(k = 0, 1, 2, \dots)$, presenting the limits of Stieltjes sums, are calculated pointwise, which correspondingly gives $V(R_\lambda)(q)$, $V(I_\lambda)(q)$, and $s_k(q)$ $(k = 0, 1, 2, \dots;\ q \in Q \setminus E_0)$. Moreover, we presume that, for each $q \in Q \setminus E_0$, the numbers $C(q)$ and $R(q)$ are the parameters of the Weyl–Hamburger circle for the sequence $\{s_k(q)\}_{k=0}^{\infty}$. For all $q \in Q \setminus E_0$ and $\varepsilon > 0$ there is some ψ in $\mathscr{P}(\mathbb{R}, \mathscr{A}, \mathbb{Q})$ such that $\psi \geqslant \alpha_1 R_\lambda + \alpha_2 I_\lambda$ and

$$U_{-1}(\psi)(q) < \alpha_1(q) \int_{\mathbb{R}} R_\lambda(u)\, d\sigma_q(u) + \alpha_2(q) \int_{\mathbb{R}} I_\lambda(u)\, d\sigma_q(u) + \varepsilon,$$

where $\sigma_q(u) = \mu_1((-\infty, u))(q)$ $(u \in \mathbb{R})$. Using $\psi(q) \in \mathscr{P}(\mathbb{R})$ and $U_{-1}(\psi)(q) = U(\psi(q))$ together with Lemma 5.6.5, we infer

$$\overline{\alpha(q)} w_1(q) = \overline{\alpha(q)} C(q) + R(q) < \int_{\mathbb{R}} \operatorname{Re} \frac{\overline{\alpha(q)}}{u-\lambda}\, d\sigma_q(u) + \varepsilon.$$

Since ε is arbitrary and $w_1(q)$ lies on the boundary of the circle $K_\infty(\lambda)(q)$, we must have

$$\operatorname{Re}\{\overline{\alpha(q)} w_1(q)\} = \int_{\mathbb{R}} \operatorname{Re} \frac{\overline{\alpha(q)}}{u-\lambda}\, d\sigma_q(u).$$

By Theorem 2.2.4 of [23], we come to the equality

$$w_1(q) = \int_{\mathbb{R}} \frac{1}{u-\lambda}\, d\sigma_q(u) \quad (q \in Q \setminus E_0).$$

This remains valid in the vector case; i.e.,

$$w_1 = \int_{\mathbb{R}} \frac{1}{u-\lambda}\, \mu_1(du).$$

An analogous construction for the vector $w_2 = C - \alpha R$ allows us to find another solution μ_2 to the moment problem for which

$$w_2 = \int_{\mathbb{R}} \frac{1}{u-\lambda}\, \mu_2(du).$$

For the vector $w = c_1 w_1 + c_2 w_2$ we now arrive at the solution $\mu = c_1\mu_1 + c_2\mu_2$, satisfying the equality (6.9).

Claims (2) and (3) of Theorem 5.6.3 are readily available from the just-proven (1). We now sketch the proof of (4).

In the above-mentioned Stone realization, to the band $\{\mathscr{D}_{n-1}\}^{\perp\perp} \cap \{\mathscr{D}_n\}^{\perp}$ there correspond the continuous functions on Q vanishing beyond some clopen subset Q_n. Moreover, there is a meager set $E_0 \subset Q_n$ such that $\mathscr{D}_{n-1}(q) > 0$ for all $q \in Q_n \setminus E_0$. In addition, $\mathscr{D}_n(q) > 0$ $(q \in Q_n)$. Consequently, there is a unique family of continuous functions $\xi_0, \xi_1, \dots, \xi_n$ on $Q_n \setminus E_0$ such that

$$\sum_{i=1}^{n} \xi_i s_{i+k} = 0, \quad \xi_n = 1 \quad (k = 0, 1, \dots, n).$$

It is elementary that the polynomial $\xi_0(q) + \xi_1(q)\lambda + \dots + \xi_n(q)\lambda^n = 0$ $(q \in Q_n \setminus E_0)$ has n distinct real roots $\lambda_1(q) < \lambda_2(q) < \dots < \lambda_n(q)$. The Rouché Theorem, well-known in complex analysis, easily ensures the continuity of the functions $\{\lambda_i(q)\}_{i=1}^{n}$ $(q \in Q_n \setminus E_0)$. Therefore, each function $\lambda_i(q)$ realizes some vector λ_i in a universal completion of the band $\pi_n F = \{\mathscr{D}_{n-1}\}^{\perp\perp} \cap \{\mathscr{D}_n\}^{\perp}$ $(i = 1, \dots, n)$. The solution degenerates of the moment problem for the projections $\{\pi_n(s_k)\}_{k=1}^{\infty}$ to this band, becoming the simultaneous equations

$$\pi_n(s_k) = \sum_{i=1}^{n} c_i^{(n)} \lambda_i^k \quad (k = 0, 1, 2, \dots)$$

for some $c_i^{(n)} \geqslant 0$ in a universal completion of the space $\pi_n F$. Consider the spectral measures $\nu_i^{(n)}$ for the elements λ_i $(i = 1, \dots, n)$ (with respect to the order-unity $\pi_n(s_0)$). By unique solvability of the moment problem in this band, we derive the equality

$$\pi_n \circ \mu_0 = \sum_{i=1}^{n} \mu_i^{(n)}, \tag{6.21}$$

where $\mu_i^{(n)} = c_i^{(n)} \nu_i^{(n)}$ $(i = 1, \dots, n)$ are also spectral measures in the sense of our definition which take values in the original space F. The disjointness of $\{\mu_i^{(n)}\}_{i=1}^n$ follows from the fact that all roots $\{\lambda_i(q)\}_{i=1}^n$ $(q \in Q_n \setminus E_0)$ are distinct. Uniqueness of the representation (6.21) (in the sense of Item (4) of Theorem 5.6.3) follows from the fact that the roots $\{\lambda_i(q)\}_{i=1}^n$ are uniquely determined to within renumbering on clopen subsets of Q_n. The proof of Theorem 5.6.3 is complete.

5.7. The Hamburger Moment Problem for Dominant Moment Sequences

Solution of the Hamburger problem in a *bo*-complete lattice normed space $(Y, |\cdot|, F)$ reduces actually to Theorem 5.6.2 in the case when this space is decomposable (in the Kantorovich sense). The situation changes drastically in the general case when we do not assume that Y is decomposable. Some technical difficulties arise invoking the extra hypothesis (7.2) in the statement of Theorem 5.7.1 which amounts to the well-known Hardy condition in the scalar case (cf. [23]).

5.7.1. Theorem. *Assume given a sequence $\{y_k\}_{k=0}^{\infty}$ in an o-complete lattice normed space Y. Suppose that the sequence $\{s_k\}_{k=0}^{\infty} \subset F$ satisfies the conditions*

$$\left| \sum_{k,l=0}^{n} y_{k+l} \sigma_k \sigma_l \right| \leqslant \sum_{k,l=0}^{n} s_{k+l} \sigma_k \sigma_l \tag{7.1}$$

$$(\{\sigma_k\}_{k=0}^{\infty} \subset \mathbb{R}, n = 0, 1, 2 \dots);$$

the positive series

$$\sum_{k=0}^{\infty} \frac{a^{2k} s_{2k}}{(2k)} \tag{7.2}$$

converges in the Kantorovich space F for some $a \in \mathbb{R}$, $a > 0$.

Then there is a unique Borel measure $\mu : \mathscr{B}(\mathbb{R}) \to Y$ having bounded vector variation and satisfying

$$y_k = \int_{\mathbb{R}} \lambda^k \mu(d\lambda) \quad (k = 0, 1, 2 \dots). \tag{7.3}$$

◁ From (7.1) it follows that $\{s_k\}_{k=0}^{\infty}$ is a positive-definite sequence. Moreover, by Theorem 5.6.2 there is a positive Borel measure $\nu : \mathscr{B}(\mathbb{R}) \to F$ satisfying

$$s_k = \int_{\mathbb{R}} \lambda^k \nu(d\lambda) \quad (k = 0, 1, 2 \dots).$$

Since the series (7.2) converges, for some $a > 0$ the function $e^{a|\lambda|}$ belongs to $\mathscr{L}_1(\nu)$. Define the operator $T : \mathscr{P}(\mathbb{R}) \to Y$ by the formulas

$$Tp = \sum_{k=0}^{n} c_k y_k, \text{ with } p(\lambda) = \sum_{k=0}^{n} c_k \lambda^k \quad (\lambda \in \mathbb{R}).$$

Using (7.1), we see that for $p \in \mathscr{P}(\mathbb{R})$, $p \geqslant 0$, the estimate holds $|Tp| \leqslant Up$, where the dominant U is the integral

$$U f = \int_{\mathbb{R}} f(\lambda)\nu(d\lambda) \quad (f \in \mathscr{L}_1(\nu)).$$

Assume that $q \in \mathscr{P}(\mathbb{R})$, $0 < \beta < a$ and $\{p_k(\lambda)\}_{k=1}^{\infty}$ is a sequence of polynomials convergent pointwise to $\cos(\beta\lambda/2)$ and satisfying the estimates $|p_k(\lambda)| \leqslant e^{\beta|\lambda|/2}$ $(k \in \mathbb{N},\ \lambda \in \mathbb{R})$. (These conditions are fulfilled, for instance, by the sequence of partial sums of the Taylor series expansion for the function $\cos(\beta\lambda/2)$.) For all $\varepsilon > 0$ and $n \in \mathbb{R}$ there is $m \in \mathbb{R}$ satisfying

$$\begin{aligned} &|q(\lambda)[(p_k(\lambda))^2 - (p_l(\lambda))^2]| \\ &\leqslant \varepsilon + \frac{\lambda^2}{n^2}[(q(\lambda))^2 + 1][(p_k(\lambda))^2 + (p_l(\lambda))^2] = r_{k,l}(\lambda) \end{aligned}$$

for all $\lambda \in \mathbb{R}$, $k, l \in \mathbb{N}$, $k \geqslant m$, $l \geqslant m$. The polynomials

$$q_{k,l}(\lambda) = r_{k,l}(\lambda) - q(\lambda)[(p_k(\lambda))^2 - (p_l(\lambda))^2]$$

are positive on $\mathbb{R}$. Consequently,

$$|T(qp_k^2) - T(qp_l^2)| \leqslant \int (q_{k,l} + r_{k,l})d\nu \leqslant 3 \int r_{k,l} d\nu$$

$$= 3\,\varepsilon \int_{\mathbb{R}} \nu(d\lambda) + \frac{6}{n}^{2} \int_{\mathbb{R}} \lambda^2[(q(\lambda))^2 + 1]e^{\beta|\lambda|}\nu(d\lambda).$$

This proves that $\{T(q p_k^2)\}_{k=1}^{\infty}$ is a *bo*-Cauchy sequence. The above arguments show also that $\{T(q q_k^2)\}_{k=1}^{\infty}$, with $q_k(\lambda) = p_k(\frac{\pi}{2\beta} - \lambda)$ $(k \in \mathbb{N})$, is an *o*-Cauchy sequence as well.

Now, given the function

$$g(\lambda) = p(\lambda) + q(\lambda)\cos\beta\lambda + r(\lambda)\sin\beta\lambda \quad (p,\, q,\, r \in \mathscr{P}(\mathbb{R})),$$

define

$$T_1 g = Tp + bo\text{-}\lim_k \{T(q(2p_k^2 - 1)) + T(r(2q_k^2 - 1))\}.$$

We have thus come to some linear operator $T_1 : \mathscr{P}_1(\mathbb{R}) \to Y$ on the space $\mathscr{P}_1(\mathbb{R})$ of functions of the form

$$g(\lambda) = p(\lambda) + q(\lambda)\cos\beta\lambda + r(\lambda)\sin\beta\lambda \quad (p, q, r \in \mathscr{P}(\mathbb{R})).$$

Let a so-presented g be positive. Then, for all $\varepsilon > 0$ and $n \in \mathbb{N}$ there is $m \in \mathbb{N}$ satisfying

$$\begin{aligned} -\varepsilon - \frac{\lambda^2}{n^2}\{[p(\lambda))^2 + 1] + [(q(\lambda)^2 + 1][2(p_k(\lambda))^2 + 1] \\ +[(r(\lambda))^2 + 1][2(q_k(\lambda))^2 + 1]\} \\ \leqslant p(\lambda) + q(\lambda)[2(p_k(\lambda))^2 - 1] + r(\lambda)[2(q_k(\lambda))^2 - 1] \end{aligned}$$

for all $k \in \mathbb{N}$, $k \geqslant m$. Analogous arguments demonstrate that

$$\begin{aligned} |T_1 g| \leqslant \int_{\mathbb{R}} g(\lambda)\nu(d\lambda) + 2\varepsilon \int_{\mathbb{R}} \nu(d\lambda) \\ +\frac{2}{n^2}\int_{\mathbb{R}} \lambda^2\{(p(\lambda))^2 + 1 + [(q(\lambda))^2 + 1](2e^{\beta|\lambda|} + 1) \\ + [(r(\lambda))^2 + 1](2e^{\frac{\pi}{2}+\beta|\lambda|} + 1)\}\nu(d\lambda). \end{aligned}$$

This implies the estimate

$$|T_1 g| \leqslant \int_{\mathbb{R}} g d\nu \quad (g \in \mathscr{P}_1(\mathbb{R}),\ g \geqslant 0).$$

Fixing $p, q, r, s, t \in \mathscr{P}(\mathbb{R})$, put

$$\begin{aligned} g_k(\lambda) = p(\lambda) \,+\, q(\lambda)\cos\beta\lambda \,+\, r(\lambda)\sin\beta\lambda \\ +s(\lambda)\{2[(p_k(\lambda))^2 - 1]\cos\beta\lambda - [2(q_k(\lambda))^2 - 1]\sin\beta\lambda\} \\ +t(\lambda)\{[4(p_k(\lambda))^2 - 2]\sin\beta\lambda\}. \end{aligned}$$

It is now easy that $\{T_1 g_k\}_{k-1}^{\infty}$ is a *bo*-Cauchy sequence. We then consider the function

$$g(\lambda) = p(\lambda) + q(\lambda)\cos\beta\lambda + r(\lambda)\sin\beta\lambda + s(\lambda)\cos 2\beta\lambda + t(\lambda)\sin 2\beta\lambda$$

and put

$$T_2 g := bo\text{-}\lim_k T_1 g_k.$$

Continuing by induction, we obtain some linear operator $T_\beta : \mathscr{P}_\beta(\mathbb{R}) \to Y$ that is define on the space $\mathscr{P}_\beta(\mathbb{R})$ of functions of the form

$$s(\lambda) = \sum_{k=0}^{n} [c_k(\lambda)\cos k\beta\lambda + d_k(\lambda)\sin k\beta\lambda]$$

$$\big(\{c_k\}_{k=1}^n,\ \{d_k\}_{k=1}^n \subset \mathscr{P}(\mathbb{R}),\ n \in \mathbb{N}\big)$$

and extends T, satisfying the estimate

$$|T_\beta s| \leqslant \int_{\mathbb{R}} s\,d\nu \quad (s \in \mathscr{P}_\beta(\mathbb{R}),\ s \geqslant 0). \tag{7.4}$$

Consider the sequence $\beta_n = \beta/2^{n-1}$ $(n \in \mathbb{N})$ for some $0 < \beta < a$. Put $T_n = T_{\beta_n}$, $\mathscr{P}_n = \mathscr{P}_{\beta_n}(\mathbb{R})$. Clearly, $\mathscr{P}_n \subset \mathscr{P}_{n+1}$ $(n \in \mathbb{N})$.

It can be shown that the operator T_{n+1} extends T_n for all $n \in \mathbb{N}$. We omit the proof since it utilizes exactly the same technique. There is a unique linear operator $T_\infty : \mathscr{P}_\infty \to Y$ on the space $\mathscr{P}_\infty = \bigcup_n \mathscr{P}_n$ such that for all $n \in \mathbb{N}$ the restriction of T_∞ to $\mathscr{P}_n$ agrees with T_n. In $\mathscr{P}_\infty$ consider the linear subspace $\mathscr{P}_0$ of trigonometric polynomials of the form

$$s(\lambda) = \sum_{k=1}^{m} \left\{ c_k \cos\left(\frac{k\lambda}{2^{n-1}}\right) + d_k \sin\left(\frac{k\lambda}{2^{n-1}}\right) \right\}$$

$$\big(\{c_k\}_{k=1}^m,\ \{d_k\}_{k=1}^m \subset \mathbb{R},\ m, n \in \mathbb{N}\big).$$

Denote by T_0 the restriction of T_∞ to $\mathscr{P}_0$. By uniform continuity, the operator T_0 extends uniquely to the linear operator $\overline{T}_0 : \overline{\mathscr{P}}_0 \to Y$ on the uniform closure $\overline{\mathscr{P}}_0$ of the space $\mathscr{P}_0$. Obviously, $\overline{\mathscr{P}}_0$ includes the subspace $\mathscr{P}_*$ of continuous periodic functions with periods $2^n\pi/\beta$ $(n \in \mathbb{N})$. The main implication of the above lengthy considerations is the fact that $\mathscr{P}_*$ is a function vector lattice. Let T_* stand for the restriction of $\overline{T}_0$ to the space $\mathscr{P}_*$. From (7.4) is immediate that T_* is a dominated operator. Its sequential *bo*-continuity is straightforward from the sequential *bo*-continuity of the dominant U. There is a unique Borel measure $\mu : \mathscr{B}(\mathbb{R}) \to Y$ such that

$$T_* s = \int_{\mathbb{R}} s d\mu \quad (s \in \mathscr{P}_*).$$

Moreover, the vector variation $|\mu|$ is less than or equal to ν. Show that μ enjoys (7.3). Given an even $k \in \mathbb{N}$ and a natural n, put

$$s_{k,n}(\lambda) = \left(\lambda - \frac{\pi \cdot m \cdot 2^n}{\beta}\right)^k,$$

if

$$\left|\lambda - \frac{\pi \cdot 2^{n-1}}{\beta}\right| \leqslant \frac{\pi \cdot 2^{n-1}}{\beta} \quad (m \in \mathbb{Z}).$$

Given an odd $k \in \mathbb{N}$ and a natural n, put

$$s_{k,n}(\lambda) = (-1)^m \left(\lambda - \frac{\pi \cdot m \cdot 2^n}{\beta}\right)^k$$

for λ satisfying the previous inequality. All functions $s_{k,n}$ $(k, n \in \mathbb{N})$ are continuous and periodic with period $\pi \cdot 2^n/\beta$. For all $k, n \in \mathbb{N}$ and $\varepsilon > 0$ there is a trigonometric polynomial $t_{k,n} \in \mathscr{P}_0$ satisfying

$$|s_{k,n}(\lambda) - t_{k,n}(\lambda)| < \varepsilon \quad (\lambda \in \mathbb{R}).$$

This means that

$$|\lambda^k - t_{k,n}(\lambda)| \leqslant \varepsilon + \frac{\lambda^2 \beta^2}{\pi^2 \cdot 2^{2n}} (\lambda^{2n} + 1).$$

We thus have the inequality

$$|T_\infty(f_k) - T_\infty(t_{k,n})| \leqslant 3\varepsilon \int_{\mathbb{R}} \nu(d\lambda) + \frac{3\beta^2}{\pi^2 \cdot 2^{2n-2}} \int_{\mathbb{R}} \lambda^2(\lambda^{2k} + 1)\nu(d\lambda),$$

where $f_k(\lambda) = \lambda^k$ $(\lambda \in \mathbb{R})$. Since $T_\infty(f_k) = y_k$, it follows that

$$\left|y_k - \int \lambda^k \mu(d\lambda)\right|$$
$$\leqslant |y_k - T_\infty(t_{k,n})| + \left|\int (t_{k,n} - f_k) d\nu\right|$$
$$\leqslant 4\varepsilon U(1) + \frac{4\beta^2}{\pi^2 \cdot 2^{2n}} \int_{\mathbb{R}} \lambda^2(\lambda^{2k} + 1)\nu(d\lambda).$$

Since $\varepsilon > 0$ and $n \in \mathbb{N}$ are arbitrary, we immediately derive (7.3). The proof of uniqueness for μ repeats the above constructions in many points and so we omit it. ▷

As examples of nondecomposable lattice normed space we list countably-normed spaces as well as arbitrary locally convex spaces. We take as the vector norm of a vector the numeric family of the values of all seminorms at this vector. The *bo*-convergence in this case amounts to the topological convergence for bounded nets. Another example is an ordered vector space having a strong order-unity. The norm lattice in this case is a Dedekind completion of the original space.

5.7.2. Theorem. *Let Y be a monotonically complete ordered vector space. Assume given a positive-definite sequence $\{s_k\}_{k=0}^{\infty} \subset Y$ such that the series (7.2) converges for some $a > 0$. Then there is a unique positive Borel measure $\nu : \mathscr{B}(\mathbb{R}) \to F$ satisfying*

$$s_k = \int_{\mathbb{R}} \lambda^k \nu(d\lambda) \quad (k = 0, 1, 2 \dots).$$

◁ Let F stand for a Dedekind completion of the order ideal $Y(s)$, with s the sum of the series (7.2). Clearly, F is a Kantorovich space. There is a unique Borel measure $\nu : \mathscr{B}(\mathbb{R}) \to F$ solving our moment problem (we apply Theorem 5.6.2 with $Y = F$). Our aim is to show that $\nu(\mathscr{B}(\mathbb{R})) \subset Y$. By condition, for all $s \in \mathscr{P}(\mathbb{R})$ we have

$$\int_{\mathbb{R}} s(\lambda)\nu(d\lambda) \in Y.$$

We will demonstrate that this inclusion remains valid for every trigonometric polynomial $s \in \mathscr{P}_0$. As an example we take the function $\sin \beta\lambda$ $(0 < \beta < a)$. The Taylor series expansion of $\cos(\beta\lambda/2)$ is the difference of two positive series. Therefore,

$$\int_{\mathbb{R}} \sin \beta\lambda \nu(d\lambda) = \int_{\mathbb{R}} \left\{ 2\cos^2\left(\frac{\pi}{4} - \frac{\beta\lambda}{2}\right) - 1 \right\} \nu(d\lambda) \in Y.$$

Since each continuous periodic function on $\mathbb{R}$ is the uniform limit of a monotone sequence of trigonometric polynomials, the required inclusion holds for this new class of functions. Using only the limits of monotone sequences, we can obtain an arbitrary characteristic function χ_B $(B \in \mathscr{B}(\mathbb{R}))$. Consequently, $\nu(B) \in Y$ $(B \in \mathscr{B}(\mathbb{R}))$. ▷

It is well known that the space of bounded selfadjoint operators in Hilbert space is monotonically complete with respect to the natural order relation. We thus obtain the following (cf. [38]):

5.7.3. Corollary. *The Hamburger moment problem is uniquely solvable for every positive-definite sequence of bounded selfadjoint operators in Hilbert space for which the series (7.2) converges in the weak operator topology (with some $a > 0$).*

5.8. Dominated Mappings

Let F be a universally complete Kantorovich space with order-unity $\mathbf{1}$. Furnish F with multiplication so that $\mathbf{1}$ becomes a ring-unity (see [8]). Denote the complexification of F by $F_{\mathbb{C}}$. Assume that Y is a vector space over $\mathbb{C}$. A *lattice norm* or *F-norm* is a mapping $|\cdot| : Y \to F$ satisfying the axioms $|x| = 0 \leftrightarrow x = 0$, $|x+y| \leqslant |x|+|y|$, $|cx| = |c||x|$ (x, $y \in Y$, $c \in \mathbb{C}$). The triple $(Y, |\cdot|, F)$ is a *complex lattice normed space* (cf. [9]). We view $F_{\mathbb{C}}$ as a lattice normed space with lattice norm $|a + ib| = (a^2 + b^2)^{1/2}$ (a, $b \in F$). Moreover, we put $C_0(\mathfrak{X}, \mathbb{C}) := C_0(\mathfrak{X})_{\mathbb{C}}$, $C_{00}(\mathfrak{X}, \mathbb{C}) := C_{00}(\mathfrak{X})_{\mathbb{C}}$.

5.8.1. REMARK. If is possible to take as a norm lattice of Y an arbitrary Archimedean vector lattice. However, such a lattice may be naturally embedded into a universally complete Kantorovich space.

Let G stand for an arbitrary group.

5.8.2. DEFINITION. A mapping $\psi : G \to F_{\mathbb{C}}$ is *positive-definite* provided that all elements of the form

$$\sum_{j,k=1}^{n} \overline{c_j} c_k \psi(g_j^{-1} g_k)$$

belong to F_+ for all $n \in \mathbb{N}$, $g_1, \dots, g_n \in G$, $c_1, \dots, c_n \in \mathbb{C}$.

5.8.3. DEFINITION. A mapping $\varphi : G \to Y$ is *dominated* if there is a positive-definite mapping $\psi : G \to F_{\mathbb{C}}$ satisfying

$$\left| \sum_{j,k=1}^{n} \overline{c_j} c_k \varphi(g_j^{-1} g_k) \right| \leqslant \sum_{j,k=1}^{n} \overline{c_j} c_k \psi(g_j^{-1} g_k)$$

for all $n \in \mathbb{N}$, $g_1, \dots, g_n \in G$, and $c_1, \dots, c_n \in \mathbb{C}$. In this event, ψ is a *dominant* for φ.

5.8.4. Theorem. *Suppose that a mapping $\varphi : G \to Y$ has some dominant $\psi : G \to F_{\mathbb{C}}$. The following inequality holds:*

$$\left| \sum_{j,k=1}^{n} \overline{c_j} d_k \varphi(g_j^{-1} g_k) \right|^2$$

$$\leqslant 4\Big(\sum_{j,k=1}^{n} \overline{c_j} c_k \psi(g_j^{-1} g_k)\Big)\Big(\sum_{j,k=1}^{n} \overline{d_j} d_k \psi(g_j^{-1} g_k)\Big)$$

for all $n \in \mathbb{N}$, $g_1, \dots, g_n \in G$, *and* $c_1, \dots, c_n, d_1, \dots, d_n \in \mathbb{C}$. *If* G *is a non-Abelian locally compact group then there is a mapping* $\varphi : G \to \mathbb{C}$ *and a continuous dominant* $\psi : G \to \mathbb{C}$ *for* φ *such that for all* $\varepsilon > 0$ *the reverse inequality holds:*

$$\Big|\sum_{j,k=1}^{n} \overline{c_j} d_k \varphi(g_j^{-1} g_k)\Big|^2$$

$$> (4-\varepsilon)\Big(\sum_{j,k=1}^{n} \overline{c_j} c_k \psi(g_j^{-1} g_k)\Big)\Big(\sum_{j,k=1}^{n} \overline{d_j} d_k \psi(g_j^{-1} g_k)\Big)$$

for all $n \in \mathbb{N}$, $g_1, \dots, g_n \in G$, *and* $c_1, \dots, c_n, d_1, \dots, d_n \in \mathbb{C}$.

$\lhd$ Take $n \in \mathbb{N}$ and $g_1, \dots, g_n \in G$. Consider the sesquilinear form $\Phi : \mathbb{C}^n \times \mathbb{C}^n \to Y$ determined from φ by the rule:

$$\Phi(c, d) = \sum_{j,k=1}^{n} \overline{c_j} d_k \varphi(g_j^{-1} g_k),$$

where $c = (c_1, \dots, c_n) \in \mathbb{C}^n$ and $d = (d_1, \dots, d_n) \in \mathbb{C}^n$. By analogy, the mapping $\psi : G \to F_{\mathbb{C}}$ generates the sesquilinear form $\Psi : \mathbb{C}^n \times \mathbb{C}^n \to F_{\mathbb{C}}$. The definition of domination implies the inequality $|\Phi(c,c)| \leqslant \Psi(c,c)$ $(c \in \mathbb{C}^n)$. For all $\alpha \in \mathbb{C}$ it now follows that

$$|\Phi(c,c) \pm (\alpha\Phi(c,d) + \overline{\alpha}\Phi(d,c)) + |\alpha|^2 \Phi(d,d)|$$
$$\leqslant \Psi(c,c) \pm (\alpha\Psi(c,d) + \overline{\alpha}\Psi(d,c)) + |\alpha|^2 \Psi(d,d).$$

Sum up these inequalities with the plus and minus signs to obtain

$$|\alpha\Phi(c,d) + \overline{\alpha}\Phi(d,c)| \leqslant \Psi(c,c) + |\alpha|^2 \Psi(d,d).$$

We first put $\alpha = t$; second, $\alpha = it$ $(t \in \mathbb{R})$; summing up the results again, we then deduce that

$$t|\Phi(c,d)| \leqslant \Psi(c,c) + t^2 \Psi(d,d).$$

Solving this inequality for t in F, obtain

$$|\Phi(c,d)|^2 - 4\Psi(c,c)\Psi(d,d) \leqslant 0 \quad (c, d \in \mathbb{C}^n).$$

To within notation, this is the sought relation.

Let G stand for a non-Abelian locally compact group. Prove that the constant 4 is unimprovable in the last inequality. Take two noncommuting elements $a, b \in G$ and a symmetric compact neighborhood U of the unity for which $ab \notin Uba$. In G there is a σ-compact clopen subgroup G_1 including the neighborhood U and the elements a and b (see [39, Theorem 5.14]). In G_1 there is a compact normal subgroup H_1 such that $H_1 \subseteq U$ and the factor-group $\widehat{G}_1 = G_1/H_1$ is metrizable and separable (see [39, Theorem 8.7]). Denote by p the canonical homomorphism of G_1 onto $\widehat{G}_1$. Consider the Hilbert space $L^2(\widehat{G}_1, \lambda)$, with λ the left Haar measure on the group $\widehat{G}_1$. The regular left representation π_r, acting in $L^2(\widehat{G}_1, \lambda)$, admits the direct integral decomposition of irreducible unitary representations with respect to some measure space (Ω, Σ, μ), i.e.,

$$\pi_r = \int_{\Omega}^{\oplus} \pi_\omega \mu(d\omega)$$

(see [40, Ch. 8, Theorem 3]). Since $\widehat{G}_1$ is a non-Abelian group, the dimension of the representation π_ω is greater than 1 for at least one $\omega \in \Omega$. Assume that such a representation π_ω acts in the Hilbert space $\mathscr{H}_\omega$. Consider the von Neumann algebra $\mathscr{A}$ generated by the set of operators $\pi_\omega(\widehat{G}_1)$. Irreducibility implies in fact that $\mathscr{A}$ is a von Neumann factor.

We now demonstrate the following: there are two orthonormal vectors $e_1, e_2 \in \mathscr{H}_\omega$ and two operators $Q, P \in \mathscr{A}$ satisfying

$$Qe_1 = e_2, \quad Qe_2 = 0, \quad Pe_1 = 0, \quad Pe_2 = e_2.$$

The two cases are possible:

(1) $\mathscr{A}$ is a type I factor. Since $\dim \mathscr{H}_\omega > 1$, there are two minimal nonzero subspaces $\mathscr{H}_1$ and $\mathscr{H}_2$ orthogonal to one another and adjoint to $\mathscr{A}$ (see [40]). These subspaces are equivalent with respect to $\mathscr{A}$; consequently, there is a partial isometry Q of the space $\mathscr{H}_1$ onto $\mathscr{H}_2$. We now take a unit vector $e_1 \in \mathscr{H}_1$ and put $e_2 = Qe_1$. As P we take the orthoprojection to $\mathscr{H}_2$.

(2) $\mathscr{A}$ is a type II or type III factor (this happens for the so-called "wild" groups). On these cases we may also find two nonzero orthogonal subspaces $\mathscr{H}_1$ and $\mathscr{H}_2$ adjoint to $\mathscr{A}$ and of the same relative dimension. We then choose e_1, e_2, Q, and P in exactly the same manner as in Case (1).

The operators Q and P belong to the closure of the linear span of $\pi_\omega(\widehat{G}_1)$ with respect to the strong operator topology. Therefore, for each $\delta > 0$ there are elements $g_1, \dots, g_n \in G$ and $c_1, \dots, c_n, d_1, \dots, d_n \in \mathbb{C}$ such that

$$\Big\| Qe_1 - \sum_{j=1}^{n} c_j \pi_\omega(g_j) e_1 \Big\| < \delta, \quad \Big\| \sum_{j=1}^{n} c_j \pi_\omega(g_j) e_2 \Big\| < \delta,$$

$$\Big\| Pe_2 - \sum_{j=1}^{n} c_j \pi_\omega(g_j) e_2 \Big\| < \delta, \quad \Big\| \sum_{j=1}^{n} d_j \pi_\omega(g_j) e_1 \Big\| < \delta.$$

Put $\widehat{\varphi}(g) = (\pi_\omega(g)e_1, e_2)$ $(g \in \widehat{G}_1)$. The relations

$$\Big| \sum_{j,k=1}^{n} \overline{c_j} c_k \varphi(g_j^{-1} g_k) \Big| = \Big| \Big(\sum_{j,k=1}^{n} \overline{c_j} c_k \pi_\omega(g_j^{-1}) \pi_\omega(g_k) e_1, e_2 \Big) \Big|$$

$$= \Big| \Big(\sum_{k=1}^{n} c_k \pi_\omega(g_k) e_1, \sum_{j=1}^{n} c_j \pi_\omega(g_j) e_2 \Big) \Big|$$

$$\leqslant \frac{1}{2} \sum_{j,k=1}^{n} \overline{c_j} c_k [(\pi_\omega(g_k)e_1, \pi_\omega(g_j)e_1) + (\pi_\omega(g_k)e_2, \pi_\omega(g_j)e_2)]$$

imply that the function

$$\widehat{\psi}(g) = \frac{1}{2}[(\pi_\omega(g)e_1, e_1) + (\pi_\omega(g)e_2, e_2)] \quad (g \in \widehat{G}_1)$$

is a dominant for $\widehat{\varphi}$. From the inequalities

$$\Big| \sum_{j,k=1}^{n} \overline{c_j} d_k (\pi_\omega(g_j^{-1} g_k) e_1, e_2) \Big| \geqslant (Qe_1, Pe_2) - 2\delta - \delta^2,$$

$$\Big| \sum_{j,k=1}^{n} \overline{c_j} c_k \widehat{\psi}(g_j^{-1} g_k) \Big| \leqslant \frac{1}{2}[(Qc_1, Qc_1) + 2\delta + \delta^2],$$

$$\Big| \sum_{j,k=1}^{n} \overline{d_j} d_k \widehat{\psi}(g_j^{-1} g_k) \Big| \leqslant \frac{1}{2}[(Pe_2, Pe_2) + 2\delta + 2\delta^2]$$

the claim of the theorem is immediate for a metrizable separable group $\widehat{G}_1$, on taking δ sufficiently small. To demonstrate the theorem for the initial group G, it suffices to put

$$\varphi(g) = \begin{cases} \widehat{\varphi} \circ p(g) & (g \in G_1), \\ 0 & (g \in G \setminus G_1), \end{cases} \quad \psi(g) = \begin{cases} \widehat{\psi} \circ p(g) & (g \in G_1), \\ 0 & (g \in G \setminus G_1). \end{cases} \ \triangleright$$

Using the Schur Lemma, we easily derive

5.8.5. Corollary. *If G has a finite-rank irreducible unitary representation of dimension greater than 1 (for instance, if G is a non-Abelian almost periodic locally compact group; cf. [41]); then there are a continuous function $\varphi : G \to \mathbb{C}$ and a continuous dominant $\psi : G \to \mathbb{C}$ of φ such that*

$$\left(\sum_{j,k=1}^{n} \overline{c_j} d_k \varphi(g_j^{-1} g_k) \right)$$

$$= 4 \left(\sum_{j,k=1}^{n} \overline{c_j} c_k \psi(g_j^{-1} g_k) \right) \left(\sum_{j,k=1}^{n} \overline{d_j} d_k \psi(g_j^{-1} g_k) \right) > 0$$

for all $n \in \mathbb{N}$, $g_1, \dots, g_n \in G$, $c_1, \dots, c_n, d_1, \dots, d_n \in \mathbb{C}$.

The situation is completely different for the Abelian groups. Namely, we may substitute 1 for 4 in the last inequality, thus arriving at a dominant analog of the Cauchy–Bunyakovskiĭ–Schwarz inequality.

5.8.6. Theorem. *Assume that G is an Abelian group and a mapping $\varphi : G \to Y$ has a dominant $\psi : G \to F_{\mathbb{C}}$. For all $n \in \mathbb{N}$, $g_1, \dots, g_n \in G$, $c_1, \dots, c_n, d_1, \dots, d_n \in \mathbb{C}$ the inequality holds:*

$$\left| \sum_{j,k=1}^{n} \overline{c_j} d_k \varphi(g_k - g_j) \right|^2$$

$$\leqslant \left(\sum_{j,k=1}^{n} \overline{c_j} c_k \psi(g_k - g_j) \right) \left(\sum_{j,k=1}^{n} \overline{d_j} d_k \psi(g_k - g_j) \right).$$

$\lhd$ Fix some $g_1, \dots, g_n \in G$ and consider the subgroup H of G generated by these elements. Let p be a homomorphism of the free group $\mathbb{Z}^n$ onto H. Clearly, the mapping $\psi \circ p$ is a dominant for $\varphi \circ p$. If we prove the theorem for the mappings $\varphi \circ p$ and $\psi \circ p$, then it appears valid for the original mapping φ and ψ as a corollary. We thus loss no generality in assuming that $G = \mathbb{Z}^n$ from the very beginning.

Further, we may solve the trigonometric moment problem for the mappings φ and ψ on the dual group of $\mathbb{Z}^n$. Observe that this way presumes the order completeness of Y. We therefore give another proof.

Denote by $\mathscr{P}^n$ the space of trigonometric polynomials of the form

$$p(x) = \sum_{k=1}^{l} c_k e^{i\langle g_k, x \rangle}$$

where $c_1, \dots, c_l \in \mathbb{C}$, $g_1, \dots, g_l \in \mathbb{Z}^n$, and $\langle \cdot , \cdot \rangle$ is the usual inner product in $\mathbb{R}^n$. The degree of this polynomial is

$$\deg p = \max_{1 \leqslant k \leqslant l} \sum_{j=1}^{n} |g_k(j)|.$$

Consider the n-multiple Fejér kernel

$$\Delta_m^{(n)}(x) = \frac{1}{[2m(m+1)]^n} \prod_{j=1}^{n} \frac{\sin^2(mx_j/2)}{\sin^2(x_j/2)}.$$

Given $p \in \mathscr{P}^n$, put

$$p_{(m)}(x) = \sum_{k_1=-(m+1)}^{m} \cdots \sum_{k_n=-(m+1)}^{m} p\left(\frac{\pi k}{m+1}\right) \Delta_m^{(n)}\left(x - \frac{\pi k}{m+1}\right),$$

where $k = (k_1, \dots, k_n) \in \mathbb{Z}^n$ and $x = (x_1, \dots, x_n) \in \mathbb{R}^n$, while $m \in \mathbb{N}$ is odd. Clearly, if

$$p(x) = \sum_{k=1} c_k e^{i\langle g_k , x \rangle},$$

then for $m > \deg p$ we have the equality

$$p_{(m)}(x) = \sum_{k=1}^{l} c_k e^{i\langle g_k , x\rangle} \prod_{j=1}^{m} \left(1 - \frac{|g_k(j)|}{m}\right) \quad (x \in \mathbb{R}^n),$$

i.e., $\deg p_{(m)} = \deg p$. If $p \in \mathscr{P}^n$ has the above representation then we put

$$T_\varphi(p) = \sum_{k=1}^{l} c_k \varphi(g_k), \quad T_\psi(p) = \sum_{k=1}^{l} c_k \psi(g_k).$$

We have thus constructed two linear operators $T_\varphi : \mathscr{P}^n \to Y$ and $T_\psi : \mathscr{P}^n \to F_{\mathbb{C}}$. Note that

$$|T_\varphi(p - p_{(m)})| \leqslant \sum_{k=1}^{l} |c_k| \left[1 - \prod_{j=1}^{n} \left(1 - \frac{|g_k(j)|}{m}\right)\right] |\varphi(g_k)|.$$

Consequently, for each $\varepsilon > 0$ there is an odd m satisfying

$$|T_\varphi(p - p_{(m)})| \leqslant \varepsilon \sum_{k=1}^{l} |\varphi(g_k)|. \tag{8.1}$$

An analogous inequality is valid for the operator T_ψ. Since $\Delta_m^{(n)}$ is the square of a trigonometric polynomial, from the definition of domination it follows that the following equality holds:

$$\left|T_\varphi(\Delta_m^{(n)})\right| \leqslant T_\psi(\Delta_m^{(n)}).$$

Consider another trigonometric polynomial:

$$q(x) = \sum_{k=1}^{l} d_k e^{i\langle g_k, x\rangle}.$$

We have the following inequalities:

$$|T_\varphi(\overline{p}q)_{(m)}|^2 \leqslant \left| \sum_{k=(k_1,\dots,k_n)} \overline{p}\left(\frac{\pi k}{m+1}\right) q\left(\frac{\pi k}{m+1}\right) T_\psi(\Delta_m^{(n)}) \right|^2$$

$$\leqslant \left(\sum_{k=(k_1,\dots,k_n)} \left| p\left(\frac{\pi k}{m+1}\right)\right|^2 T_\psi(\Delta_m^{(n)}) \right)$$

$$\times \left(\sum_{k=(k_1,\dots,k_n)} \left| q\left(\frac{\pi k}{m+1}\right)\right|^2 T_\psi(\Delta_m^{(n)}) \right) = T_\psi\big(|p|^2_{(m)}\big) \cdot T_\psi\big(|q|^2_{(m)}\big). \qquad (8.2)$$

Given $\varepsilon > 0$, we may find m so that (8.1) becomes valid on replacing p with $\overline{p}q$, $|p|^2$, $|q|^2$. Therefore, passing to the r-limit in (8.2), we obtain

$$|T_\varphi(\overline{p}q)|^2 \leqslant T_\psi(|p|^2) \cdot T_\psi(|q|^2). \ \triangleright$$

5.8.7. Remark. Theorem 5.8.6 holds also for the dominated mapping on a Kreĭn block-algebra.

5.8.8. *Corollary.* *If $\varphi : G \to Y$ has a dominant $\psi : G \to F_{\mathbb{C}}$ o-continuous at zero then φ is an order bounded uniformly bo-continuous mapping.*

$\triangleleft$ The second inequality of Theorem 5.8.6 implies $|\varphi(g)| \leqslant 2\psi(0)$ $(g \in G)$ on putting $n = 1$ and $c_1 = 1$. The same inequality with $n = 2$ and $c_1 = -c_2 = 1$ provides the estimate

$$|\varphi(g_1) - \varphi(g_2)|^2 \leqslant 8\psi(0)(\psi(0) - \operatorname{Re}\psi(g_1 - g_2)) \quad (g_1, g_2 \in G),$$

implying uniform bo-continuity at a moment's thought φ. $\triangleright$

Let $\lambda : \mathscr{B}(G) \to \overline{\mathbb{R}}^+$ be a positive measure finite at compact subsets of G.

5.8.9. Theorem. *If $\varphi : G \to Y$ has a dominant $\psi : G \to F_{\mathbb{C}}$ order continuous at zero then the following integral inequalities hold:*

$$\Bigg| \int\limits_{G_1 \times G_2} \varphi(g-h)u(g)\overline{v(h)}\lambda(dg)\lambda(dh)\Bigg|^2$$

$$\leqslant \Bigg(\int\limits_{G\times G} \psi(g-h)u(g)\overline{u(h)}\lambda(dg)\lambda(dh)\Bigg)\Bigg(\int\limits_{G\times G} \psi(g-h)v(g)\overline{v(h)}\lambda(dg)\lambda(dh)\Bigg),$$

$$\Bigg| \int\limits_{G} \varphi(g)u(g)\lambda(dg)\Bigg|^2 \leqslant \psi(0)\Bigg(\int\limits_{G\times G} \psi(g-h)u(g)\overline{u(h)}\lambda(dg)\lambda(dh)\Bigg).$$

◁ Note first that the mappings φ and ψ are uniformly *bo*-continuous. Therefore, all integrals make sense (they are understood to be integrals over some compact set including the supports of the functions u and v).

Given a neighborhood $U \in \mathscr{U}_0$, put

$$h_U = \sup\{8\psi(0)(\psi(0) - \psi(g_1 - g_2)) : g_1, g_2 \in G,\ g_1 - g_2 \in U\}.$$

Let $V \in \mathscr{U}_0$ be a symmetric zero neighborhood such that $V + V \subseteq U$. Considering the compact set $K = (\operatorname{supp} u) \cup (\operatorname{supp} v)$, look at the partition of unity $f_1, \dots, f_n \in C_{00}(G)_+$ such that $\operatorname{supp} f_k - \operatorname{supp} f_k \subseteq V$ for all $k = 1, \dots, n$. Choose an element g_k in $\operatorname{supp} f_k$ $(k = 1, \dots, n)$. Putting

$$c_k = \int\limits_G u(g)f_k(g)\lambda(dg), \quad d_k = \int\limits_G v(g)f_k(g)\lambda(dg),$$

we may write down the following inequality:

$$\Bigg| \int\limits_{G_1 \times G_2} \varphi(g-h)u(g)\overline{v(h)}\lambda(dg)\lambda(dh)\Bigg|$$

$$\leqslant \Bigg| \sum_{j,k=1}^{n} \varphi(g_j - g_k)c_j\overline{d_k}\Bigg| + h_U^{1/2}\|u\|_1\|v\|_1,$$

where the L_1-norm $\|\cdot\|_1$ is defined with respect to the measure λ. An analogous estimate is valid for the mapping ψ. Therefore, the first inequality of Theorem 5.8.6

implies the estimate

$$\begin{aligned}\Big|\int\limits_{G\times G}&\varphi(g-h)u(g)\overline{v(h)}\lambda(dg)\lambda(dh)\Big|^2\\ \leqslant&\Big(\int\limits_{G\times G}\psi(g-h)u(g)\overline{u(h)}\lambda(dg)\lambda(dh)\Big)\\ \times&\Big(\int\limits_{G\times G}\psi(g-h)v(g)\overline{v(h)}\lambda(dg)\lambda(dh)\Big)\\ +&(\|u\|_1+\|v\|_1)\big(2\psi(0)h_U^{1/2}+h_U\big).\end{aligned}$$

Since the net $\{h_U : U \in \mathscr{U}_0\}$ vanishes in order; passing to the limit with respect to $U \in \mathscr{U}_0$, we obtain the first of the desired inequalities. The second is proven in much the same way. ▷

5.9. The Bochner Theorem for Dominated Mappings

5.9.1. Suppose that G is a locally compact Abelian group and $\mathfrak{X}$ is the dual group of G.

Theorem. *For a mapping $\varphi : G \to Y$ the following are equivalent:*

(1) *φ has a dominant o-continuous at zero;*

(2) *there is a unique measure $\mu \in \operatorname{qca}(\mathfrak{X}, Y)$ satisfying*

$$\varphi(g) = \int\limits_{\mathfrak{X}} \chi(g)\mu(d\chi) \quad (g \in G).$$

◁ (2)→(1): Assume that the mapping $\varphi : G \to Y$ is represented by some measure $\mu \in \operatorname{qca}(\mathfrak{X}, Y)$ as claimed in (2). Put

$$\psi(g) = \int\limits_{\mathfrak{X}} \chi(g)|\mu|(d\chi) \quad (g \in G).$$

The mapping ψ is a dominant for φ as follows from the usual integral inequalities (cf. [39]). Prove that ψ is o-continuous at zero. Take $\varepsilon > 0$. To each compact set $K \in \mathscr{K}(\mathfrak{X})$ there is a zero neighborhood U in the group G such that $|1 - \chi(g)| < \varepsilon$ $(\chi \in K,\ g \in U)$. Therefore, for $g \in U$ we have the estimate

$$|\psi(0) - \psi(g)| \leqslant \varepsilon\psi(0) + 2|\mu|(\mathfrak{X} \setminus K).$$

Whence it follows in view of the quasi-Radon property of ν that

$$\inf_{U\in\mathscr{U}_0}\sup\{|\psi(0)-\psi(g)| : g\in U\}=0.$$

(1)→(2): Let $\psi : G \to F_{\mathbb{C}}$ be a dominant for φ which is o-continuous at zero. In the sequel we let $\lambda : \mathscr{B}(G) \to \overline{\mathbb{R}}_+$ stand for the invariant Haar measure on G. Given $u, v \in C_{00}(G, \mathbb{C})$, note that the convolution $u * v$ with respect to the Haar measure λ belongs to $C_{00}(G, \mathbb{C})$ too. For all $u, v \in C_{00}(G, \mathbb{C})$ we have the equality

$$\int\limits_{G\times G}\varphi(g-h)u(g)\overline{v(h)}\lambda(dg)\lambda(dh)=\int\limits_{G}\varphi(g)(u*\breve{v})(g)\lambda(dg),$$

where $\breve{v}(g) = \overline{v(-g)}$ $(g \in G)$ is the *involution* on G. This is immediate from the possibility of changing the order of integration (Theorem 5.4.6) together with the formula

$$\int\limits_{G}\varphi(g-h)\overline{v(h)}\lambda(dh)=\int\limits_{G}\varphi(h)\overline{v(g-h)}\lambda(dh),$$

which is valid since λ is translation-invariant. Putting $Y = F_{\mathbb{C}}$ and $\varphi = \psi$, arrive at the same equality for the mapping ψ. Consider the two linear operators

$$\Phi(u)=\int\limits_{G}\varphi(g)u(g)\lambda(dg),\quad \Psi(u)=\int\limits_{G}\psi(g)u(g)\lambda(dg)\quad (u\in C_{00}(G,\mathbb{C})).$$

The just-indicated equalities, together with Theorem 5.8.9, imply the estimates

$$|\Phi(u*\breve{u})|\leqslant\Psi(u*\breve{u}),\quad |\Phi(u)|^2\leqslant 4\psi(0)\Psi(u*\breve{u})\quad (u\in C_{00}(G,\mathbb{C})).$$

Consider the Fourier transforms of the operators Φ and Ψ; i.e., $\widehat{\Phi}(f) := \Phi(\widehat{f})$ and $\widehat{\Psi}(f) = \Psi(\widehat{f})$ for all $\widehat{f} \in C_{00}(G, \mathbb{C})$. Here, $\widehat{f}$ stands for the Fourier transform on the dual group $\mathfrak{X}$ of the function $f \in C_0(\mathfrak{X}, \mathbb{C})$. The operators $\widehat{\Phi}$ and $\widehat{\Psi}$ are given on the subspace $\mathscr{L} \subset C_0(\mathfrak{X}, \mathbb{C})$ which is (the image under) the Fourier transform of the space $C_{00}(G, \mathbb{C})$. Therefore, $\mathscr{L}$ is uniformly dense in $C_0(\mathfrak{X}, \mathbb{C})$. The above inequalities imply

$$|\widehat{\Phi}(|f|^2)|\leqslant\Psi(|f|^2),\quad |\widehat{\Phi}(f)|^2\leqslant 4\psi(0)\widehat{\Psi}(|f|^2)\quad (f\in\mathscr{L}).$$

The proof is now completed along the standard lines. Arguing as in [39, Corollary 21.21], for all $n \in \mathbb{N}$, $f \in \mathscr{L}$ we derive the estimates

$$|\widehat{\Phi}(f)|\leqslant 4\psi(0)\|(|f|^{2^n})^\wedge\|_1^{2^{-n}},\quad |\widehat{\Psi}(f)|\leqslant 4\psi(0)\|(|f|^{2^n})^\wedge\|_1^{2^{-n}}$$

(with $\|\cdot\|_1$ standing for the L^1-norm with respect to the Haar measure λ). Passing in these estimates to the r-limit as $n \to \infty$ and using the well-known Gelfand Theorem, infer that

$$|\widehat{\Phi}(f)| \leqslant 4\psi(0)\|f\|_\infty, \quad |\widehat{\Psi}(f)| \leqslant 4\psi(0)\|f\|_\infty \quad (f \in \mathscr{L}).$$

Consequently, we may use r-continuity (with regulator $\psi(0)$) to extend the operators $\widehat{\Phi}$ and $\widehat{\Psi}$ to the whole space $C_0(\mathfrak{X}, \mathbb{C})$. Preserve the available notations for these extensions. By adapting the proof of Theorem 33.2 in [39], it is easy to show the inequality

$$|\widehat{\Phi}(f)| \leqslant \widehat{\Psi}(f) \quad (f \in C_0(\mathfrak{X})_+).$$

We may now use Theorem 5.3.7 which claims existence of a unique measure $\mu \in \operatorname{qca}(\mathfrak{X}, Y)$ such that

$$\widehat{\Phi}(f) = \int_{\mathfrak{X}} f(\chi)\mu(d\chi) \quad (f \in C_0(\mathfrak{X})).$$

If $f \in \mathscr{L}$ then $\widehat{f} = u$ for some $u \in C_{00}(G, \mathbb{C})$. Considering the integral representation for $\widehat{\Phi}$, write down the chain of equalities

$$\Phi(\widehat{f}) = \int_G u(g)\varphi(g)\lambda(dg) = \widehat{\Psi}(f) = \int_{\mathfrak{X}} f(\chi)\mu(d\chi)$$

$$= \int_{\mathfrak{X}} \Big(\int_G \chi(g)u(g)\lambda(dg) \Big) \mu(d\chi)$$

$$= bo\text{-}\lim_{K \in \mathscr{K}(\mathfrak{X})} \int_K \Big(\int_G \chi(g)u(g)\lambda(dg) \Big) \mu(d\chi)$$

$$= bo\text{-}\lim_{K \in \mathscr{K}(\mathfrak{X})} \int_G \Big(\int_K \chi(g)\mu(d\chi) \Big) u(g)\lambda(dg).$$

The last equality is valid by the vector version of the Fubini Theorem 5.4.5. Thus,

$$\int_G u(g)\varphi(g)\lambda(dg) = \int_G \Big(\int_{\mathfrak{X}} \chi(g)\mu(d\chi) \Big) u(g)\lambda(dg)$$

for all $u \in C_{00}(G, \mathbb{C})$. Denote by ρ the mapping from G to Y given by the formula

$$\rho(g) = \varphi(g) - \int\limits_{\mathfrak{X}} \chi(g)\mu(d\chi) \quad (g \in G).$$

From the definition it follows that ρ is a uniformly *bo*-continuous and order-bounded mapping satisfying

$$\int\limits_{G} \rho(g)u(g)\lambda(dg) = 0$$

for all $u \in C_{00}(G)$. Show that $\rho \equiv 0$. To this end, to each $g \in G$ and an arbitrary neighborhood $U \in \mathscr{U}_0$ we put into correspondence the function $\omega_{U,g} \in C_{00}(G)_+$ supported by $U + g$ and satisfying

$$\int\limits_{G} \omega_{U,g}(h)\lambda(dh) = 1.$$

We have the equality

$$\rho(g) = \int\limits_{G} \omega_{U,g}(h)(\rho(g) - \rho(h))\lambda(dh),$$

and so

$$|\rho(g)| \leqslant \sup\{|\rho(g) - \rho(h))| : h \in G,\ h - g \in U\}.$$

Passing to the *o*-limit along the net $U \in \mathscr{U}_0$ in this inequality, derive $\rho(g) = 0$ $(g \in G)$. Thus, the sought integral representation is valid for the mapping φ.

Demonstrate uniqueness. Let $\mu \in \mathrm{qca}(\mathfrak{X}, Y)$ be another representing measure for φ. Denote by $\mathrm{Lin}(G)$ the span of the set of continuous characters of the dual group. By condition

$$\int\limits_{\mathfrak{X}} f(\chi)\mu(d\chi) = \int\limits_{\mathfrak{X}} f(\chi)\mu_1(d\chi) \quad (f \in \mathrm{Lin}(G)).$$

Let $u\text{-}\mathrm{Lin}_{\mathbb{R}}(G)$ stand for the uniform closure of the space of real functions in $\mathrm{Lin}(G)$. It is easy that $u\text{-}\mathrm{Lin}_{\mathbb{R}}(G)$ is a function vector lattice, and so the previous inequality is valid for all $f \in u\text{-}\mathrm{Lin}_{\mathbb{R}}(G)$. Assume that K belongs to $\mathscr{K}(\mathfrak{X})$ and some decreasing net $(f_\alpha)_{\alpha \in \mathrm{A}}$ in $u\text{-}\mathrm{Lin}_{\mathbb{R}}(G)$ tends to 1_K. Since μ_1 and μ are quasi-Radon measures; therefore,

$$\mu(K) = bo\text{-}\lim_{\alpha \in \mathrm{A}} \int\limits_{\mathfrak{X}} f_\alpha(\chi)\mu(d\chi) = bo\text{-}\lim_{\alpha \in \mathrm{A}} \int\limits_{\mathfrak{X}} f_\alpha(\chi)\mu_1(d\chi) = \mu_1(K).$$

By σ-additivity we further extend this equality to all sets in $\mathscr{B}(\mathfrak{X})$. ▷

The unique representing measure $\mu \in \mathrm{qca}(\mathfrak{X}, Y)$ for φ, in accord with Theorem 5.9.1, is called the *Fourier transform* of φ and denoted by $\widehat{\varphi}$.

5.9.2. Theorem. *For a mapping $\psi : G \to F_{\mathbb{C}}$ the following are equivalent:*

(1) *ψ is positive-definite and o-continuous at zero;*

(2) *there is a unique measure $\nu \in \mathrm{qca}(\mathfrak{X}, F)_+$ satisfying*

$$\psi(g) = \int_{\mathfrak{X}} \chi(g)\nu(d\chi) \quad (g \in G).$$

$\lhd$ This is immediate from Theorem 5.9.1. $\rhd$

Theorems 5.9.1 and 5.9.2 straightforwardly imply the following results on isomorphism:

5.9.3. Theorem. *The Fourier transform is a linear and order isomorphism between the space $\mathrm{qca}(\mathfrak{X}, F_{\mathbb{C}})$ and the space $\mathscr{M}_0(G, F_{\mathbb{C}})$ (with the ordering cone $\mathscr{M}_0(G, F_{\mathbb{C}})_+$). In particular, $\mathscr{M}_0(G, F_{\mathbb{C}})$ is a complex Kantorovich space.*

From Theorem 5.9.3 it ensues that to each mapping $\varphi \in \mathscr{M}(G, Y)$ there corresponds its least dominant $\psi \in \mathscr{M}_0(G, F_{\mathbb{C}})^+$ referred to as the *norm* of φ and denoted by $|\varphi|$.

5.9.4. Theorem. *The Fourier transform is an isometry between the spaces $\mathrm{qca}(\mathfrak{X}, Y)$ and $\mathscr{M}_0(G, Y)$. In particular, $\mathscr{M}_0(G, Y)$ is a bo-complete lattice normed space and $|\varphi|^\wedge = |\widehat{\varphi}|$.*

Note also that if Y is a Banach–Kantorovich space then so is $\mathscr{M}_0(G, Y)$.

5.9.5. Remark. The above theorem differs from an analogous result of [42] in the following details: First, [42] discusses only a special Kantorovich space of commuting selfadjoint operators. Second, the definition of positive-definiteness in [42] involves the operator coefficients c_j. Third, the representing measures in $\mathrm{qca}(\mathfrak{X}, F)$ are simpler than the measures $M^{(m)}(\mathfrak{X})$ in [42]. Indeed, for $\mu \in \mathrm{qca}(\mathfrak{X}, F)_+$ the integral I_μ is a dominated positive operator from $C_0(\mathfrak{X})$ to F. This operator extends to the vector functions $F \otimes C_0(\mathfrak{X})$ on letting $\otimes I_\mu = \mathrm{id}_F \otimes I_\mu$. Further, we may use the o-denseness of the subspace $F \otimes C_0(\mathfrak{X})$ of $C_0^{(m)}(\mathfrak{X})$ to extend the operator to the positive operator $\otimes\mu : C_0^{(m)}(\mathfrak{X}) \to F_{\mathbb{C}}$. In this event $\otimes\mu$ belongs to $M^{(m)}(\mathfrak{X})$, and the correspondence $\mu \mapsto \otimes\mu$ is a bijection between $\mathrm{qca}(\mathfrak{X}, F_{\mathbb{C}})$ and $M^{(m)}(\mathfrak{X})$.

5.10. Convolution

The main result of the previous section, Theorem 5.9.1, enables us to furnish the space of quasi-Radon vector measures with convolution and to obtains the spectral decomposition for unitary representations of a locally compact Abelian group in a complex Kantorovich space. In this event Theorem 5.9.2 may be extended to the case of monotonically complete ordered vector spaces.

Assume that Y is furnished with a bilinear mapping $\odot : Y \times Y \to Y$, enjoying the domination condition $|y_1 \odot y_2| \leqslant |y_1| \cdot |y_2|$ $(y_1, y_2 \in Y)$. Consider two measures $\mu_j \in \operatorname{qca}(\mathfrak{X}, Y)$ $(j = 1, 2)$. Their tensor product $\mu = \mu_1 \times \mu_2$ with respect to $\odot$ is also a quasi-Radon σ-additive measure by Theorem 5.4.2. Define the linear operator $T_\mu : C_0(\mathfrak{X}) \to Y$ by the formula

$$T_\mu(f) = \int\limits_{\mathfrak{X}\times\mathfrak{X}} f(\chi_1 \cdot \chi_2)\mu(d\chi_1 d\chi_2).$$

Clear, this is a dominated operator and $|T_\mu| \leqslant T_{|\mu|}$. Moreover, we have the inequality $|T_\mu(f)| \leqslant |\mu|(\mathfrak{X} \times \mathfrak{X})\|f\|_\infty$. By Theorem 5.3.7 there is a unique measure $\mu_1 * \mu_2 \in \operatorname{qca}(\mathfrak{X}, Y)$ satisfying

$$T_\mu(f) = \int\limits_{\mathfrak{X}} f(\chi)(\mu_1 * \mu_2)(d\chi).$$

The representing measure $\mu_1 * \mu_2$ is natural to be called the *convolution* of μ_1 and μ_2. Since $|\mu_1 \times \mu_2| \leqslant |\mu_1| \times |\mu_2|$ by Theorem 5.4.2; therefore, $|\mu_1 * \mu_2| \leqslant |\mu_1| * |\mu_2|$.

The lattice normed space $\operatorname{qca}(\mathfrak{X}, Y)$ is a lattice normed algebra under the convolution $*$. Note that the convolution of scalar Radon measures on a locally-compact group is defined for instance in [39].

By $\varphi_1 \odot \varphi_2$ we denote the product of the mappings $\varphi_j : G \to Y$ $(j = 1, 2)$; i.e., $(\varphi_1 \odot \varphi_2)(g) = \varphi_1(g) \odot \varphi_2(g)$ $(g \in G)$.

5.10.1. *Theorem.* *If $\varphi_j \in \mathscr{M}_0(G, Y)$ $(j = 1, 2)$ then $\varphi_1 \odot \varphi_2 \in \mathscr{M}_0(G, Y)$ and the mapping $|\varphi_1| \cdot |\varphi_2|$ is an o-continuous dominant for $\varphi_1 \odot \varphi_2$ maintaining the equality*

$$(\varphi_1 \odot \varphi_2)^\wedge = \widehat{\varphi}_1 * \widehat{\varphi}_2.$$

$\lhd$ By Theorem 5.9.1 there are unique measures $\mu_j \in \operatorname{qca}(\mathfrak{X}, Y)$ $(j = 1, 2)$ satisfying

$$\varphi_j(g) = \int\limits_{\mathfrak{X}} \chi(g)\mu_j(d\chi) \quad (j = 1, 2,\ g \in G).$$

By the vector version of the Fubini Theorem 5.4.5, we have

$$\varphi_1(g) \odot \varphi_2(g) = \int\limits_{\mathfrak{X}\times\mathfrak{X}} (\chi_1 \cdot \chi_2)(g)(\mu_1 \times \mu_2)(d\chi_1 d\chi_2) = \int\limits_{\mathfrak{X}} \chi(g)(\mu_1 * \mu_2)(d\chi).$$

Consequently, $\varphi_1 \odot \varphi_2 \in \mathscr{M}_0(G, Y)$ and $(\varphi_1 \odot \varphi_2)^\wedge = \widehat{\varphi}_1 * \widehat{\varphi}_2$. By definition

$$|\varphi_1 \odot \varphi_2|^\wedge = |\widehat{\varphi}_1 * \widehat{\varphi}_2| \leqslant |\widehat{\varphi}_1| * |\widehat{\varphi}_2| = |\varphi_1|^\wedge * |\varphi_2|^\wedge = (|\varphi_1| \cdot |\varphi_2|)^\wedge.$$

Therefore, $|\varphi_1| \cdot |\varphi_2| - |\varphi_1 \odot \varphi_2| \in \mathscr{M}_0(G, F_{\mathbb{C}})_+$. $\rhd$

5.10.2. Corollary. *The lattice normed space $\mathscr{M}_0(G,Y)$ with multiplication $\odot$ is a lattice normed algebra, and the Fourier transform $\varphi \mapsto \widehat{\varphi}$ is an isomorphism between the algebras $\mathscr{M}_0(G,Y)$ and* $\operatorname{qca}(\mathfrak{X},Y)$.

5.10.3. REMARK. We may consider a more general situation:

Let Y, Y_1, and Y_2 be three complex lattice normed spaces with norm lattice a Kantorovich space F. We further assume given a bilinear mapping $\odot : Y_1 \times Y_2 \to Y$ such that $|y_1 \odot y_2| \leqslant |y_1| \cdot |y_2|$.

The respective convolution becomes a bilinear mapping from $\operatorname{qca}(\mathfrak{X},Y_1) \times \operatorname{qca}(\mathfrak{X},Y_2)$ to $\operatorname{qca}(\mathfrak{X},Y)$. Moreover, the following formulas remain valid:

$$(\varphi_1 \odot \varphi_2)^\wedge = \widehat{\varphi}_1 * \widehat{\varphi}_2,$$

$$|\varphi_1| \cdot |\varphi_2| = |\varphi_1 \odot \varphi_2| \in \mathscr{M}_0(G,F_{\mathbb{C}})_+.$$

Consider the spectral decomposition problem for representations of the group G under study. Assume that F contains an order-unity $\mathbf{1}$. An element $u+iv$ in $F_{\mathbb{C}}$ is *unitary* if $u^2+v^2 = |u+iv|^2 = \mathbf{1}$. The set $\mathscr{U}(\mathbf{1})$ of unitary elements makes a group under the order multiplication in $F_{\mathbb{C}}$ having $\mathbf{1}$ as the ring-unity.

A homomorphism $\pi : G \to \mathscr{U}(\mathbf{1})$ is a *unitary representation* of G. It is easy to see that every representation is positive-definite.

By $\mathscr{B}(\mathbf{1})$ we denote the Boolean algebra of fragments of the order-unity $\mathbf{1}$. A measure $\nu : \mathscr{B}(\mathfrak{X}) \to F$ is *spectral* if the image of ν lies in $\mathscr{B}(\mathbf{1})$.

5.10.4. Theorem. *To every o-continuous unitary representation $\pi : G \to \mathscr{U}(\mathbf{1}) \subset F_{\mathbb{C}}$ there is a unique spectral measure $e : \mathscr{B}(\mathfrak{X}) \to \mathscr{B}(\mathbf{1}) \subset F_{\mathbb{C}}$ satisfying*

$$\pi(g) = \int\limits_{\mathfrak{X}} \chi(g)e(d\chi) \quad (g \in G).$$

$\lhd$ By Theorem 5.9.2 there is a unique representing measure $e \in \operatorname{qca}(\mathfrak{X},F)_+$ for π. Denote by $\operatorname{Lin}(G)$ the span of the character space of the group $\mathfrak{X}$. Let K belong to $\mathscr{K}(\mathfrak{X})$ and let some bounded net of functions $(f_\alpha)_{\alpha\in\mathrm{A}} \subset \operatorname{Lin}(G)$ converge pointwise to 1_K. If

$$f_\alpha(\chi) = \sum_{j=1}^{n_\alpha} c_{\alpha,j}\chi(g_{\alpha,j})$$

for some $c_{\alpha,j} \in \mathbb{C}$, $g_{\alpha,j} \in G$ ($\alpha \in \mathrm{A}$, $j = 1, \dots, n_\alpha$) then

$$e(K)^2 = o\text{-}\lim_{\alpha \in \mathrm{A}} \Big(\int_{\mathfrak{X}} f_\alpha(\chi) e(d\chi) \Big) \Big(\int_{\mathfrak{X}} \overline{f_\alpha(\chi)} e(d\chi) \Big)$$

$$= o\text{-}\lim_{\alpha \in \mathrm{A}} \sum_{j,k=1}^{n_\alpha} c_{\alpha,j} \overline{c_{\alpha,k}} \pi(g_{\alpha,j}) \overline{\pi(g_{\alpha,k})}$$

$$= o\text{-}\lim_{\alpha \in \mathrm{A}} \sum_{j,k=1}^{n_\alpha} \int_{\mathfrak{X}} \chi(g_{\alpha,j} - g_{\alpha,k}) c_{\alpha,j} \overline{c_{\alpha,k}} e(d\chi)$$

$$= o\text{-}\lim_{\alpha \in \mathrm{A}} \int_{\mathfrak{X}} |f_\alpha(\chi)|^2 e(d\chi) = e(K).$$

By σ-additivity, $e(B)^2 = e(B)$ ($B \in \mathscr{B}(\mathfrak{X})$). Consequently, $e(\mathscr{B}(\mathfrak{X})) \subseteq \mathscr{B}(\mathbf{1})$. $\triangleright$

5.10.5. Remark. The Stone Theorem on unitary representation of a locally compact group (see [43]) is a particular case of Theorem 5.10.4. Indeed, if $\mathscr{H}$ is a complex Hilbert space and $B(\mathscr{H})$ is the endomorphism space of $\mathscr{H}$ then as $F_\mathbb{C}$ we must take the commutative von Neumann algebra generated by $\pi(G)$.

Assume now that E is a monotonically complete ordered vector space. We preserve the same definition of positive definiteness for a mapping $\psi : G \to E_\mathbb{C}$ (cf. Definition 5.8.2). A mapping $\omega : G \to E$ is *o-continuous at a point* $g_0 \in G$ if there is a net $\{h_U : U \in \mathscr{U}_0\} \subset E_+$, decreasing in order to zero and such that $-h_U \leqslant \omega(g) - \omega(g_0) \leqslant h_U$ for all $g \in U$ ($U \in \mathscr{U}_0$). A mapping $\psi : G \to E_\mathbb{C}$ is *o-continuous at a point* $g_0 \in G$, if $\operatorname{Re}\psi$ and $\operatorname{Im}\psi$ are o-continuous at g_0. By analogy we define uniform o-continuity for E-valued and $E_\mathbb{C}$-valued mappings.

For a positive-definite mapping $\psi : G \to E_\mathbb{C}$ the mappings $\operatorname{Re}\psi$ are $\operatorname{Im}\psi$ order bounded and take values in the ideal generated by $\psi(0)$. We denote by $\operatorname{qca}(\mathfrak{X}, E)_+$ the set of σ-additive quasi-Radon measures $\nu : \mathscr{B}(\mathfrak{X}) \to E_+$ (the definition of a quasi-Radon measure remains the same as before since it involves only monotone nets).

5.10.6. ***Theorem.*** *For a mapping $\psi : G \to E_\mathbb{C}$ the following are equivalent:*

(1) *ψ is positive-definite and o-continuous at zero;*

(2) *there is a unique measure $\nu \in \operatorname{qca}(\mathfrak{X}, E)_+$ satisfying*

$$\psi(g) = \int_{\mathfrak{X}} \chi(g) \nu(d\chi) \quad (g \in G).$$

◁ (1)→(2): Let the o-continuity of ψ at zero be checked for the net $\{h_U : U \in \mathscr{U}_0\}$. Take $U_0 \in \mathscr{U}_0$ and $h_0 = h_{U_0}$. In E consider the ideal $E(\mathbf{1})$ generated by $\mathbf{1} = \psi(0) + h_0$. Following [33], look at a Dedekind completion F of this ideal. The above implies that $\psi(g) \in E(\mathbf{1})_{\mathbb{C}} \subset F_{\mathbb{C}}$ $(g \in G)$. Since $h_0 \in F$, the mapping $\psi : G \to E_{\mathbb{C}}$ is o-continuous at zero too. By Theorem 5.9.2 there is a unique measure $\nu \in \operatorname{qca}(\mathfrak{X}, E)^+$ satisfying

$$\psi(g) = \int_{\mathfrak{X}} \chi(g)\nu(d\chi) \quad (g \in G).$$

We are left with checking that the measure ν has range in the original space E. Let $K \in \mathscr{K}(\mathfrak{X})$. There is a net $(f_\alpha)_{\alpha \in \mathrm{A}}$ in $\operatorname{Lin}(G)$ decreasing pointwise to 1_K. Observe that

$$\nu(K) = \inf_{\alpha \in \mathrm{A}} \int_{\mathfrak{X}} f_\alpha(\chi)\nu(d\chi) \in E_+.$$

From σ-additivity of ν it follows that $\nu(B) \in E_+$ $(B \in \mathscr{B}(\mathfrak{X}))$. ▷

5.10.7. REMARK. We list as instances of monotonically complete ordered vector spaces the so-called O^*-algebras [44]. A nontrivial example of an O^*-algebra is the algebra of measurable operator with respect to some von Neumann algebra (see [44]). The Bochner Theorem for positive-definite operator valued mappings was also treated in the article [45] by M. Christensen.

5.11. Boolean Valued Interpretation of the Wiener Lemma

We now turn to the case of a compact group G. The dual group $\mathfrak{X}$ of G is discrete; and Theorem 5.9.1 asserts in this event that a mapping $\varphi : G \to Y$ expands in an absolutely convergent Fourier series in the characters of the group G if and only if φ has an o-continuous dominant $\psi : G \to F_{\mathbb{C}}$. The Fourier coefficients may be written down by the conversion formula

$$y_\chi = \int_G \varphi(g)\overline{\chi(g)}dg = \widehat{\varphi}(\{\chi\}),$$

with dg the normalized Haar measure on G.

The next assertion about the group S^n, with S the circumference of radius 1, is known as the *Wiener Lemma* (see [46, Lemma 11.6] and [39, Vol. 2, Theorem 39.31]).

5.11.1. Theorem. *Assume that $\varphi \in \mathscr{M}_0(G, \mathbb{C})$ and φ does not vanish on G. Then $1/\varphi \in \mathscr{M}_0(G, \mathbb{C})$.*

$\triangleleft$ Given $\psi \in \mathscr{M}_0(G, \mathbb{C})$, define the scalar norm $\|\psi\|_\wedge = |\psi|(0) = |\widehat{\psi}|(\mathfrak{X})$. The space $\mathscr{M}_0(G, \mathbb{C})$ becomes a Banach algebra under this norm and pointwise multiplication. If $\varphi \in \mathscr{M}_0(G, \mathbb{C})$ then we also have $\overline{\varphi} \in \mathscr{M}_0(G, \mathbb{C})$, and by Theorem 5.10.1 for $Y = \mathbb{C} = F_\mathbb{C}$ we obtain $|\varphi|^2 = \varphi \cdot \overline{\varphi} \in \mathscr{M}_0(G, \mathbb{C})$. Put

$$M = \sup\{|\varphi(g)|^2 : g \in G\}, \quad m = \inf\{|\varphi(g)|^2 : g \in G\}.$$

For the uniform norm of $\psi = M \cdot 1_G - |\varphi|^2$ we obviously have $\|\psi\|_\infty = M - m$. Therefore, the spectral radius of the element ψ in the Banach algebra $\mathscr{M}_0(G, \mathbb{C})$ equals $M - m$ (cf. Theorem C.24 as well as the proof of Theorem 23.13 in [39, Vol. 1]). Consequently, the element $M \cdot 1_G - \psi = |\varphi|^2$ is invertible in the algebra $\mathscr{M}_0(G, \mathbb{C})$. Appealing to Theorem 5.10.1 once again, note that the element $\varphi^{-1} = \overline{\varphi} \cdot (|\varphi|^2)^{-1}$ belongs to the algebra $\mathscr{M}_0(G, \mathbb{C})$. $\triangleright$

Given a locally-compact Abelian group G, we consider in the Banach algebra $\mathscr{M}_0(G, \mathbb{C})$ the Banach subalgebra $\mathscr{M}_{ad}(G, \mathbb{C})$ of the functions $\varphi \in \mathscr{M}_0(G, \mathbb{C})$ such that the measure $\widehat{\varphi} \in \operatorname{qca}(\mathfrak{X}, \mathbb{C})$ has no singular part, i.e., the Lebesgue decomposition for $\widehat{\varphi}$ has the form $\widehat{\varphi} = \widehat{\varphi}_a + \widehat{\varphi}_d$, where $\widehat{\varphi}_a$ is the absolutely continuous part and $\widehat{\varphi}_d$ is the discrete part of $\widehat{\varphi}$ with respect to the Haar measure on $\mathfrak{X}$.

5.11.2. Theorem. *The spectrum of each element φ of the Banach algebra $\mathscr{M}_{ad}(G, \mathbb{C})$ is the closure in $\mathbb{C}$ of the image $\varphi(G)$. In particular, the spectral radius of $\varphi \in \mathscr{M}_{ad}(G, \mathbb{C})$ equals $\|\varphi\|_\infty$.*

$\triangleleft$ Assume that λ belongs to $\mathbb{C}$ and does not lie in the closure of $\varphi(G)$. Consider the function $\psi = \lambda 1_G - \varphi$. It has the representation: $\psi(g) = \psi_a(g) + \psi_d(g)$ $(g \in G)$, where $\widehat{\psi}_a$ is an absolutely continuous measure with respect to the Haar measure on $\mathfrak{X}$ and ψ_d has the form

$$\psi_d(g) = \sum_{n=1}^{\infty} c_n \chi_n(g) \quad (g \in G)$$

for some $\chi_n \in \mathfrak{X}$, $c_n \in \mathbb{C}$ $(n \in \mathbb{N})$. By condition $|\psi(g)| \geqslant \varepsilon_0$ $(g \in G)$ for some $\varepsilon_0 > 0$. Since

$$\lim\{\sup\{|\psi_a(g)| : g \notin K\} : K \in \mathscr{K}(G)\} = 0,$$

there is a compact set $K_0 \in \mathscr{K}(G)$, satisfying $|\psi_d(g)| \geqslant \varepsilon_0/2$ $(g \in G \setminus K_0)$. The function ψ_d is uniformly almost periodic and so admits the continuation ψ_d^* to the Baire compactification G^* of G. Assume that $\psi_d(g_0) = 0$ for some $g_0 \in G$. The set $U^* = \{g \in G^* : |\psi_d(g)| < \varepsilon_0/2\}$ is open in the group G^*. We consider the nontrivial case in which G is not compact. Then there are an element $g^* \in U^* \setminus G$ and the net $\{g_\alpha : \alpha \in \mathrm{A}\}$ lying in U and converging to g^*. Obviously, it is impossible for this net to lie within the compact set K_0. Therefore, for some $\alpha \in \mathrm{A}$ we have $g_\alpha \in U \setminus K_0$. This contradicts the fact that $|\psi_d(g_\alpha)| \geqslant \varepsilon_0/2$. Our argument implies the inequality

$$\inf\{|\psi_d(g)| : g \in G\} > 0.$$

Consequently, ψ_d^* does not vanish on G^*, and by Theorem 5.11.1 the function ψ_d^{*-1} expands in an absolutely convergent Fourier series in the character of the group $\mathfrak{X}$. The same holds for the function ψ_d^{-1}. Consequently, $\tau = \psi\psi_d^{-1} - 1_G$ belongs to $\mathscr{M}_0(G, \mathbb{C})$, and $\widehat{\tau}$ is an absolute continuous with respect to the Haar measure. The function $(1_G + \tau)^{-1}$ belongs to $\mathscr{M}_0(G, \mathbb{C})$ as well. This is proved like in Theorem 5.11.1. We thus find that ψ is invertible in $\mathscr{M}_{ad}(G, \mathbb{C})$, and so λ does not belong to the spectrum of φ. ▷

5.11.3. *Theorem.* *Let F be a universally complete Kantorovich space with ring- and order-unity* $\mathbf{1}$. *If the mappings $\varphi : G \to F_{\mathbb{C}}$, $\psi : G \to F_{\mathbb{C}}$ enjoy the conditions $\varphi(g)\psi(g) = \mathbf{1}$ $(g \in G)$ and $\varphi \in \mathscr{M}_0(G, F_{\mathbb{C}})$, then $\psi \in \mathscr{M}_0(G, F_{\mathbb{C}})$.*

◁ We sketch a proof that rests on Boolean valued interpretation of Theorem 5.11.1. Let $\mathbf{V}^{(\mathbb{B})}$ stand for the Boolean valued universe, with $\mathbb{B}$ the Boolean algebra of band projections in F. Assume that $\mathscr{G}$ is a completion of the topological group $G^\wedge$ inside $\mathbf{V}^{(\mathbb{B})}$. Then $[\![\mathscr{G}$ is a compact group$]\!] = \mathbf{1}$. Furthermore, if $\mathscr{C}$ is the field of complex numbers inside $\mathbf{V}^{(\mathbb{B})}$, then $\mathscr{C}{\downarrow}$ is a complex Kantorovich space isomorphic to $F_{\mathbb{C}}$. Without loss of generality, we may thus assume that $\mathscr{C}{\downarrow} = F_{\mathbb{C}}$. Suppose that φ and ψ meet the hypotheses of the theorem. The function $\varphi^\wedge : G^\wedge \to \mathscr{C}$ inside $\mathbf{V}^{(\mathbb{B})}$, determined by the equalities $[\![\varphi^\wedge(x^\wedge) = \varphi(x)]\!] = \mathbf{1}$ $(x \in G)$, is uniformly continuous, since φ is uniformly o-continuous. This is easy to check by simple calculations with Boolean truth-values. Denote by $\bar{\varphi}$ the continuation of φ from $G^\wedge$ to the completion $\mathscr{G}$. Then $[\![\bar{\varphi}$ is a dominated function with a dominant continuous at zero$]\!] = \mathbf{1}$. Indeed, if φ_0 is a dominant for φ which is o-continuous at zero, then there is $\bar{\varphi}_0 : \mathscr{G} \to \mathscr{C}$ such that $[\![\bar{\varphi}_0$ is a dominant for $\bar{\varphi}]\!] = \mathbf{1}$. Apply Theorem 5.11.1 to $\bar{\varphi}$. Inside $\mathbf{V}^{(\mathbb{B})}$ there is a function $\omega : \mathscr{G} \to \mathscr{C}$ having a continuous dominant at zero such that $[\![\bar{\varphi}(g)\omega(g) = 1 \ (g \in G)]\!] = \mathbf{1}$. Then the restriction ψ_0 of the mapping $\omega{\downarrow} : \mathscr{G}{\downarrow} \to F_{\mathbb{C}}$ to $G \subset \mathscr{G}{\downarrow}$ satisfies the conditions $\psi_0 \in \mathscr{M}_0(G, F_{\mathbb{C}})$ and $\varphi(g)\psi_0(g) = \mathbf{1}$ $(g \in G)$. This shows that $\psi = \psi_0$. ▷

References

1. Dinculeanu N., Vector Measures, VEB Deutscher Verlag der Wissenschaften, Berlin (1966).

2. Diestel J. and Uhl J. J., Vector Measures, Amer. Math. Soc., Providence, RI (1977). (Math. Surveys; **15**.)
3. Kantorovich L. V., Vulikh B. Z., and Pinsker A. G., Functional Analysis in Semiordered Spaces [in Russian], Gostekhizdat, Moscow and Leningrad (1950).
4. Fremlin D. M., "A direct proof of the Matthes–Wright integral extension theorem," J. London Math. Soc. (2), **11**, No. 3, 276–284 (1975).
5. Riečan B., "A simplified proof of the Daniell integral extension theorem in ordered spaces," Math. Slovaca, 32, No. 1, 75–79 (1982).
6. Hausdorff F., Set Theory, Chelsea, New York (1991).
7. Plesner A. I., Spectral Theory of Linear Operators. Vol. 1 and 2, Frederick Ungar Publishing Co., New York (1969).
8. Vulikh B. Z., Introduction to Functional Analysis, Pergamon Press, Oxford (1963).
9. Kusraev A. G., Vector Duality and Its Applications [in Russian], Nauka, Novosibirsk (1985).
10. Kusraev A. G. and Strizhevskiĭ V. Z., "Lattice normed spaces and dominated operators," in: Studies on Geometry and Functional Analysis. Vol. 7 [in Russian], Trudy Inst. Mat. (Novosibirsk), Novosibirsk, 1987, pp. 132–157.
11. Wright J. D. M., "Stone-algebra-valued measures and integrals," Proc. London Math. Soc., **19**, No. 1, 107–122 (1969).
12. Horn A. and Tarski A., "Measures in Boolean algebras," Trans. Amer. Math. Soc., **64**, No. 3, 467–497 (1948).
13. Loš J. and Marczewski E., "Extensions of measure," Fund. Math., **36**, 267–276 (1949).
14. Wright J. D. M., "The measure extension procedure for vector lattices," Ann. Inst. Fourier Grenoble, **21**, No. 4, 65–85 (1971).
15. Panchapagesan T. V. and Palled Sh. V., "On vector lattice-valued measures. I," Math. Slovaca, **33**, No. 3, 269–292 (1983).
16. Riečan J., "On the Kolmogorov consistency theorem for Riesz space valued measures," Acta Math. Univ. Comen., **48/49**, 173–180 (1986).
17. Wright J. D. M., "Vector lattice measures on locally compact spaces," Math. Z., **120**, No. 3, 193–203 (1971).
18. Kasch F., Modules and Rings, Academic Press, London and New York (1982).
19. Wright J. D. M., "Products of positive vector measures," Quart. J. Math., **24**, No. 94, 189–206 (1973).
20. Duchoň M. and Kluvanek I., "Inductive tensor product of vector valued measures," Mat. Casop., **17**, No. 2, 108–112 (1967).
21. Duchoň M., "On the projective tensor product of vector valued measures," Mat. Casop., **17**, No. 2, 113–120 (1967).

22. Kluvanek I., "On the product of vector measures," J. Austral. Math. Soc., **15**, No. 1, 22–26 (1973).

23. Akhiezer N. I., The Classical Moment Problem and Some Related Topics in Analysis, Oliver & Boyd, Edinburgh and London (1965).

24. Kreĭn M. G. and Nudel′man A. A., The Markov Moment Problem and Extremal Problems: Ideas and Problems of P. L. Chebyshëv and A. A. Markov and Their Further Development, American Mathematical Society, Providence (1977).

25. Niåstad O., "Unique solvability of an extended Hamburger moment problem," J. Math. Anal. Appl., **124**, No. 2, 502–519 (1987).

26. Alden E., On Indeterminacy of Strong Moment Problems [Preprint/Univ. Umea; No. 2], Sweden (1988).

27. Shonkwiler R., "On the solution of moment problems by reproducing kernel methods," J. Math. Anal. Appl., **130**, No. 1, 271–299 (1988).

28. F. Riesz and B. Szökefalvi-Nagy, Functional Analysis, Dover Publications, New York (1990).

29. Vorob′ëv Yu. V., "Orthogonal operator polynomials and approximate methods for determining spectra of bounded linear operators," Uspekhi Mat. Nauk, **9**, No. 1, 83–90 (1954).

30. Berezanskiĭ Yu. M., "The generalized power moment problem," Trudy Moskovsk. Mat. Obshch., **21**, 47–102 (1970).

31. Sebestyen Z., "Moment theorems for operators on Hilbert spaces," Acta Sci. Math. Szegel., **47**, No. 1–2, 101–106 (1984).

32. Schmüdgen K., "On a generalization of the classical moment problem," J. Math. Anal. Appl., **125**, No. 2, 461–470 (1987).

33. Khurana S. S., "Lattice-valued Borel measures," Rocky Mount. J. Math., **6**, No. 2, 377–382 (1976).

34. Malyugin S. A., "Quasi-Radon measures," Sibirsk. Mat. Zh., **32**, No. 5, 103–111 (1991).

35. Malyugin S. A., "On the vector Hamburger moment problem," Optimization, No. 48, 124–141 (1990).

36. Kreĭn M. G., "Infinite J-matrices and the matrix moment problem," Dokl. Akad. Nauk SSSR, **69**, No. 2, 125–128 (1949).

37. Berezanskiĭ Yu. M., Expansions in Eigenfunctions of Selfadjoint Operators, American Mathematical Society, Providence (1968).

38. Sz.-Nagy B., "A moment problem for self-adjoint operators," Acta Math. Acad. Sci. Hung., **3**, 285–293 (1952).

39. Hewitt E. and Ross K., Abstract Harmonic Analysis. Vol. 1 and 2, Springer-Verlag, New York (1994).

40. Naĭmark M. A., Normed Rings, McGraw-Hill Book Co., New York (1973).

41. Heyer H., Probability Measures on Locally Compact Groups, Springer-Verlag, Berlin etc. (1977).
42. Takeuti G., "A transfer principle in harmonic analysis," J. Symbolic Logic, **44**, No. 3, 417–440 (1979).
43. Loomis L. H., An Introduction to Abstract Harmonic Analysis, D. Van Nostrand Company Inc., Princeton (1953).
44. Sarymsakov T. A., Ayupov Sh. A., Khadziev Dzh., and Chilin V. I., Ordered Algebras [in Russian], Fan, Tashkent (1983).
45. Christensen M. J., "Extension theorems for operator-valued measures," J. Math. Phys., **20**, No. 3, 385–389 (1979).
46. Rudin W., Functional Analysis, McGraw-Hill Book Co., New York (1973).

Notation Index

Subject Index

Zeitfracht Medien GmbH
Ferdinand-Jühlke-Straße 7
99095 Erfurt, Deutschland
produktsicherheit@kolibri360.de